Cytogenetics of Livestock

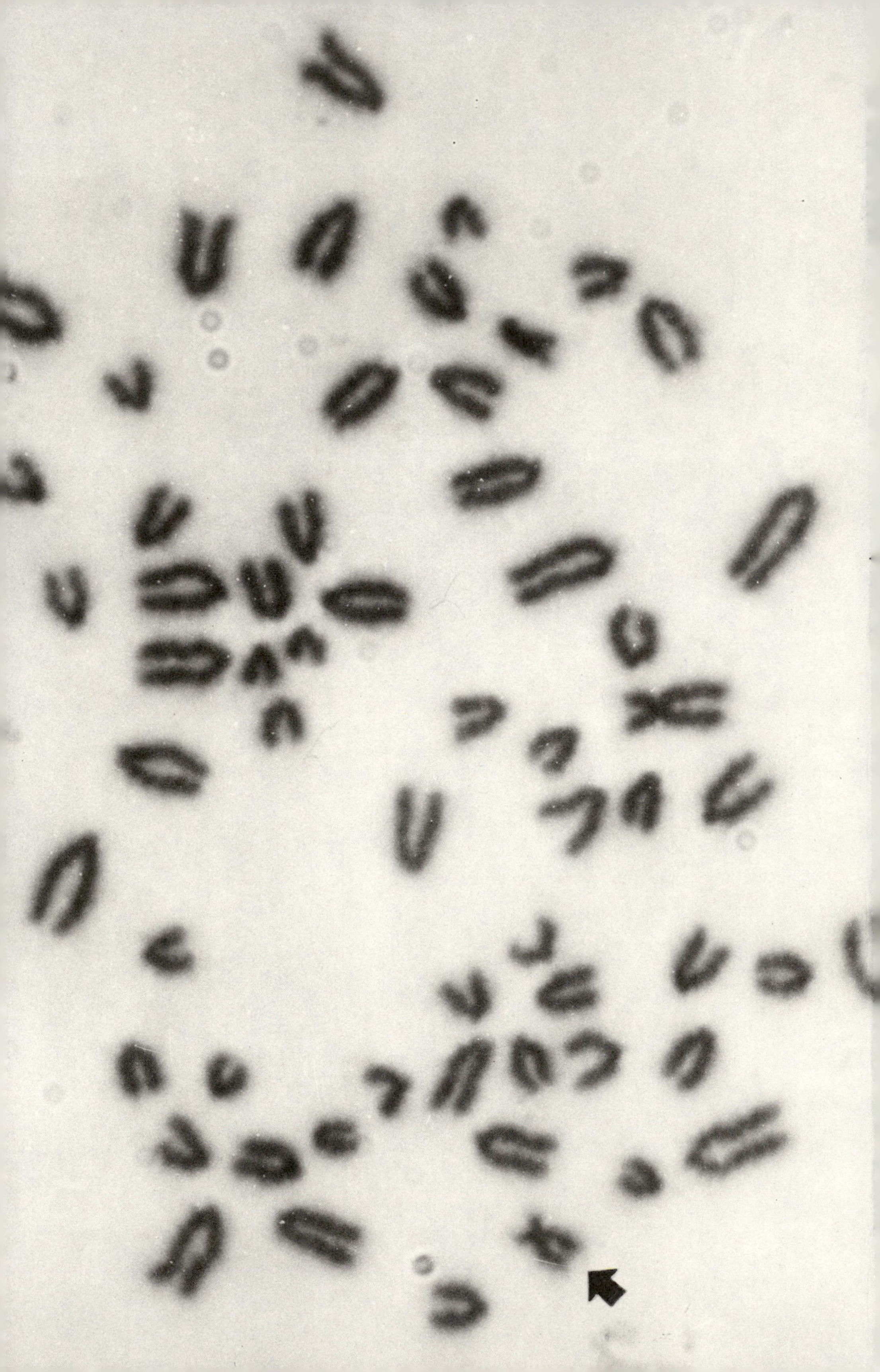

Cytogenetics of Livestock

Franklin E. Eldridge

Department of Animal Science
University of Nebraska
Lincoln, Nebraska

AVI PUBLISHING COMPANY, INC.
Westport, Connecticut

Frontispiece

Metaphase chromosomes from a Simmental bull which had a long Y chromosome. The centromeric index for the Y chromosome (arrow) is .66 as compared to Holstein and Brown Swiss of .57 and Jersey of .52.

THE AVI PUBLISHING COMPANY, INC.
P.O. Box 831
250 Post Road East
Westport, Connecticut 06881

Library of Congress Cataloging in Publication Data

Eldridge, Franklin E.
Cytogenetics of livestock.

Includes bibliographies and index.
1. Livestock—Cytology. 2. Livestock—Genetics.
3. Cytogenetics. I. Title.
SF757.25.E47 1985 599'.0873223 85-1274
ISBN 0-87055-483-2

Printed in the United States of America
A B C D 4321098765

Contents

Preface ix

1 Introduction to the Historical Development of Cytology and Cytogenetics **1**

Cytology and Cytogenetics 1
Light and Electron Microscopy 2
Early History of Cytology 4
References 6
Additional References 7

2 Cell Division: Mitosis **9**

Cell Division 9
Synthesis 10
Stages of Mitosis 13
References 15
Additional Reference 15

3 Cell Division: Meiosis **17**

Introduction 17
Stages of Meiosis 18
Gametogenesis 21
References 22

4 Chromosome Number and Morphology **23**

Introduction 23
Chromosome Morphology 23
References 27

5 Chromosomal Aberrations **29**

Introduction 29
Modification of Chromosome Numbers 30

Robertsonian Translocations 31
Centric Fission 35
Supernumerary Chromosomes 35
Modifications of Chromosome Structure 36
Reciprocal Translocations 40
References 44
Additional Reference 44

6 Banding of Chromosomes and Karyotyping 45

Introduction 45
Salivary Gland Chromosomes of *Drosophila* 45
Karyotyping 46
Karyotyping without Banding 48
Fluorescent Banding 51
Banding Techniques in Karyotyping 52
References 59
Additional References 60

7 Fertilization, Parthenogenesis, and Sex Determination 61

Fertilization 61
Parthenogenesis 63
Early Embryology 64
Sex Determination 64
References 68
Additional References 69

8 Somatic Cell Hybridization 71

References 73

9 Fertility As Affected by Chromosomes 75

Introduction 75
Factors Affecting Fertility 75
Freemartinism 79
Fertility of Males with Sex Chromosome Chimerism 82
Intersexes 83
Trisomy 85
General Fertility 86
Direct Chromosomal Effects 86
References 89

10 Laboratory Procedures for Chromosome Studies 93

Introduction 93
Development of Synthetic Media 94
Control of Contamination 95
Tissue Culture 96

Lymphocyte Cultures 99
Somatic Cell Genetics 105
Chromosomes of Blastocysts 107
Meiotic Studies 108
Squash Technique 110
References 112

11 Cattle Chromosomes 115

Historical Background 115
Robertsonian Translocations 122
X-Autosome Translocations 147
Double Translocations in Cattle 150
Other Translocations 152
Trisomy 153
Pericentric Inversion 158
Y-Chromosome Polymorphisms 160
Chimerism and Mosaicism in Cattle 161
Homology with Sheep and Goats and Other Members of the Superfamily Bovidea 163
Cattle Hybrids 164
Population Studies 167
Blastocyst and Embryo Chromosomes 170
Meiotic Studies 172
Sex Identification 178
Sister-Chromatid Exchange 179
Mitochondrial Variation in Bovine DNA 180
References 180

12 Sheep and Goat Chromosomes 189

Introduction and Early Studies 189
Robertsonian Translocations in Sheep 195
Chimerism in Sheep 197
Other Chromosomal Aberrations in Sheep 200
Other Causes of Chromosomal Aberrations 203
Early Studies on Goat Chromosomes 204
Robertsonian Translocations in Goats 207
Goat Intersexes 207
Sheep × Goat Hybrids 209
References 215

13 Swine Chromosomes 219

Establishment of Chromosome Number 219
Wild Swine 221
Banding and Karyotyping 223
Swine Intersexes 226
Embryological Studies 229

Environmentally Induced Chromosomal Aberrations 230
Chromosomal Aberrations Occurring Naturally 231
Chromosomes of Phenotypically Abnormal Swine 233
General 236
References 238

14 Chromosomes of Horses, Asses, and Mules 243

Chromosome Numbers 243
Chromosome Banding 248
Intersexuality 251
Chromosomal Aberrations Associated with Fertility Problems 257
Mules, Hinnies, and Other Hybrids 259
References 260

15 Bird Cytogenetics *by R. N. Shoffner* 263

Autosomes 264
Gonosomes 268
Meiotic Chromosomes 270
Chromosomal Analysis 271
Chromosome Variation 273
Ploidy 277
Summary 279
References 280

16 Application of Cytogenetics to Livestock Improvement 285

References 290

Index 291

Preface

This book is an outgrowth of a course entitled "Cytogenetics of Livestock," which was first taught at the National Dairy Research Institute at Karnal (Haryana), India. The course was subsequently enlarged and has been taught several times at the University of Nebraska at Lincoln, Nebraska. Obviously, the major emphasis of this book is directed toward cytogenetics of farm livestock: cattle, sheep, swine, horses, asses, mules, goats, and poultry.

Since there are numerous textbooks that detail the basic concepts and development of cytology and cytogenetics, coverage of these topics is minimal in this text. However, it seemed necessary to review, at least briefly, the concepts of mitosis and meiosis, chromosome morphology, chromosome numbers, and common aberrations with techniques of culturing, slide preparation, banding, and karyotyping. Some duplication of information is inevitable as specific situations with different livestock species are presented.

Cytogeneticists, veterinarians, and livestock scientists and producers may also find this book useful as a compilation of current knowledge in this field as of the date of publication. The field of cytogenetics of livestock is expanding rapidly. More papers are now being published monthly than were being published yearly in the recent past. Consequently, it is inevitable that some important research papers have not been cited. If errors or omissions are noted by readers, I hope they will be called to my attention.

I wish to extend my sincere thanks to Dr. R. N. Shoffner, University of Minnesota, for providing the chapter on Bird Cytogenetics. Dr. Shoffner has specialized in this field for many years. Research in cytogenetics in poultry has progressed farther than in many other livestock species because of the shorter generation interval and the more readily available ova (eggs) and other factors.

Many people have made contributions to this book, both directly and indirectly. These include graduate assistants, technical assistants, secretaries, family members, and others. Their help is gratefully acknowledged.

FRANKLIN E. ELDRIDGE

Related AVI Books

ADVANCES IN MEAT RESEARCH
Pearson and Dutson
COMMERCIAL CHICKEN PRODUCTION MANUAL, 3rd Edition
North
DIGESTIVE PHYSIOLOGY AND METABOLISM IN RUMINANTS
Ruckebusch and Thivend
FUNDAMENTALS OF DAIRY CHEMISTRY, 2nd Edition
Webb, Johnson, Alford
STATISTICAL METHODS FOR FOOD AND AGRICULTURE
Bender, Douglass, Kramer
SWINE PRODUCTION AND NUTRITION
Pond and Maner
VETERINARY PHARMACOLOGY AND TOXICOLOGY
Ruckebusch, Toutain, Koritz
WORLD FISH FARMING, 2nd Edition
Brown

Cytogenetics of Livestock

1

Introduction to the Historical Development of Cytology and Cytogenetics

Even before Aristotle's time, there was the concept of pangenesis, that is, that in the semen of the male there were perfectly formed but minute models of an adult animal. The female simply provided the environment for development to parturition. Because at that time "experimentation" itself was not possible and philosophy or "logic" was popular, based on what could be observed with the naked eye, this theory could not be proved or disproved. Curiosity was there, but the methods were inadequate. Today our research and knowledge of cytogenetics and other fields depend on methods, or techniques.

CYTOLOGY AND CYTOGENETICS

Cytology is the study of cells—their morphology, internal structures, cell membranes, metabolic functions, and other cell material. A cytologist is interested in cells for many different reasons, including heredity. Genetics is the study of hereditary variations among organisms. Cytogenetics is a science combining cytology and genetics, and because inheritance is mostly controlled by genes on the chromosomes, cytogenetic study is concentrated mostly on chromosomes. However, not all inheritance is controlled by chromosomes, Some hereditary characteristics are controlled by materials in the cell that are not on the chromosomes, a type of inheritance known as extrachromosomal. At this time much more is known about extrachromosomal inheritance in plants than in animals.

Mitochondrial inheritance in animals may explain some of the maternal effects which have been difficult to evaluate quantitatively. Laipis *et al.* (1982) found some interesting polymorphisms in cattle and evidence for shifts in mitochondrial DNA populations which were more rapid than expected between maternally related animals.

All living organisms, with the exception of virus particles, are made up of cells. This description is true of single-cell organisms such as bacteria and yeast, and of multicellular plants and animals. There are about a million billion cells in the human body. Although an animal such as a cow is much larger than a human, each cell is about the same size as a human cell. The number of cells in an animal can be roughly estimated by the ratio of its weight to that of a human. The cells make up many different specialized tissues—for example, skin, nervous tissue, blood, and bone. The specialized tissues, in turn, make up the specialized organs.

Animal cells have a number of structures, each of which has certain functions. The cell membrane surrounds the cell and forms its boundaries, but less distinctly than in plant cells, where it is usually more rigid. Within the cell are protoplasm, mitochondria, ribosomes, Golgi apparatus, centrioles, lysosomes, the nucleus, and various other organelles.

Within the nucleus are found nucleoli (which disappear during cell division), chromosomes and microtubules (which can only be distinguished under the light microscope during mitosis and meiosis), and nucleoplasm. The nucleus is surrounded by a nuclear envelope or membrane, except during mitosis when it breaks down permitting the chromosomes to move about.

A large amount of research has been directed toward the anatomy, morphology, physiology, biochemistry, and biophysics of mammalian cells. In cytogenetics the nucleus and centrioles, and within the nucleus, the chromosomes and nucleoli, have received the most attention.

LIGHT AND ELECTRON MICROSCOPY

The development of cytology and cytogenetics has been closely related to the development of the light microscope and the electron microscope. Since information about the earliest construction of a microscope has been lost, scientific historians have attempted to reconstruct its origins. The spectacle makers Hans Janssen, his son Zacharias, and Hans Lipershey in Holland were involved in the earliest documented cases. Many people who manufactured lenses for spectacles probably first used two convex lenses to magnify an object, and then took the next step of mounting these lenses in a tube, or pair of tubes, so that the object could be brought into focus. The next step in microscope evolution was apparently the light microscope, which was constructed between 1590 and 1609.

In the seventeenth century, many people used and improved the microscope. The most notable were Hooke and Leeuwenhoek. Then in the 1700s a mirror to use transmitted light and a glass table upon which a specimen could be placed, rather than fastened to a point, were introduced. In 1827 von Baer saw and described the mammalian ovum under such a "light microscope."

Achromatic lenses were the next development, important because they eliminated fuzziness and colored edges. These were perfected in the late nineteenth century.

Little additional progress was made until 1934 when F. Zernike developed, and patented with Zeiss, the phase-contrast microscope. This is a specially constructed microscope which contains a phase-shifting ring, whereby small differences in the index of refraction are made visible as intensity or contrast differences in the image. It is particularly useful for examining structural details of transparent specimens, such as living unstained cells or tissue.

Interference microscopy, an improvement over phase microscopy, was the next advance; it was developed by F. H. Smith (Bradbury 1968) in 1947.

Fluorescent microscopy grew out of the knowledge that greater resolution could be obtained with light of shorter wavelengths, although this aspect of fluorescent microscopy has not been pursued. Fluorescent microscopy has been developed, however, to show the details made visible by use of fluorescent stains. The first chromosome banding was done with fluorescent stains or dyes.

In 1928 Knoll and Ruska in Berlin started work on the electron microscope. In 1933 the first prototype was made, and in the 1940s electron microscopes were commercially produced. In somewhat reverse order to light microscopy in which surface features of specimens were observed first and thin sections were observed later, the first electron microscopes transmitted electrons through ultrathin sections. The scanning electron microscope, which defines surfaces three dimensionally, was developed afterward.

The resolving power of an electron microscope can produce sharp images of objects 1000 times smaller than those that can be seen clearly by a light microscope. This is a function of these shorter electromagnetic radiation waves. The shorter electromagnetic wavelengths are converted to images by a fluorescent screen or a photographic plate, and objects can be magnified more than is physically possible with the longer, visible light waves (Bradbury 1968).

Most cytologists or cytogeneticists using light or electron microscopes do not fully understand all of the physical principles of their operation. A microscope is a tool that permits visualization of the materials, primarily chromosomes, in which they are interested. Although it is not necessary to understand these basic physical principles, the instrument becomes more useful as understanding of its principles increases. As with many other research fields, a team approach in which one technician understands one type of equipment or techniques while another works with a different phase is valuable. For a brief description of the practical use of the light microscope, see Humason (1967) or Bourne (1964), and for a more detailed review of the development of microscopes, read Bradbury (1968).

EARLY HISTORY OF CYTOLOGY

In 1665 Robert Hooke identified cells in thin slices of cork as the basic structural unit in plants. He also noted that in green plants these cells were filled with juices (Bradbury 1968). Chromosomes were first observed by Nägeli in 1842. He called them transitory cytoblasts, a very descriptive term, since they were observable only in some cells and at certain times. Chromosomes could be seen under a light microscope only during mitosis and meiosis; they were not distinguishable at other times.

Late in the nineteenth century the threadlike chromosomes were observed, by the German anatomist Walter Flemming, to split lengthwise and migrate to separate poles of the nucleus, a process called mitosis (*mitos* meaning thread). In 1888 Waldeyer gave the name chromosome (*chromo* meaing stained and *some* meaning body) to these bodies. Each species was found to have a characteristic number of chromosomes and this number was the same in each cell nucleus. Qualitative and morphological differences were noted between chromosomes, so that in several species individual chromosomes could be identified.

In the early 1890s a special type of cell division in sperm and egg cells in animals (and in spores of plants), called meiosis, was recognized. Meiosis differs from mitosis in producing daughter cells with *half* the chromosome number of the parent cell. When an egg and a sperm unite, the total number of chromosomes is again achieved. In this way, chromosome numbers are kept constant between generations.

At the turn of the century homologous chromosomes at meiosis were observed to be in pairs. These discoveries were recognized to be parallel to Mendel's laws, in which genes had been postulated to occur in pairs, and firmly established the relationship between genetics, newly rediscovered, and cytology.

McClung in 1902 and then Wilson in 1905 showed that sex was determined by a chromosomal mechanism. This was the first hereditary characteristic positively associated with a chromosome, thus providing a great impetus to the concept that the chromosomes were the carriers of hereditary material.

Cytological studies in the nineteenth century also established the fact that cells only arise from preexisting cells and that spontaneous generation of cells was not an acceptable hypothesis. During this century fertilization was determined to be a process in which one gamete from each of two parents united to form the basic zygote. However, many scientists still clung tenaciously to the concept of pangenesis for several years.

The concepts of crossing-over during meiosis and the consequent exchange of genetic material between two homologous chromosomes were established in the early years of the twentieth century, and were confirmed both cytologically and genetically. Linkage maps of genes arranged in linear order along the length of the chromosomes became realities.

Two types of mutations were recognized early in the twentieth century: one type was the point mutation of a single gene, which may be called a genetic mutation; the other type, which also was genetic in nature since it resulted in hereditary variation among organisms, was the result of chromosomal modifications which could be observed microscopically. Polyploidy and aneuploidy also were discovered early in the twentieth century.

In the 1930s a flurry of new activity accompanied Müller's discovery of the effects of X-rays in inducing mutations. These efforts also focused attention and interest on the ultrastructure of chromosomes and on the biochemical nature of genes and chromosomes. In 1944 Avery, MacLeod, and McCarty added another step in molecular genetics when they demonstrated that bacterial DNA (deoxyribonucleic acid) appeared to be the material that was the determinant of the hereditary characteristic, permitting transformation of R mutants of pneumococcus to the S wild type. This discovery was met with considerable skepticism, but also helped lay the groundwork leading to the later development of the Watson-Crick model of the DNA structure of chromosomes. Tremendous strides have occurred in the last 10 years in the further understanding of DNA. DNA sequences have been identified, genes have been modified and spliced, genetic material has been inserted into microorganisms to change their function, and most recently genes have been inserted into mammalian genomes (Abelson 1983). Cytogenetics contributed much to the earlier understanding of chromosomes in relation to heredity, and it has now become one of the many techniques to be combined with the newer techniques of molecular biology.

The electron microscope has added much to our knowledge of the ultrastructure of chromosomes, but to date has been used only occasionally in cytogenetics of livestock. Brown (1972) has presented a short summary of these historical concepts, from the early light microscope studies through electron microscopy.

REFERENCES

ABELSON, P. H. 1983. Biotechnology: An overview. Science *219,* 611–613.

AVERY, O. T., MACLEOD, C. M., and MCCARTY, M. 1944. Studies on the chemical nature of the substance inducing transformation of pneumococcal types. Induction of transformation by a desoxyribonucleic acid fraction isolated from pneumococcus Type III. J. Exp. Med. *79,* 137.

BOURNE, G. H. 1964. Cytology and Cell Physiology, 3rd Edition. Academic Press, New York.

BRADBURY, S. 1968. The Microscope, Past and Present. Pergamon Press, Oxford.

BROWN, W. V. 1972. Textbook of Cytogenetics. Mosby, St. Louis, MO.

HUMASON, G. L. 1967. Animal Tissue Techniques. Freeman, San Francisco.

LAIPIS, P. J., WILCOX, C. J., and HAUSWIRTH, W. W. 1982. Nucleotide sequence variation in mitochondrial deoxyribonucleic acid from bovine liver. J. Dairy Sci. *65,* 1655–1662.

ADDITIONAL REFERENCES

BROWN, W. V., and BERTKE, E. M. 1974. Textbook of Cytology, 2nd Edition. Mosby, St. Louis, MO.

CLARK, G. L. 1961. The Encyclopedia of Microscopy. Rheinhold Publishing, New York.

DEROBERTIS, E. D. P., NOWINSKI, W. W., and SAEZ, F. A. 1968. Cell Biology. Saunders, Philadelphia, PA.

MEEK, G. A. 1976. Practical Electron Microscopy for Biologists. Wiley, London.

MOORE, J. A. 1972. Readings in Heredity and Development. Oxford University Press, New York.

SCHULTZ-SCHAEFFERT, J. 1980. Cytogenetics: Plants, Animals, Humans. Springer-Verlag, New York.

STENT, G. 1971. Molecular Genetics, Freeman, San Francisco.

WHITE, M. I. D. 1973. Animal Cytology and Evolution, 3rd Edition. Cambridge University Press, Cambridge, England.

WILSON, S. D. 1967. Applied and Experimental Microscopy. Burgess Publishing, Minneapolis, MN.

2

Cell Division: Mitosis

Through the process of mitosis plants and animals are able to grow by multiplying their somatic cells while still maintaining the typical number of chromosomes characteristic of the species in each cell. Somatic cells are all the cells in the body other than those in the germinal tissues, the testis or ovary. This process assures that in each of the cells at the time of division, identical, or nearly identical, hereditary material will be transmitted into each cell.

CELL DIVISION

Cell division is a cyclical and repeatable process which can be described diagrammatically as a circle (Fig. 2.1). For convenience in description this circle can be broken arbitrarily so that the first part of the cell division cycle is designated as interphase. Sometimes this phase is called the resting stage, but that is a misleading term, for "resting" intimates that the cell is inactive, which is not true. Most of the life of a cell is spent in interphase and considerable metabolic activity is characteristic of the cell at this time. Only a small part of the lifetime of a cell is spent in mitosis. Some cells live for a year or considerably longer before going through cell division while other types of cells divide much more frequently. Interphase can be described as the major time of metabolic or physiological activity in the cell. During interphase new chromosomes are being synthesized and each cell is proceeding with the specific activity characteristic of the organ, or site which it composes (for example, the cells lining the alveoli of the mammary gland are secreting milk if the female is

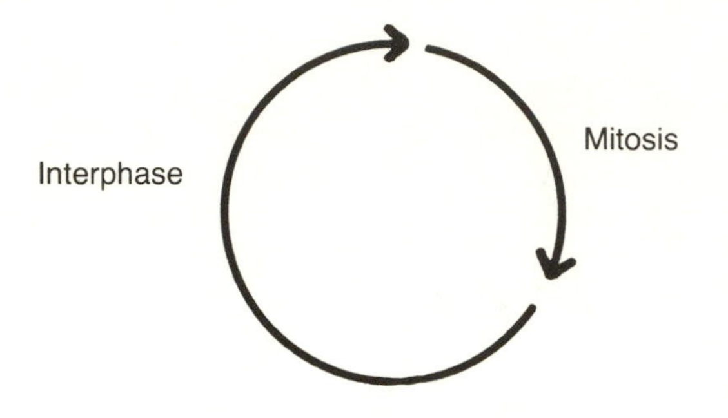

FIG. 2.1. Schematic illustration of the sequence of events occurring in the life of a cell, resulting in two identical cells at the completion of mitosis.

lactating; the muscle cells, as well as some other types of cells, are metabolizing carbohydrates into energy). During interphase the nucleotides, which are the building blocks of the chromosomes, are being synthesized and collected in the nucleus in preparation for the formation of the new chromatids, the two daughter strands of a chromosome. The simple circle illustrated in Figure 2.1 can be further subdivided as shown in Figure 2.2.

SYNTHESIS

The G_1 stage is the time immediately after completion of mitosis when the chromatids are still single but are made up of two sub-

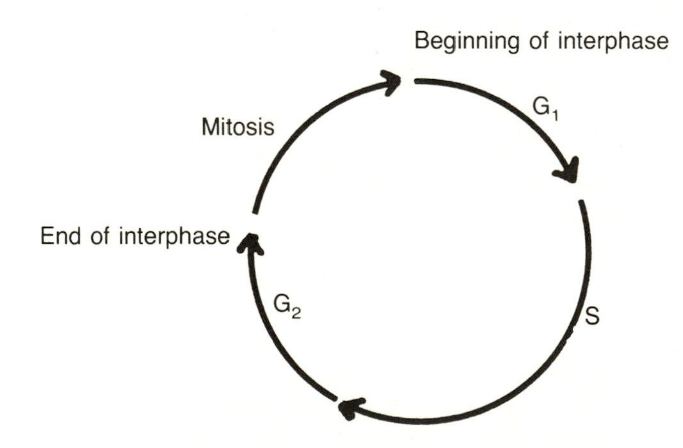

FIG. 2.2. Schematic illustration of the cell cycle including Gap 1(G_1), S (DNA synthesis), and Gap 2 (G_2)

units. In Figure 2.2 G_1 designates the "gap" of time between telophase and synthesis of nucleotides and the S stands for synthesis (DNA synthesis), where each subunit of the chromatid acts as a template, or model, against which the two new subunits are actually formed.

Synthesis starts simultaneously at many points along the length of each chromatid during the S period of interphase. Subunits of the newly formed chromatid then are joined so that by the completion of synthesis two new chromatids, identical to the original, have been formed. During the following mitosis the two original chromatids of the chromosome, each with its newly formed chromatid, separate to form the "arms," which are visible through the light microscope, and then finally separate at the centromere as they migrate to the opposite poles. The sequence of steps in the synthesis of the new chromatid is described in detail in textbooks on biochemical genetics [e.g., see Stent (1971) and Swanson and Webster (1977)].

The time period following the completion of synthesis (or S stage) until the beginning of mitosis is called G_2; or "Gap 2." Metabolic activities other than synthesis of the new chromatids presumably continue at this stage, similar to those activities occurring during G_1. The length of time the cell spends in each of these stages varies between organisms and between different types of cells within organisms, and in cell culture is also influenced by temperature, nutrient availability, and other environmental variables. Cells in vitro (in tissue culture outside an organism) may have different time schedules than cells in vivo (inside a living organism) for the stages that occur during interphase.

How can these stages—G_1, S, and G_2—be determined? The amount of DNA in a diploid cell has been measured at increasing time intervals after telophase, the final stage of mitosis. For cells from any specific source, the amount of DNA will be constant for a time, 2C, which is the designation for this amount. During the time that the amount of DNA remains constant at the 2C level, the nucleus is in the G_1 stage. The S period begins when the amount of DNA abruptly starts to increase. When the 4C level of DNA is reached, indicating the doubling of DNA in the cell, the S period is completed. This level then remains constant throughout the G_2 stage, until mitosis begins. In order to make these measurements in culture, cell division is first synchronized so that large numbers of cells can be obtained at each stage for measuring DNA quantity (Brown and Bertke 1974).

The major basic protein of the chromosome is histone. There is a constant quantitative relationship between DNA and histone in any specific type of cell in culture. During the S period there is also a doubling of the amount of histone.

Another method for observing the S stage is by autoradiographic methods. Radioactive thymidine (tritiated thymidine) can be added to the culture media at different stages of culture (Swanson and Webster 1977, pp. 173–178). Microscope slides are then made by the usual method. By placing photographic film over the slides which have had cells placed upon them and leaving the slides for a period of time, up to several days, the film will record the radiation emitted from the chromatid which has been synthesized utilizing the radioactive thymidine. The same slides then can be stained and photographed.

The following chart shows the amount of time that mouse cells in culture spend in each stage.

G_1	S	G_2	Prophase	Metaphase	Anaphase	Telophase	Total
9.5 hr	7.5 hr	1 hr	18 min	6 min	8 min	8.5 min	18.7 hr

The beginning of mitosis marks the end of the G_2 phase. Mitosis itself is divided into four stages: prophase, metaphase, anaphase, and telophase. To assist in remembering these stages of mitosis, the Greek root words are helpful: *pro* means prior to; *meta* means be-

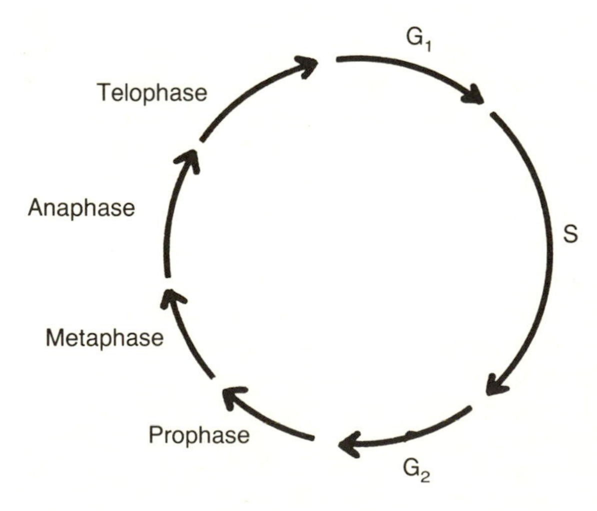

FIG. 2.3. Schematic illustration of the cell cycle including the four phases of cell division, mitosis, resulting in two cells following telophase.

tween or middle; *ana* means upward, toward the poles; *telo* means the last or final.

STAGES OF MITOSIS

The first step after interphase, or the first stage of mitosis, is called prophase, the time when the very elongated chromatin material starts to condense and individual chromosomes can begin to be recognized. The precise time of beginning of prophase is difficult to establish. Each chromosome has split into two parts, the old "chromosome" and the newly synthesized "chromosome." Since the term "chromosome" is used to designate a single entity, these two parts are called chromatids. They cannot be distinguished as being double at prophase under a light microscope, but together look like one chromosome. In early prophase the chromosomes are so long and tangled that individual chromosomes cannot be identified. By late prophase the chromosomes have shortened enough that at least some can be seen to be separate, but they are still long strands. In species with very small numbers of chromosomes, each chromosome may be distinguished, but in cattle, with 60 chromosomes, or other livestock (all of which have many chromosomes), even in late prophase there may be much overlapping and intertwining of chromosomes.

The time at which prophase ends and metaphase begins is not a clearly defined break since mitosis is a continuous process. When the chromosomes start to line up on the equatorial plate of the nucleus, metaphase has started. If a cell nucleus happens to be at a right angle the chromosomes can be observed to line up across the center, or the median plate of the nucleus. This can be seen in a time-lapse, phase-contrast film of mitosis. In prophase and in very early metaphase the chromosomes are scattered throughout the nucleus with no particular arrangement. As metaphase is approached the chromosomes appear to be bouncing around in the nucleus. They then line up like a squad of soldiers that have been called to attention. They arrange themselves on the equatorial plate and remain there with some random motion, but they are accurately and carefully lined up on the plate.

Outside of the nucleus in animal cells is a body called the centrosome, which at some time during interphase divides into two bodies, each of which is called a centriole. At prophase the centrosome can be seen to be divided into two centrioles and as mitosis progresses each centriole travels around the nucleus so there is one centriole at each pole. A line drawn from one centriole to the other would be at a right angle to the equatorial plate. In the region of the centriole, spindle fibers (microtubules) begin to develop. The spindle fibers apparently do not come from the centrioles but are organized by them, although it is not known exactly how this occurs. About this time the nuclear membrane also disappears. As the spindle fibers develop into the nucleus they apparently cause the chromosomes to become organized into the metaphase plate. Each of the spindle fibers becomes attached to the centromere—the point on the chromosome where the two chromatids remain attached. The spindle fibers appear to be pulling against each other since each centromere has not separated into two bodies at this time. The nucleus is prepared to proceed into the next stage of division after the chromosomes are lined up on the metaphase plate. The separation of each centromere into two bodies signals the end of metaphase.

The next stage is anaphase. If the centromere is located near one end of the chromosome the chromosomes will appear to be a J shape during anaphase. If it is located near the middle it tends to be a V shape. If it is located at the end, the chromosomes during anaphase are rod-shaped. Ultrastructural studies of chromosomes have not explained fully the mechanical forces under which the chromosomes move to the poles during anaphase. It is not certain whether the centromere provides the force which causes each half to go toward the poles from the equatorial plate or whether the spindle fiber pulls. Some force causes the separation, a tendency either for a centromeres to migrate in opposite directions or for the spindle fibers to pull the centromeres apart, with the remainder of the chromosome following.

Each of the split chromosomes (chromatids) goes to opposite poles. Telophase is the term used to describe the stage of mitosis when the chromosomes have reached the poles. The chromatin material becomes clumped near each pole so that individual chromosomes cannot be distinguished. A "pinching in" of the middle of the nucleus then begins and the final step of cell division, the formation of two cells, occurs. This final stage is called cytokinesis.

The formation of a new cell membrane between the two cells completes the cell division. From telophase each of the two cells goes into the G_1 stage of interphase.

REFERENCES

BROWN, W. V., and BERTKE, E. M. 1974. Textbook of Cytology, 2nd Edition. Mosby, St. Louis, MO.

STENT, G. S. 1971. Molecular Genetics. Freeman, San Francisco.

SWANSON, C. P., and WEBSTER, P. H. 1977. The Cell, 4th Edition. Prentice-Hall, Englewood Cliffs, NJ.

ADDITIONAL REFERENCE

SCHULZ-SCHAEFFER, J. 1980. Cytogenetics: Plants, Animals, Human. Springer-Verlag, New York.

3

Cell Division: Meiosis

INTRODUCTION

Meiosis is a specialized form of cell division that differs in several respects from mitosis. Mitosis is the process of cell division that occurs in the somatic cells of the body. Mitosis preserves the $2n$ or diploid number of chromosomes as the cells divide. Meiosis, on the other hand, provides for a reduction of the number of chromosomes in a cell from the $2n$ or diploid number to the n or haploid number. Meiosis may be defined simply as two successive nuclear divisions accompanied, or preceded, by one replication of the chromosomes. A reduction division is necessary because at fertilization the female gamete and male gamete fuse into one cell with one nucleus. If the number of chromosomes in each gamete were to remain the same as in the somatic cells, then every generation would have double the number of chromosomes of the previous generation. It is a mechanism that has led to much thought in terms of evolutionary development, and there are many factors that are still unknown. It is not possible to recover the organisms which were the earliest forms of life, or the early descendants of these primary forms. Therefore, the evolutionary development of the mechanical and biochemical forces occurring during meiosis can only be hypothesized.

Van Beneden in 1889–1894 was the first person to demonstrate that an equal number of chromosomes is contributed by each parent. This preceded the rediscovery of Mendel's work. Weissman in 1887 developed the hypothesis that a reduction division had to occur someplace in the germinal tissue.

From the standpoint of cytogenetics all cells in the body can be divided into two classes: somatic and germinal. Somatic means all of the cells in the body other than those in the specialized tissue that creates the gametes. Germinal cells are the reproductive cells involved in gamete production, in either the ovary or the testis. Somatic cells have the $2n$ or diploid number of chromosomes. The germinal tissue has both cells with $2n$ and n chromosomes since both mitosis and meiosis are occurring.

STAGES OF MEIOSIS

In prophase of meiosis several complex changes occur which distinguish meiosis from mitosis. The stages within prophase of the first meiotic division, M-I, are leptotene (chromosomes appearing as thin threads), zygotene (yolked thread), pachytene (thick thread), diplotene (double thread), and diakinesis, the terminal stage. These stages are sometimes referred to as leptonema, zygonema, pachynema, and diplonema. Prophase is followed sequentially by metaphase, anaphase, and telophase, similar to mitosis. Then, after the nucleus proceeds into a stage called interkinesis the second division proceeds with prophase II, metaphase II, anaphase II, and telophase II. In the male, four spermatids result from meiosis. In the female, however, the first meiotic division results in nuclei of unequal size, the smaller being designated as a polar body. The second meiotic division in the female results in one more polar body and the ovum. The first polar body may or may not divide at the second meiotic division (Sybenga 1975).

Leptotene is the stage when the chromatin network starts to become more distinctive chromosome threads. Frequently, small, darkly stained dots, chromomeres, can be seen along these threads. Some cytologists have identified a stage just preceding leptotene as preleptotene. The synthesis of the new chromatids occurs at this time and/or during late interphase, but, different from mitosis, some DNA synthesis may be occurring as late as pachytene. The chromosomes at leptotene, however, usually appear to be single strands, so their replication has only been determined biochemically. The chromosomes at this stage are not scattered randomly in the nucleus, but seem to be attached by their telomeric ends to the nuclear envelope to form what is described as a bouquet.

The beginning of zygotene and the end of leptotene is the time when the homologous chromosomes begin to form pairs, a process called synapsis. This usually starts at one end of the chromosomes, and the process has been described to be similar to zipping a zipper. The pairing is very precise, occurring between homologous parts of the chromosomes. Synapsis is a very intimate pairing of the chromosomes, and it can be seen in considerable detail with an electron microscope. The region of intimate pairing of the chromosomes is the synaptinemal complex.

As soon as synapsis is complete, the next stage, pachytene, begins. Condensation of the chromosomes, which causes them to become shorter, continues through this stage. The chromosomes are now called bivalents, since each body consists of two chromosomes. It is during this stage that breaks occur in the chromosomes followed by fusion between the homologous chromosomes, so that any one chromosome may consist of parts of the maternal and parts of the paternal chromosomes. This is called crossing-over. The places where fusion has occurred can be observed under the light microscope as chiasmata, along the length of the homologous pair. Each chiasma is a cross-over point. Each bivalent along its entire length is composed of four chromatids. Exchanges may occur between two nonsister chromatids at this time, but the cytological evidence is apparent when chiasmata are clearly seen at anaphase I.

Diplotene starts when the chromosomes start to separate. Several configurations occur at this time, influenced by the crossing-over that has occurred, the cross-overs being visible as chiasmata. They may be X-shaped, or rings, or in longer chromosomes like chains in links. The bivalents continue to shorten in the diplotene stage, but in early diplotene the outlines of the chromosomes usually are quite fuzzy. As diplotene progresses the outlines of the bivalents become clearer and sharper. The number of chiasmata decreases as the nucleus proceeds through diplotene because of terminalization of the chiasmata. Terminalization appears to be the moving of the chiasmata away from the centromere toward the telomeres as the synaptinemal complex disintegrates. This is a mechanical process that allows the chromatids with cross-overs to become disentangled by slipping out over the ends of the chromosomes.

Diakinesis is the endpoint of diplotene. The bivalents reach their maximum shortening and may move out to the periphery of the nucleus.

The first metaphase starts with the breakdown of the nuclear membrane and the beginning of the formation of a spindle. The bivalents become oriented on the metaphase plate in such a way that the two centromeres for each are located on opposite sides of the equator so that a spindle fiber from one pole becomes attached to one centromere and a spindle fiber from the other pole is attached to the homologous centromere. The nucleolus also disappears at this stage. By electron microscopy it has been observed that each chromatid is formed of a spiral thread, and that each thread is also in a spiral form. These may be called major and minor spirals, and conform to the Watson-Crick model of DNA structure. Near the centromere the chromosome (sometimes called a dyad) frequently shows the effect of the "pulling force" of the spindle fiber by becoming quite long and thin. The forces that result in the separation of the dyads from each other are not fully understood, but this stretched appearance leads to the hypothesis that the spindle fibers do indeed pull, as contrasted with a "repelling" force between the homologous centromeres sometimes postulated for this separation.

At anaphase I the undivided centromeres for each dyad are moved toward opposite poles. The separate chromatids in each dyad become quite evident at this stage. The acrocentric chromsomes have two arms and appear as V's, and the metacentric chromosomes have four arms extending from the centromeres in the shape of X's. The chiasmata become completely terminalized as the separation occurs. This division can be referred to as the reductional division with respect to the centromeres and chromatids immediately adjacent. The area of the chromatid immediately after the first chiasma, moving away from the centromere, would actually be an equational division since non-sister chromatids have been exchanged. Therefore, the entire process of the two meiotic divisions cannot be identified as reductional or equational, only the centromere and adjacent chromatid being equational. Since the orientation of the centromeres of the bivalents on the metaphase plate is random with respect to paternal or maternal origin, anaphase results in a random distribution of chromsomes to the resulting cells. Crossing-over increases the randomness of this assortment of blocks of genes with respect to maternal or paternal origin.

When the dyads have been completely separated from their homologues, telophase has been reached. In animal cells, generally telophase I may proceed directly into prophase II, but in some species the dyads may enter a phase called interkinesis in which the chromatin material loses its condensed appearance, becoming more diffuse. A major difference between mitosis and meiosis is that there is no synthesis of new chromatids in interkinesis.

Metaphase II, anaphase II, and telophase II proceed as in mitosis.

GAMETOGENESIS

Spermatogenesis occurs within the seminiferous tubules of the testis. The process starts at puberty and is continuous throughout the reproductive life of the animal. Lying along the basal membrane of the seminiferous tubules are type A spermatogonia which reproduce by mitosis to produce additional type A spermatogonia and type B spermatogonia. The type B spermatogonia by cell division produce cells which are primary spermatocytes and which are found nearer the lumen of the tubule. The first meiotic division occurs in the primary spermatocytes, resulting in secondary spermatocytes. The secondary spermatocytes divide during the second meiotic division to produce spermatids. The spermatids, possessing a haploid set of chromosomes, proceed through a maturation process into spermatozoa which then move out of the seminiferous tubules into the epididymis for further maturation before ejaculation and fertilization of the ovum. The total time from primary spermatocyte to ejaculation in the bull is approximately 48 days (Salisbury *et al.* 1978).

In mammalian females the proliferation of the oogonia, the cells that produce the oocytes, is completed during intrauterine life. All of the ova produced are derived from oocytes which are already present at birth (Balinsky 1975).

The number of oocytes in mammals actually decreases with age. In the 3-month-old heifer there are about 75,000, but by 1½ to 3 years of age the number has been reduced to 21,000, and in old cows to 2500.

The precise timing of the meiotic divisions in relationship to the maturation of the oocyte, ovulation, and fertilization varies among species of livestock (Hafez 1974; Salisbury *et al.* 1978). The oocyte is usually in pachytene, diplotene, or diakinesis of prophase I during diestrus. In cattle, sheep, and swine at ovulation the first polar body has already been formed. The horse and dog are in the process of the first meiotic division, but the first polar body has not yet been formed at ovulation. Thus, the horse and dog ovulate primary oocytes, but cattle, sheep, and swine ovulate secondary oocytes. As the ovum matures in the graafian follicle the nucleus becomes enlarged, with an increase of nucleoplasm to form the germinal vesicle, a name describing the enlarged nucleus. The level of luteinizing hormone (LH) is associated with the completion of the first meiotic division. There are some important biochemical reactions that occur in the oocyte as the nuclear membrane disappears and the nucleoplasm becomes mixed with the cytoplasm at metaphase I.

The chromosomes migrate to one edge of the oocyte where the spindle is formed vertical to the oocyte cell membrane. As the chromosomes (bivalents) proceed through metaphase I, anaphase I, and telophase I, a bulge appears in the oocyte membrane and one set of chromosomes (dyads) is extruded into this bulge. As it pinches off, it becomes the first polar body.

The second meiotic division is initiated by fertilization which occurs in the lower portion of the ampulla of the oviduct (Hafez 1974). The second polar body is then formed and extruded, the remaining haploid set of chromosomes becoming the maternal pronucleus. The single sperm that has penetrated the zona pellucida and vitellus, which surrounds the oocyte, then becomes the paternal pronucleus. In the pronuclei the chromosomes elongate and form a chromatin network. The two pronuclei enlarge and after about 15 hours the two pronuclei coalesce. Prophase actually begins before coalescence, so the two sets of chromosomes can be seen. The two groups then become one and the first mitotic division of the new zygote proceeds.

REFERENCES

BALINSKY, B. I. 1975. An Introduction to Embryology, 4th Edition. Saunders, Philadelphia, PA.

HAFEZ, E. S. E. 1974. Reproduction in Farm Animals, 3rd Edition. Lea & Febiger, Philadelphia, PA.

SALISBURY, G. W., VANDEMARK, N. L., and LODGE, J. R. 1978. Physiology of Reproduction and Artificial Insemination of Cattle, 2nd Edition. Freeman, San Francisco.

SYBENGA, J. 1975. Meiotic Configurations. Springer-Verlag, New York.

4

Chromosome Number and Morphology

INTRODUCTION

Each animal species normally has a characteristic number of chromosomes and each pair has a characteristic form or shape. Variations from this numerical and morphological pattern occur, but such aberrations have a low frequency in the general population. Hsu and Benirschke (1967) have published an atlas of the normal number and morphology of chromosomes of a large number of mammalian species.

"Chromosome numbers (haploid) range from one in the horse round worm *Parascaris equorum* var. *univalens* to about $n = 630$ in the fern *Ophioglossum reticulatum*" (White 1973A). In animals the lycaenid butterfly *Lysandra atlantica* has the largest number, $n = 223$.

Among mammals the Indian muntjac deer *Muntiacus muntjac* has the smallest $2n$ number, 7 and 6, with the Y chromosomes being divided into two. Among common mammals, the dog *Canis familiaris* has one of the larger numbers, $2n = 78$ (White 1973A). Chromosome numbers of other common domestic animals are listed in Table 4.1.

CHROMOSOME MORPHOLOGY

There are two major types of chromosomes based upon morphology at mitotic metaphase. Acrocentric chromosomes are those

TABLE 4.1. Chromosome Numbers of Domestic Animals

Animal	Number
Cattle (*Bos taurus, Bos indicus*)	60
Swine (*Sus scrofa*)	38
Sheep (*Ovis aries*)	54
Horse (*Equus caballus*)	64
Ass (*Equus asinus*)	62
Mule (hinny, hybrids of horse and ass)	63
Goat (*Capra hircus*)	60
Human (*Homo sapiens*)	46
Chicken (*Gallus domesticus*)	78
Turkey (*Meleagris gallopavo*)	82
Pigeon (*Columba livia*)	80
Duck (*Anas platyrhyncha*)	80
Rabbit (*Oryctolagus cuniculus*)	44
Dog (*Canis familiaris*)	78
Cat (*Felis catus*)	38
Murrah buffalo (*Bubalis bubalis*)	50
Asiatic swamp buffalo (*Bubalis bubalis*)	48
Mink (*Mustela vison*)	30
Elephant (*Elephas maximus, Loxodonta africana*)	56
Camel (*Camelus bactrianus*)	74
Reindeer (*Rangifer tarandus*)	70
Musk ox (*Oribus muschatus*)	48

in which the centromere is located very near the end of the chromosome. The term telocentric is used to indicate that the centromere is located right at the end of the chromosome. There is some argument among cytogeneticists about whether telocentric chromosomes really exist. There is evidence to indicate that telocentric chromosomes do exist, but even though the centromere appears to be located at the end there may be a minute amount of chromatin on the other side of the centromere. It may be such a small amount that it is very difficult to detect whether it is actually there. White (1973B) presented a detailed discussion of this controversy, and John and Freeman (1975) reported a divergent view on this same topic. The term acrocentric may include telocentric chromosomes. If there is not enough chromatin opposite the long arms to see any separation of chromatids, it is called acrocentric. If any separation of chromatids occurs in this material as it is seen under a light microscope, then it may be called subtelocentric (see Fig. 4.1).

The other major type of chromosome is metacentric, *meta* meaning that the centromere is located in the middle. Since it is not always located directly in the middle, this classification is some-

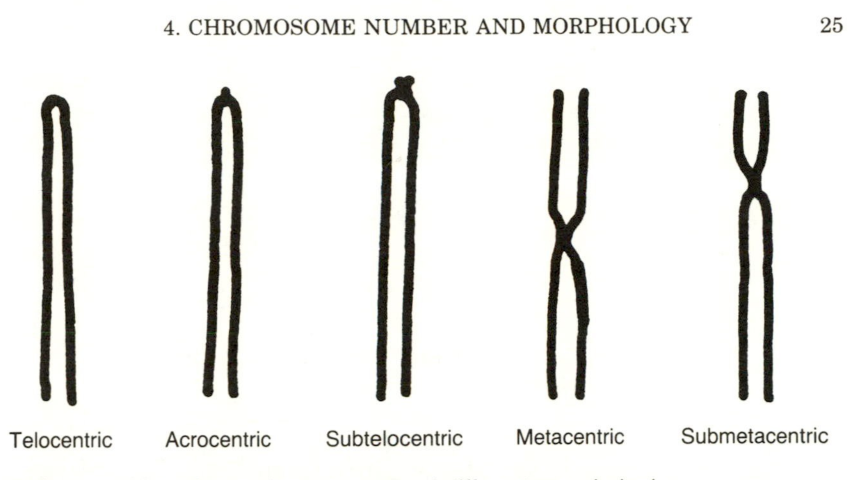

FIG. 4.1. Metaphase chromosomes of different morphologies.

times modified to include the term submetacentric, which means that the centromere is almost in the middle. Usually the terms metacentric, submetacentric, and acrocentric are sufficient to identify the major types of animal chromosomes by morphology as they appear in mitotic metaphase. When chromosomes are cut from a photograph and arranged in pairs in a karyotype, the submetacentric chromosomes are placed so that the short arms, designated p, are at the top, and the long arms, designated q, are at the bottom. Usually in livestock the metacentric and submetacentric chromosomes are placed first, and may or may not be subdivided into the two groups in the top lines of chromosomes; the acrocentric chromosomes follow. In both groups, the chromosomes are arranged from longest to shortest, and are numbered from 1 as the longest metacentric or submetacentric through the last number, followed by the sex chromosomes. See Chapter 6 for karyotypes.

The appearance of chromosomes at anaphase was described in Chapter 2. The J-shaped chromosomes would be submetacentric, the V-shaped would be metacentric, and the rod-shaped acrocentric. In early cytological work these shapes at anaphase were most important because observations were based almost totally on microtome sections and the characteristic shapes could only be seen in early anaphase.

When observing cultured cells at metaphase through a microscope the location of the centromere is easily seen, except at early metaphase when the chromsomes are very long and slender (see Fig.

FIG. 4.2. Acrocentric chromosome at metaphase, with satellite.

4.1). The centromeric region also is called the primary constriction. The terminology may vary depending upon the specific cells used for illustration.

A secondary constriction is found in many chromosomes and is recognized very widely in humans, where it permits the identification of certain chromosomes. Human chromosomes 13, 14, 15, 20, and 21 have secondary constrictions, which usually are located near the ends of the chromosomes. The knob located distally to the constriction is called a satellite. Depending upon how the chromosome preparations look, a small thread connects the satellite with the chromosome. The satellites are seldom broken loose, so they are definitely associated with certain chromosomes (see Fig. 4.2). They also are recognized in humans to be the nucleolar organizer regions (NOR) on the chromosomes. The nucleolus in the nucleus of the cell is attached to, and formed by, the NOR on the chromosomes. The NOR is also involved in the synthesis of the nucleic acid portion of the nucleolus and the formation of ribosomes. In livestock, the NORs are generally found at the telomeric end of the chromosomes. In livestock, cattle, sheep, goats, asses, pigs, and horses, satellites are not normally found, except for one chromosome in the horse (the horse's chromosome 21 might be described as having a satellite).

The number 1 chromosome in cattle, the longest one, is rather easily distinguished. It is longer than each of the other acrocentric chromosomes, such that it can usually be identified under the microscope without karyotyping. In this chromosome the centromere is located near the end, so near that it may be telocentric, but usually all 58 autosomes are called acrocentric. Also, at a point two thirds of the length from the centromere on the number 1 chromosome of cattle there is a slight secondary constriction. This term, secondary constriction, does not carry the same connotation as in the human chromosomes because it is not a NOR, nor is the remainder of the chromosome a satellite.

After locating the centromere, the next characteristic used to distinguish chromosomes in terms of morphology is length. A karyotype of the chromosomes of a cell is made by arranging the photographic cut-outs of the chromosomes into accepted groups. After dividing them into groups by centromeric location, they are arranged in pairs starting with the longest. In cattle, *Bos taurus,* only the sex chromosomes, X and Y, are submetacentric, all autosomes being acrocentric. In *Bos indicus* the Y chromosome also is acrocentric. The autosomes, which are all the chromosomes other than the sex chromosomes, are then paired up by lengths and arranged in pairs from the longest to the shortest. The exactness of the pairing of nonbanded chromosomes of cattle is questionable because many of the pairs are so near the same length. Precise pairing can be done only with banded chromosomes, which will be discussed in Chapter 6.

Meiotic chromosomes differ in appearance from mitotic chromosomes; their morphology has been discussed in Chapter 3.

REFERENCES

HSU, T. C., and BENIRSCHKE, K. 1967. An Atlas of Mammalian Chromosomes. Springer-Verlag, New York.

JOHN, B., and FREEMAN, M. 1975. Causes and consequences of Robertsonian exchanges. Chromosoma *52* (2), 123–136.

WHITE, M. J. D. 1973A. The Chromosomes. Chapman & Hall, London.

WHITE, M. J. D. 1973B. Animal Cytology and Evolution. Cambridge University Press, Cambridge, England.

5

Chromosomal Aberrations

INTRODUCTION

One of the most remarkable attributes of biological organisms is their stability from one generation to the next, considering the numerous opportunities for deviations to occur. In spite of the tremendously complex nature of mammals, physiologically and genetically, the offspring resulting from reproduction are generally normal; that is, they are capable of living a normal physiological life including the reproduction of another generation, and have the usual set of chromosomes and genes. However, if all organisms were normal, or identical within species, there would be no scientific discipline of genetics or cytogenetics. The deviations from normal provide the basic data upon which research can proceed to uncover the reasons for occurrence. These deviations have also provided the basic building blocks for evolutionary change.

The morphology of chromosomes, mitosis, meiosis, and chromosome numbers (discussed in the earlier chapters) represent the "normal" situation. Aberrant cytological situations can be divided into two groups:

Modifications of	
Chromosome number	Chromosome structure
Euploid variations	Deletions or deficiencies
Haploidy	Duplications
Polyploidy	Inversions
Aneuploid variations	Translocations
Monosomy	
Trisomy	
Centric fusion or fission	
Supernumerary (B) chromosome	

Each of these aberrations has been found in some plants, microorganisms, insects, lower forms of vertebrates, and humans. Eventually all of these may be found in livestock, but to date only a few types have been identified. These will be discussed in later chapters.

MODIFICATION OF CHROMOSOME NUMBERS

When the somatic cells have chromosome numbers that are exact multiples of the haploid number for the species, the individual is termed euploid. This includes haploids, n; diploids, $2n$; triploids, $3n$; tetraploids, $4n$; and higher multiples. Triploids and above collectively are referred to as polyploids. The term heteroploid sometimes is used to include all types of euploid cells or organisms from haploids to the highest levels of polyploids.

Triploidy has been found in the adult chicken (Bloom 1970) (see Chapter 15) and in the adults of several of the lower forms of vertebrates. Polyploidy and haploidy have been found in early embryos of many species of vertebrates including cattle, but except for poultry have not been reported yet in livestock at birth. There are many examples of polyploidy in the plant kingdom, one of the most common being hexaploidy, which is typical of bread wheat varieties. In wheat, polyploidy has permitted the use of monosomics and nullisomics as useful cytological tools for chromosome mapping and breeding. Hexaploid plants missing one or two chromosomes can still be viable since each plant actually possesses three full sets, genomes, of chromosomes. The absence of one chromosome in a *diploid* species nearly always results in death, or a major anatomical or physiological disturbance.

Because tetraploidy in flowering plants often causes larger blossoms it has been widely used in breeding ornamental plants. Since triploidy usually results in sterility, this characteristic has been used in producing seedless watermelons. Diploid and tetraploid varieties of watermelon are cross-fertilized to produce the sterile triploid fruit. Although living polyploid cattle have not yet been reported, nearly all lymphocyte cultures have a few polyploid cells. These have sometimes been thought to be the result of colchicine treatment, but the presence of tetraploid cells in testicular tissue, untreated with colchicine, leads to the conclusion that in cattle they are natural phenomena. Perhaps one reason polyploid individuals are not found in mammals is that sexual reproduction is the exclusive process for producing offspring. If polyploidy occurred the animals would have various combinations of sex chromosomes (XXXX, XXXY, XXYY, XYYY, YYYY), and sterility or inviability almost always accompanies these combinations of sex chromosomes in animals.

When the cell nuclei of an organism have the number of chromosomes which is not an exact multiple of the haploid number for the species they are called aneuploid. An animal that has one more chromosome than usual ($2n + 1$), or one less ($2n - 1$), would be aneuploid. It would be called aneuploid for any number differing from exact multiples of the haploid, generally stated as $2n \pm x$. If for some reason the two homologues do not synapse during zygotene, and reach metaphase I of meiosis unpaired, then both homologues may go to one pole. This would result in one pair of gametes having one extra chromosome and the other pair missing one chromosome. A zygote formed from such a gamete which united at fertilization with a gamete containing the normal haploid number would be monosomic ($2n - 1$), having only one of one pair of chromosomes, or trisomic ($2n + 1$), having three homologous chromosomes.

ROBERTSONIAN TRANSLOCATIONS

Another type of aneuploidy occurs as the result of centric fusion or its opposite, centric fission. If a break occurs near the centromere of two nonhomologous acrocentric or telocentric chromosomes in one nucleus, the two may become fused to each other at these breaks. The result is a metacentric or submetacentric chromosome, depending upon the relative length of the two acrocentric chromosomes which were involved in the fusion. If this fusion occurred in the primordial germ cells or during meiosis some of the gametes which resulted would have one chromosome less than the n number. If these were successful in fertilization the resulting offspring would have one less chromosome than normal, but would have essentially all of the chromatin material. This type of anomaly can be called a balanced translocation. Centric fusion is the process for forming the type of chromosome known as a Robertsonian translocation.

There are three different ways in which Robertsonian translocations (centric fusions) have been postulated to occur (Fig. 5.1).

One way (A) in which fusion could occur would be for one of the break points to be on the long-arm side of the centromere, the other break point to be as in the previous example. This would result in one long submetacentric or metacentric chromosome with a single centromere (monocentric), located to one side of the fusion point. The other centromere with a minute amount of chromatin would probably become lost or discarded in subsequent cell divisions, with presumably no serious loss of genetic material if the animal remained viable.

A second was for a break to occur in each of two nonhomologous acrocentric chromosomes at a point opposite the long arm of the chromosome but leaving the centromeres in each chromosome intact (B). Fusion of these two chromosomes at their break points would result in a dicentric chromosome which would be metacentric or submetacentric. The acentric fragments would be lost.

The third way for centric fusion to occur would be for the break points to occur within the centromeric regions (C). Fusion of the long arms would result in a metacentric or submetacentric chromosome with a single centromere (monocentric) probably larger than normal, and extending across the point of fusion. The small fragments would be lost (Eldridge and Balakrishnan 1977; John and Freeman 1975).

The Robertsonian translocation is the most frequent type of chromosomal anomaly found in cattle and sheep. At least 24 different Robertsonian translocations have been found in cattle, and numerous other species have had Robertsonian translocations reported. Specific cases are discussed in the chapters on chromosomes of different livestock species. There is considerable speculation concerning the role of Robertsonian translocations in the evolutionary development of species.

The term bivalent is used to describe the synapsed chromosomes at zygotene and later stages of prophase I in meiosis. In animals heterozygous for a Robertsonian translocation three chromosomes synapse into one body which is then called a trivalent. Heterozygous in this case means an animal in which all of the nuclei have one Robertsonian translocation chromosome and the two nonfused homologues. In anaphase I of animals heterozygous for a Robertsonian translocation, three types of segregation of the trivalent are possi-

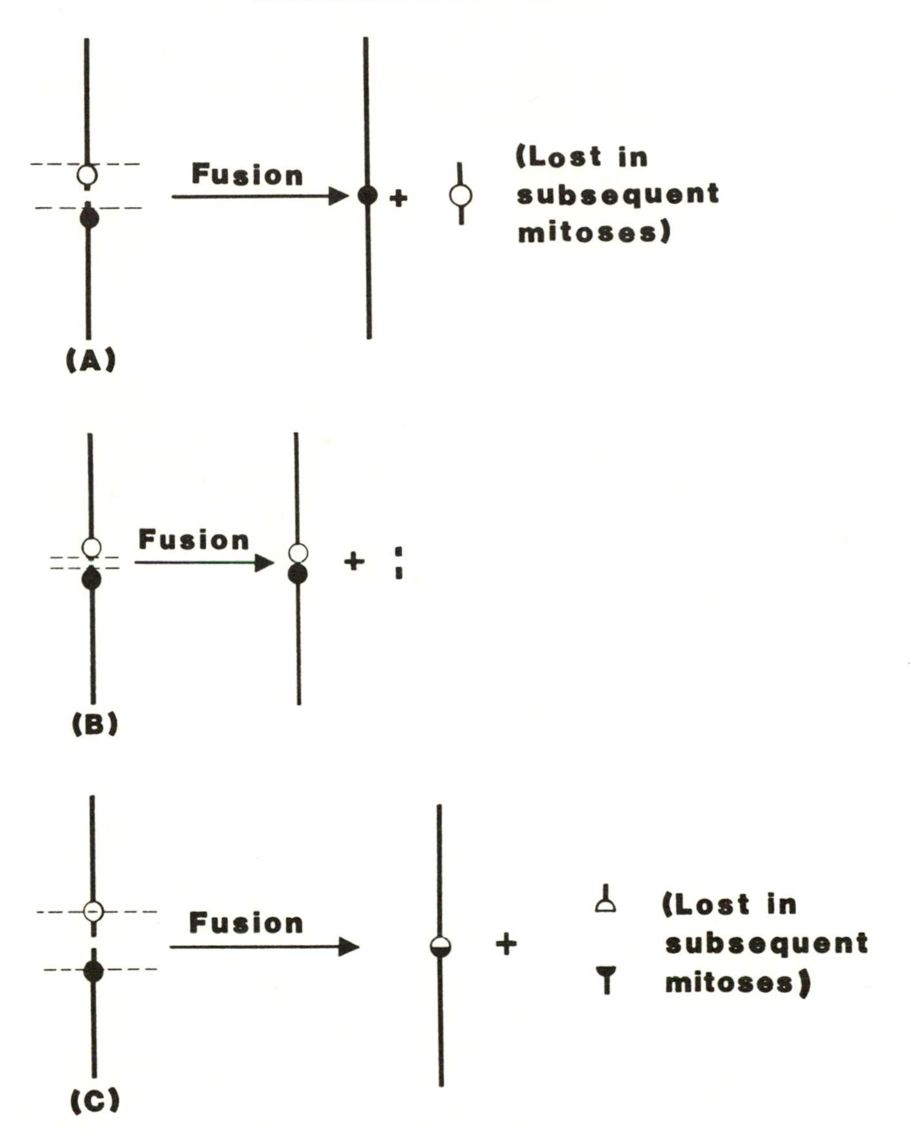

FIG. 5.1. Three mechanisms postulated to explain the formation of a Robertsonian translocation. (Dashed lines indicate the locations of breaks.) (A) The breaks as indicated result in one metacentric chromosome with one centromere and one very small piece with a centromere that is lost. (B) A dicentric chromosome occurs, but with the centromeres so close that they act as one in mitosis. Two fragments are soon lost. (C) The centromeres apparently fuse, perhaps resulting in a larger than normal centromere, and the two fragments are lost.

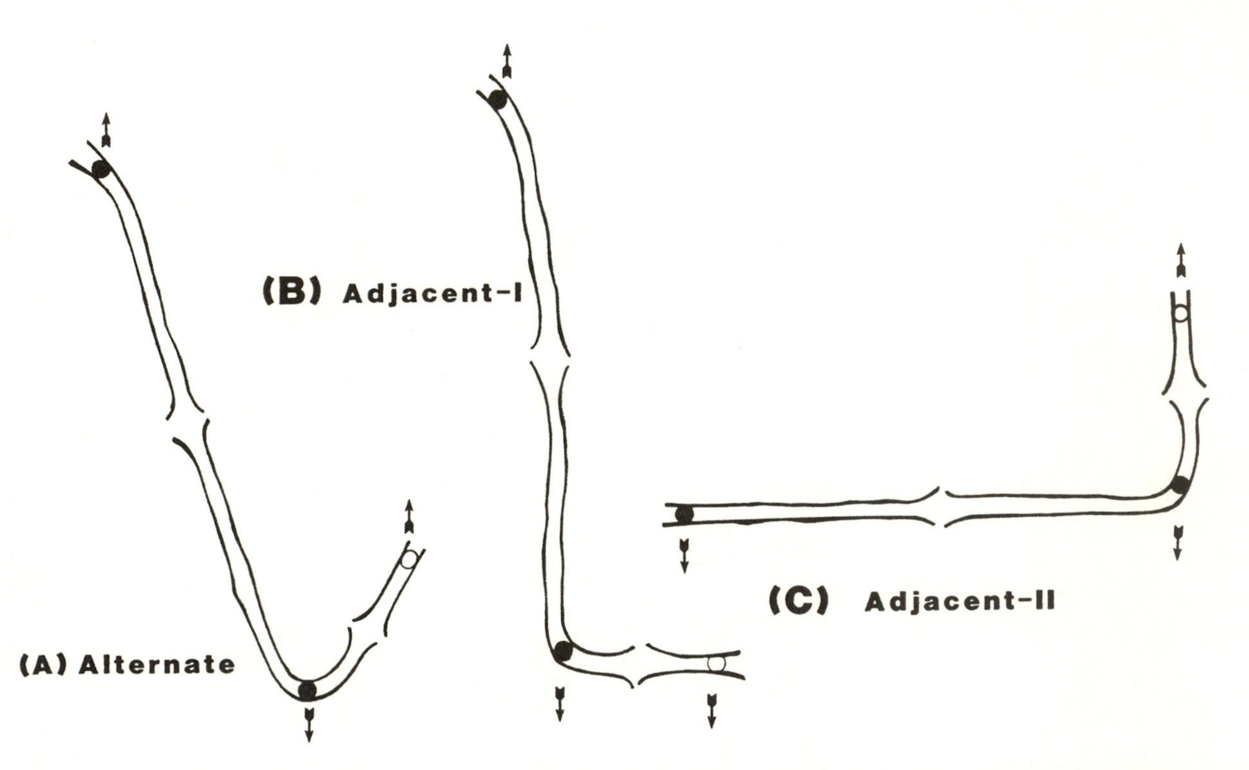

FIG. 5.2. Illustration of anaphase I in animals heterozygous for a Robertsonian translocation. In alternate segregation A the products are balanced and in the male four balanced gametes will be produced. In either adjacent segregation I or II B or C the translocation chromosome goes to the same pole as one of the independent chromosomes in the trivalent, and the resulting cells are either n with a duplication or $n - 1$ with a deficiency. The frequency of occurrence of the different metaphase orientations varies from one Robertsonian translocation to another.

ble. The frequency of each of the three types of segregation depends upon the orientation of the trivalent on the metaphase plate (Fig. 5.2). When the Robertsonian translocation chromosome goes to one pole during anaphase I and the two homologues go to the other pole, four balanced (normal) gametes will be produced (A). If, however, one of the homologues goes to the same pole as the Robertsonian translocation, then in the male two gametes will be n, including a duplication in each, and two gametes will be $n - 1$ with a deficiency in each (B or C). The term nondisjunction is used to describe this situation. If these gametes are involved in fertilization with a gamete from a normal animal, the zygotes will be $2n$ or $2n - 1$. The $2n - 1$ embryo with a deficiency is nearly always inviable, and the $2n$ which has a duplication is frequently inviable also, but the effect of a deficiency upon viability is usually more severe than a duplication. In humans some cases have been reported (Chapman *et al.* 1973) where nondisjunction has occurred involving chromosome 21, and the offspring with a duplication has been affected with Down's syndrome (mongolism). To date in livestock no calves have been found with duplications or deficiencies from heterozygous Robertsonian translocation parents. It is presumed that the embryos die before birth, which phenotypically results in longer calving intervals. Since apparently little or no chromatin material has been lost in the formation of Robertsonian translocations in animals, no phenotypic effects, other than lowered fertility, have been found associated with this type of balanced translocation.

CENTRIC FISSION

The opposite of centric fusion is centric fission. If two acrocentric or telocentric chromosomes can fuse at the centromeric regions, then a break, splitting the centromere in a metacentric or submetacentric chromosome, should also be possible. This has been found in plants and some lower animals but not in mammals.

SUPERNUMERARY CHROMOSOMES

In several species of plants, and a very limited number of animals, supernumerary, or B, chromosomes are found. These are small chromosomes which apparently have centromeres but are not distributed evenly at anaphase I. Some phenotypic effects are associated with them, particularly with the number transmitted, but the effects are variable. Such chromosomes have not been identified in livestock, other than in poultry (see Chapter 14).

MODIFICATIONS OF CHROMOSOME STRUCTURE

Deletions or deficiencies happen when a break occurs in a chromosome and a portion without a centromere is lost. This could be a segment at the telomeric end of the chromosome, but it is generally thought that a broken end of a chromosome cannot function without a telomere. Therefore deletions are usually thought to occur by the loss of an internal segment of a chromosome as the result of two breaks. One break may be extremely close to the telomere. This type of structural change in chromosomes was postulated originally by genetic studies. Its occurrence has been proven by cytological studies with *Drosophila,* maize and other species of plants, and lower forms of animal life. In the giant salivary gland chromosomes of drosophila, deletions can be detected by the absence of certain bands in one half of the chromosome and the lack of pairing at that specific region. It can also be seen at pachytene in maize. Viable deletions usually involve very short segments, since longer segments may contain genes which must occur in duplicate for life and reproduction of the organism. Deletions have been identified in humans (Van Kempen 1975). In livestock, deletions could not be identified prior to the development of banding techniques, which occurred around 1970, and there have not been enough cattle studied with this technique as yet to identify deletions clearly.

A duplication is the result of a small part of a chromosome sequence of genetic loci being repeated. This may be within a chromosome, or a piece may be broken off to produce an additional chromosome. Within a chromosome the piece may be in tandem,

A—B—C—D—E—F—G   A—B—C—D—C—D—E—F—G

or reverse tandem,

A—B—C—D—E—F—G   A—B—C—D—D—C—E—F—G

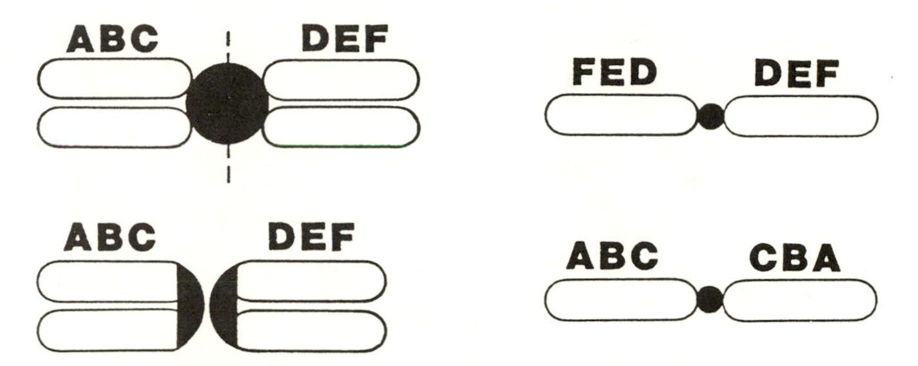

FIG. 5.3. Isochromosome formation. A transverse split in a centromere can result in two isochromosomes with identical arms.

These are difficult to identify cytologically, and usually have been discovered genetically (Swanson *et al.* 1967). It is postulated that these may occur at the time of replication with a short segment being duplicated, or by a single break in one chromatid and a double break in the sister or homologous chromatid, which results in a duplication in one and a deficiency in the opposite member of the pair. This process may have been the method by which new "genes" were created in the evolutionary development of organisms.

Occasionally a metacentric, or submetacentric, chromosome may split transversely instead of longitudinally at the centromere and each segment may then duplicate its one arm into two, forming metacentrics in which the two arms are identical but in reverse order (Fig. 5.3). Such chromosomes are called isochromosomes.

Inversions occur when a chromosome breaks in two places and the section of chromosome between the two breaks is "healed" back into the chromosome in reverse order. This could happen if a chromosome developed a loop and then broke as shown in Figure 5.4. An inversion can be detected at pachytene or in *Drosophila* salivary chromosomes by a loop as the comparable loci on the homologues have paired (Fig. 5.5). When an inversion has occurred on one side of the centromere, as in Figure 5.5, it is called a paracentric inversion. Such a chromosome can proceed through meiosis with no problem unless a cross-over occurs within the loop (Fig. 5.6). If a cross-over does occur, then the following problems develop.

Because the centromere has such a vital role in mitosis and meiosis, crossing-over within the inverted section leads to problems.

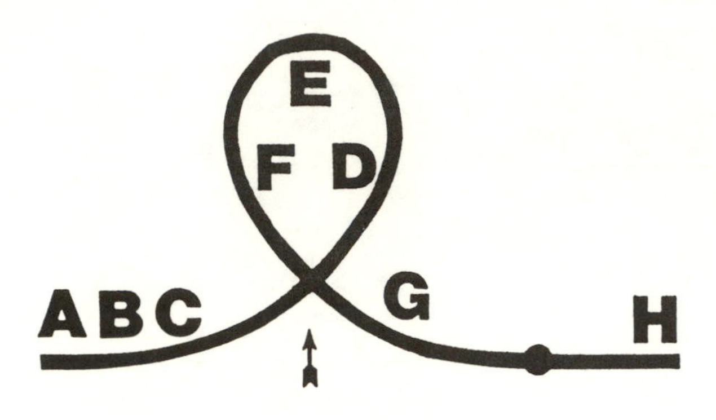

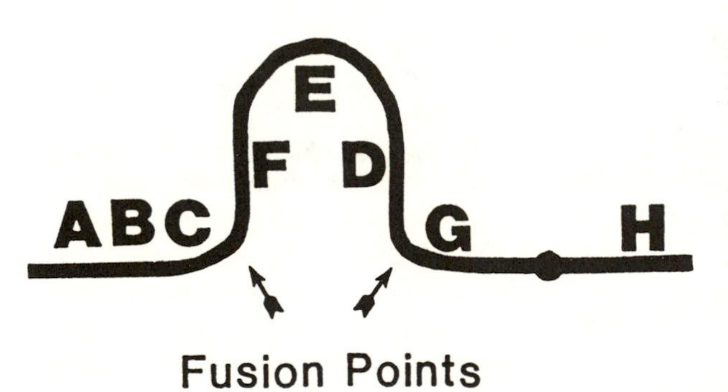

FIG. 5.4. Formation of a paracentric inversion. A chromosome may develop a loop and two breaks occur. This will result in an inversion of one segment, but if the centromere is outside the inversion little problem is encountered at meiosis, even though loops are normally formed as the chromosomes synapse.

The chromosome without a centromere, called acentric, has no process for orientation on the metaphase plate, and is composed of a duplicated pair of chromosome sections. The chromosome with two centromeres, called dicentric, may have each centromere being pulled toward opposite poles, which produces a cytological bridge, leading eventually to a random break of the chromosome. Such chromosomes are composed of duplicate chromosome segments so that duplications or deficiencies would result.

If both centromeres of the dicentric chromosomes went to one pole and the acentric chromosome went in the same direction, then the other nucleus could conceivably proceed through the second meiotic

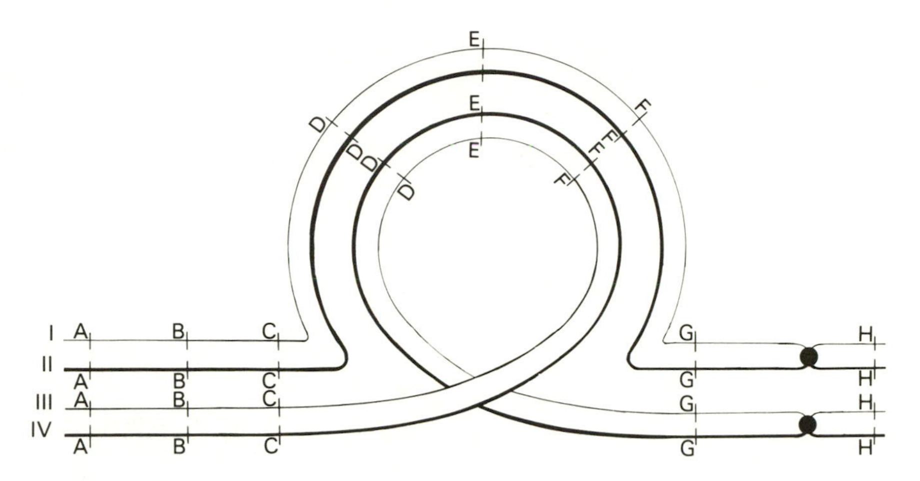

FIG. 5.5. At meiosis a characteristic loop can be observed when a paracentric inversion has occurred. When no cross-over has occurred within the loop all four resulting chromosomes will have only one centromere and they will duplicate the original and inversion chromosomes.

division normally and produce two normal gametes, one with the inversion. The second meiotic division of the other half, however, would still have problems and would almost invariably produce unbalanced gametes that would be inviable or produce lethal embryos. Theoretically, 50% would be the highest level of fertility from such an organism.

If the gamete with the inversion were to fertilize another gamete with the same inversion, full fertility could be restored unless position effects were deleterious. Thus, a new line might be formed that eventually could lead to a new species, since interfertility would be greatly reduced.

If an inversion occurs in such a way as to include the centromere, the result is called a pericentric inversion (Figs. 5.7 and 5.8). Each resulting chromosome has only one centromere, so the difficulty of meiotic division resulting from dicentric chromosomes and their consequent bridges and random breaks is avoided. However, the duplications and deficiencies would reduce viability and fertility in half of the gametes. As with paracentric inversions, if a gamete with the inversion fertilized a similar gamete, full fertility could be restored in such a line, and such a modification could lead to the evolutionary development of a new line, and possibly a new species.

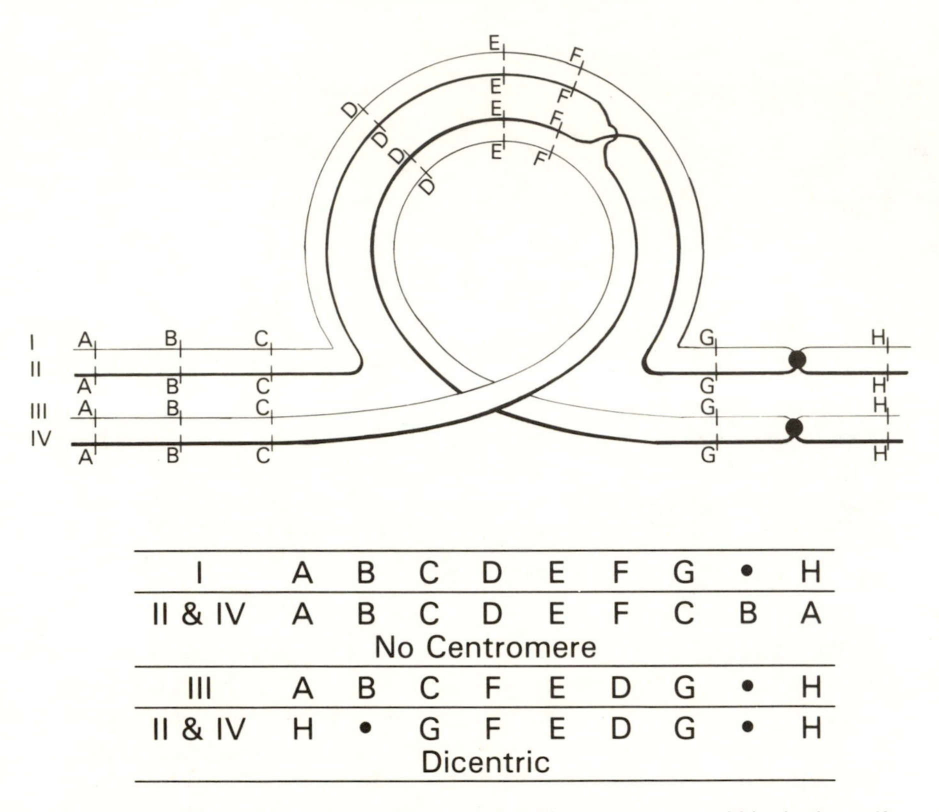

I	A	B	C	D	E	F	G	•	H
II & IV	A	B	C	D	E	F	C	B	A
	No Centromere								
III	A	B	C	F	E	D	G	•	H
II & IV	H	•	G	F	E	D	G	•	H
	Dicentric								

FIG. 5.6. Paracentric inversion at meiosis with a cross-over within the loop. If a cross-over occurs within the loop caused by the inversion the resulting chromosomes will consist of two normal chromosomes, one with the original sequence and one with the inversion, plus one dicentric chromosome and one acentric chromosome as illustrated.

Two reports have been published on pericentric inversions in cattle (Short *et al.* 1969; Popescu 1976).

RECIPROCAL TRANSLOCATIONS

When two nonhomologous chromosomes break simultaneously and each of the telomeric ends which have broken off becomes attached to the other chromosome a reciprocal translocation has oc-

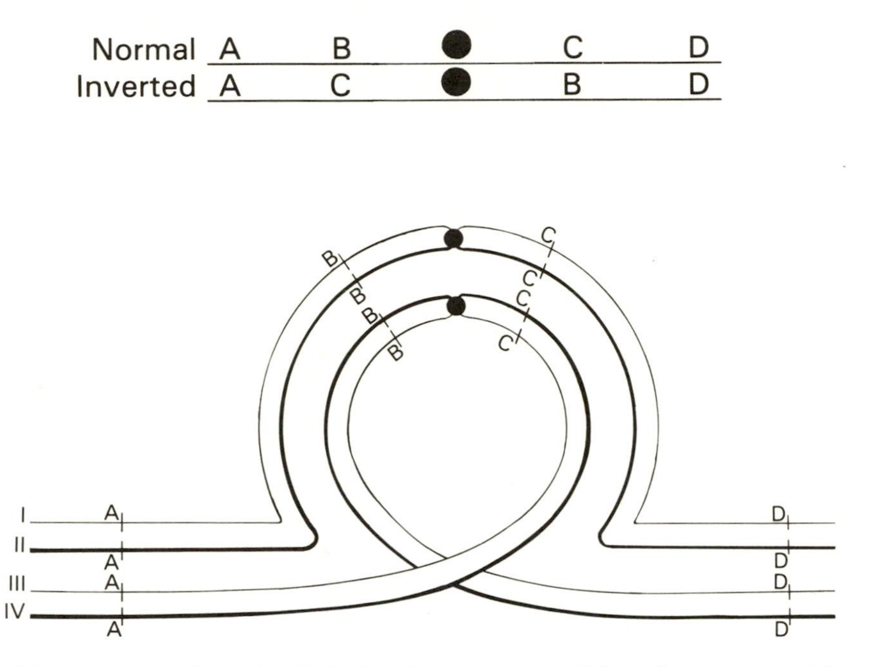

FIG. 5.7. If an inversion includes the centromere it is called a pericentric inversion. If no cross-over occurs, then at meiosis the two original and two inverted chromosomes result.

curred (Fig. 5.9). Only broken ends of chromosomes can reunite. The unbroken telomeric ends of chromosomes resist reunion with the broken end of a chromosome. Broken ends, however, seem to have a high degree of affinity for another broken end. This type of translocation is referred to as reciprocal because no chromatin material is lost. Other types of translocations may occur, but because of loss of viability due to loss of genetic material they do not survive for observation, Sometimes a translocation can be an insertion into another chromosome, but this requires four break points, two in each of two nonhomologous chromosomes.

If the segments which are exchanged are of equal lengths the translocations are not detectable except by the study of banded chromosomes or meiotic configurations. However, if the segments are of different lengths, especially if they include submetacentric chromosomes that can be individually identified, then Giemsa-stained kar-

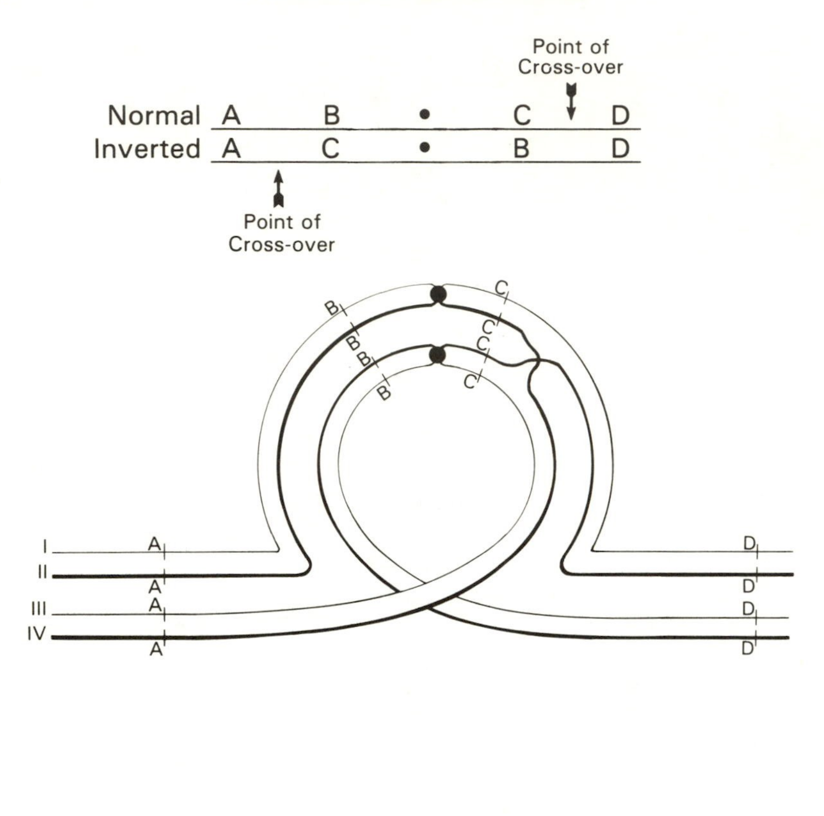

FIG. 5.8. Pericentric inversion with a cross-over in the loop. When the centromere is located within the inverted sequence, and a cross-over occurs at meiosis, the results are four chromosomes with one centromere each. One is the original sequence, one is the inverted sequence, and two chromosomes each contain a region of duplication and a deficiency.

yotypes may be sufficient to indicate that a translocation has occurred. Banding is still necessary to identify clearly the chromosomes involved.

At the zygotene of first meiotic division the homologous portions of the chromosomes will pair and at pachytene they will form an X-

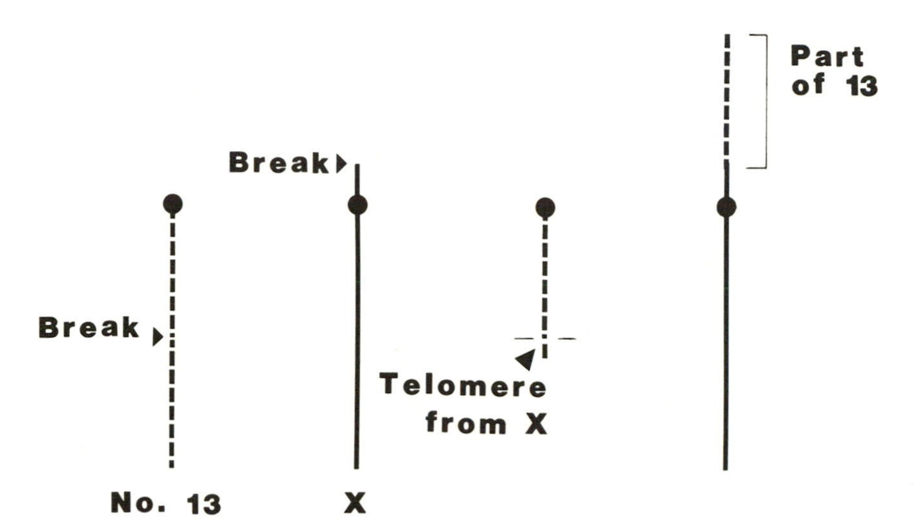

FIG. 5.9. If two breaks occur in two nonhomologous chromosomes in one nucleus and the broken ends fuse, a reciprocal translocation is the result. Illustrated are breaks in the middle of chromosome 13 and the telomeric end of the short arm of the X chromosome. The result is a short number 13 and a metacentric, long X chromosome. This is called an X-autosome reciprocal translocation.

or +-shaped configuration. The body formed by these chromosomes can now be called a quadrivalent since it contains four parts. All of the homologous portions of these two chromosomes are in the nucleus, but none of the four chromosomes is now a homologue. There are two ways in which the chromosomes can orient themselves on the metaphase plate. If one type of orientation occurs the two resulting cells will each have a portion of one chromosome duplicated and will have a deficiency for another portion. This type of orientation leads to what is called adjacent segregation. If the other type of orientation occurs the two resulting cells will be balanced for the chromatin in each chromosome. This is designated as alternate segregation. As a consequence, fertility is reduced in organisms heterozygous for a translocation. The reduction in fertility varies but would be expected to be 50% or less.

When the translocations involve chromosomes other than telocentric, rings are formed at diplotene, when the chromosomes have separated except at their telomeric ends. If more than one translocation occurs in one chromosome, rings of 6 chromosomes can be formed. In some species of plants up to 14 chromosomes form one ring.

Translocations have been found in many types of organisms, including livestock and humans. They have been studied extensively in plants.

REFERENCES

CHAPMAN, C. J., GARDNER, R. J. M., and VEALE, A. M. O. 1973. Segregation analysis of a large t (219229) family. J. Med. Genet. *10,* 362–366.

ELDRIDGE, F. E., and BALAKRISHNAN, C. R. 1977. C-band variation in Robertsonian translocations in cattle. Nucleus *20,* 28–30.

JOHN, B., and FREEMAN, M. 1975. Causes and consequences of Robertsonian exchanges. Chromosoma *52,* 123–136.

POPESCU, C. P. 1976. New data on pericentric inversions in cattle (*Bos taurus* L.). Ann. Genet. Sel. Anim. *8,* 443–448.

SHORT, R. V., *et al.* 1969. Cytogenetic and endocrine studies of a freemartin heifer and its bull co-twin. Cytogenetics *8,* 369–388.

SWANSON, S. P., MERZ, T., and YOUNG, W. J. 1967. Cytogenetics. Prentice-Hall, Englewood Cliffs, NJ.

VAN KEMPEN, C. 1975. A patient with congenital anomalies and a deletion of the long arm of chromosome 4 [46,XY del(4)(931)]. J. Med. Genet. *12*(2), 204–207.

ADDITIONAL REFERENCE

BLOOM, S.E. 1970. Trisomy-3,4 and triploidy (3A-ZZW) in chick embryos: Autosomal and sex chromosomal nondisjunction in meiosis. Science *170,* 457–458.

6

Banding of Chromosomes and Karyotyping

INTRODUCTION

After chromosomes were discovered, one of the primary needs was the identification of individual chromosomes. Much of the early work with chromosomes was devoted to determination of the characteristic number of chromosomes and description of the morphology of the chromosomes for each species. The size and shape of each chromosome at anaphase of mitosis or meiosis were the first characteristics used to identify individual chromosomes. Primary constrictions were found at the centromere in metacentric and submetacentric chromosomes, and secondary constrictions were also observed on some chromosomes. Individual chromosomes were identified by these characteristics. Many of the early studies on animals were done with testicular tissue since cell division is relatively rapid here compared with cells from somatic tissue, with some exceptions such as in bone marrow. Characterization of number and morphology reflected both meiosis and mitosis in testicular tissue.

SALIVARY GLAND CHROMOSOMES OF DROSOPHILA

Some species were selected for study by early cytologists because they had a relatively small number of chromosomes. For example, *Drosophila melanogaster* has a $2n$ number of 8. Two pairs are similar in morphology, but the others are distinctly different. The small number of chromosomes and the morphological differences in the chromosomes, as well as the ease and low cost of rearing drosophila led to considerable early work on this species. The discovery by Painter in 1934 of the giant, banded, salivary chromosomes in drosophila gave a tremendous impetus to cytological work with many *Drosophila* species. Not only could a chromosome be individually identified, but the bands on those extra-large chromosomes permitted identification of segments on each chromosome. Furthermore, the somatic synapsis of the chromosomes, making them appear like chromosomes in the early pachytene stage of meiosis I, permitted observation of the pairing of each small area of the chromosomes. Deletions, inversions, and translocations could be seen, and the characteristic loops, etc., could be related to these chromosomal aberrations. A great flurry of activity followed as chromosome maps were made and specific genes were assigned to known bands of these chromosomes. In 1935 Bridges published one of the first detailed chromosome maps using these special chromosomes. An article by McKusick and Ruddle (1977) presented a description of the chromosome map for humans, using the banding patterns from the 1971 Paris Conference (1973).

The enormous size of the salivary gland chromosome compared with normal chromosomes from a somatic cell can be seen in Figure 6.1. The fine detail, including bands, which can be seen in the salivary gland chromosomes, makes studies of species in this genus very easy compared to mammalian species. Many papers have been published concerning the evolution of *Drosophila* species—postulated from comparison of banding patterns. Crossing of the male of a species with females which have known gene arrangements yields configurations which identify chromosomal rearrangements (Kastritsis and Crumpacker 1967). The discovery of these bands on salivary gland chromosomes in drosophila led to an unsuccessful search for similar banding in mammalian species, since the banding permitted individual chromosome identification as well as identification of inversions, deletions, and other chromosomal rearrangements.

KARYOTYPING

A karyotype is a full set of metaphase chromosomes from one cell of an organism arranged in subsets of pairs for a normal diploid cell.

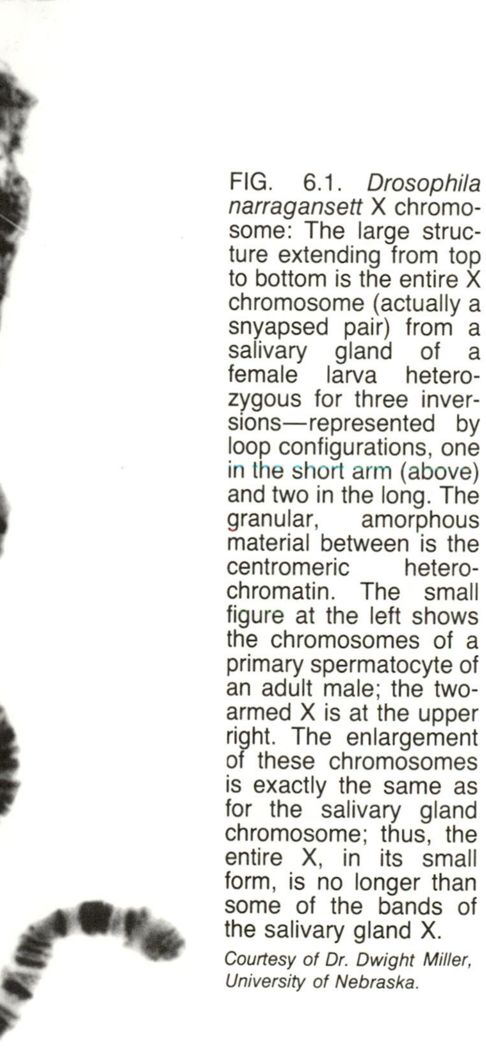

FIG. 6.1. *Drosophila narragansett* X chromosome: The large structure extending from top to bottom is the entire X chromosome (actually a snyapsed pair) from a salivary gland of a female larva heterozygous for three inversions—represented by loop configurations, one in the short arm (above) and two in the long. The granular, amorphous material between is the centromeric heterochromatin. The small figure at the left shows the chromosomes of a primary spermatocyte of an adult male; the two-armed X is at the upper right. The enlargement of these chromosomes is exactly the same as for the salivary gland chromosome; thus, the entire X, in its small form, is no longer than some of the bands of the salivary gland X.

Courtesy of Dr. Dwight Miller, University of Nebraska.

The karyotype may be characteristic of a normal cell for the species from which it was obtained, or it may show an abnormal chromosome or abnormal chromosomal arrangement. A karyotype is usually made from an enlarged photomicrographic print of the full set of chromosomes from one cell. Each chromosome is cut out and the pairs of chromosomes are identified. Banding makes karyotyping much more precise. The pairs of chromosomes are then arranged on light-colored cardboard starting with the longest and continuing through the shortest. The short arms of the submetacentric chromosomes are placed on top and the long arms below. By convention some species which have both acrocentric and metacentric or submetacentric chromosomes have the chromosomes divided into these groups and then arranged by length. The sex chromosomes are usually placed in a pair by themselves rather than arranging them by length. Chapter 15 of Brown (1972) presents a good review of karyotyping.

The Paris Conference (1973) is the accepted guide for preparing banded karyotypes for humans while the Reading Conference (see Ford *et al.* 1980) is the accepted guide to this subject for livestock.

KARYOTYPING WITHOUT BANDING

In many species two or more pairs of homologous chromosomes resemble each other so closely that it is impossible to distinguish one pair of homologues from another in standard Giemsa-stained preparations. In humans the resemblance between the chromosomes within some groups of homologues was such that a scheme of classification developed in which the chromosomes were placed in seven groups of similar size and shape, A through G. Group A included three pairs of metacentrics, Group B two pairs of submetacentrics, Group C seven pairs of smaller submetacentrics, Group D three pairs of acrocentrics, Group E the three pairs of smallest submetacentrics, Group F two pairs of the smallest metacentrics, and Group G two pairs of the smallest acrocentrics. The X chromosome is also similar to Group C and the Y chromosome is similar to Group G. Following the discovery of G-banding (to be discussed later in this chapter), individual chromosomes could be identified more accurately.

In cattle, goats, and sheep large numbers of acrocentric chromosomes form a continuum from the longest to the shortest with the

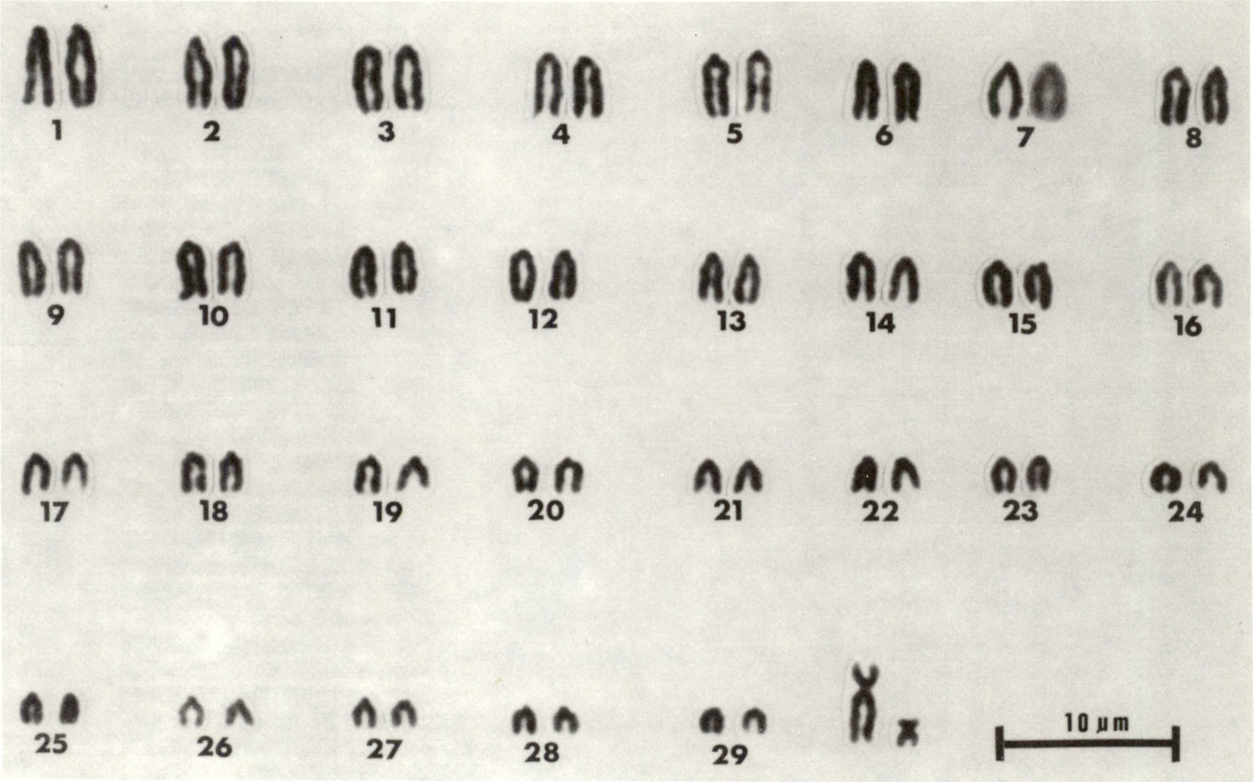

FIG. 6.2. Giemsa-stained karyotype of bovine Guernsey male. Pairing of chromosomes is unquestionably inaccurate due to the closely similar lengths of the autosomes and absence of other morphological differences.

Photograph by Eldridge.

differences in length so small from one pair to the next that it is not possible, with any desirable degree of certainty, to arrange the chromosomes in pairs from standard Giemsa-stained preparations (Fig. 6.2). In *Bos taurus,* for example, both sex chromosomes are submetacentric and can be easily identified. The longest acrocentric pair of autosomes is enough longer than the others and possesses a slight secondary constriction which permits individual identification of this pair of chromosomes in most cells. The other 28 pairs of autosomes, as seen in the karyotype in Figure 6.2, have rather small differences from one to the next. The idiogram for cattle, Figure 6.3, was based upon measurement of length only, from 10 cells of 2

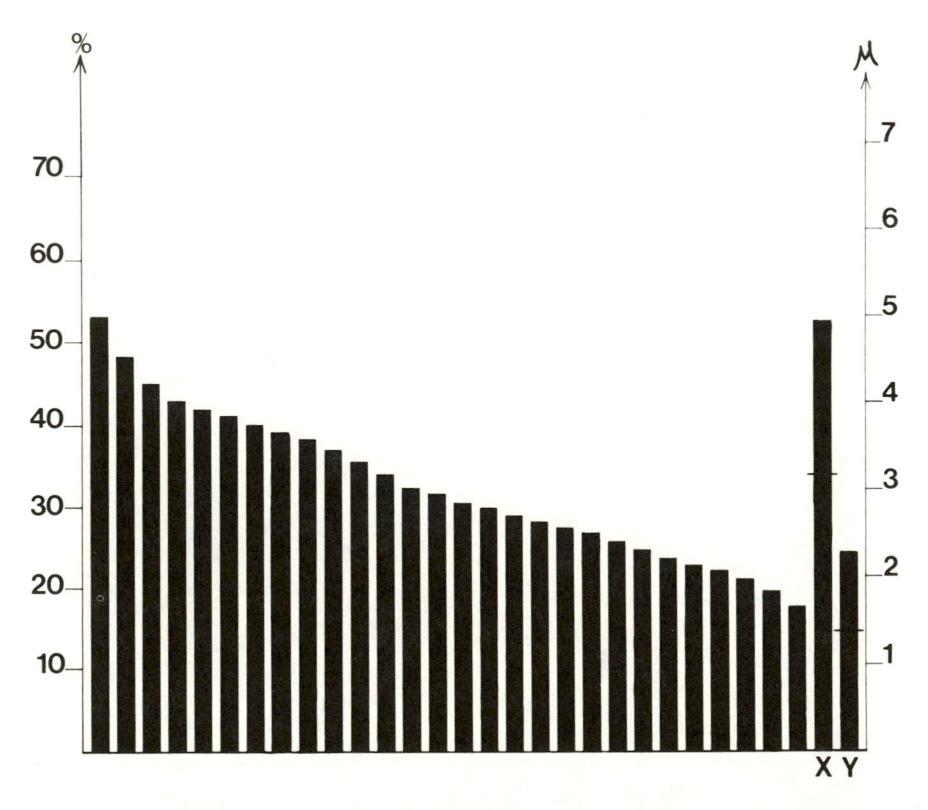

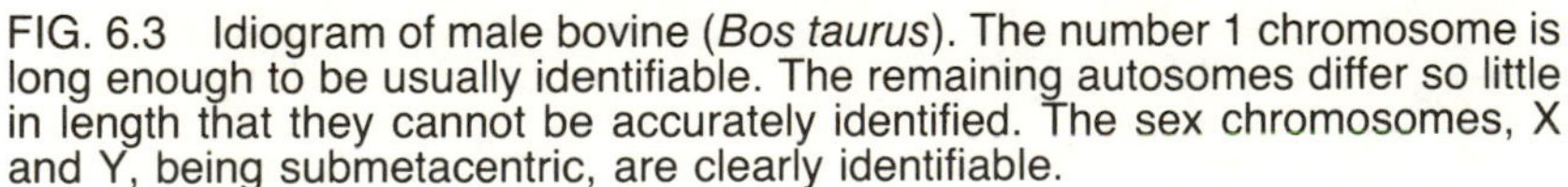
FIG. 6.3 Idiogram of male bovine (*Bos taurus*). The number 1 chromosome is long enough to be usually identifiable. The remaining autosomes differ so little in length that they cannot be accurately identified. The sex chromosomes, X and Y, being submetacentric, are clearly identifiable.

individuals. In morphologically distinct chromosomes in cattle, such as the X chromosome, the difference in length between the two members of a homologous pair varies. Prior to their attachment to the glass slide the chromosomes apparently have some amount of elasticity. Length measurements must be recognized as approximate. When cattle chromosomes are paired by length alone, it is probable, for example, that a chromosome which should be in pair 18 might actually measure long enough to be put in place 17, 16, or 15.

Goat chromosomes pose an even greater problem since the X chromosome is also acrocentric and therefore is not as easily identified as in cattle. The Y chromosome is quite small and submetacentric. In excellent Giemsa-stained preparations, in which the chromosomes are quite long, the X chromosomes can be identified but with less certainty than cattle chromosomes.

In sheep, with 54 chromosomes ($2n$), three pairs appear to be the result of centric fusion of nonhomologous autosomes. The 23 pairs of remaining autosomes form a continuum similar to those in cattle and goats (Evans *et al.* 1973).

Domestic swine have a $2n$ number of 38 chromosomes. Twenty-six chromosomes are metacentric or submetacentric, including the X and Y. Twelve are acrocentric. The difference in morphology and length alone permits rather easy pairing of most chromosomes, although it is difficult to identify the X chromosome accurately without banding.

FLUORESCENT BANDING

In 1968 Caspersson *et al.* published studies of chromosome preparations treated with quinacrine mustard, in which distinctive fluorescent bands were found across the chromosomes of plants and Chinese hamster. In 1970 Caspersson *et al.* reported that this type of banding permitted accurate identification and pairing of all human chromosomes. The fluorescent banding method has been named Q-banding after the quinacrine mustard stain. These discoveries in human and other mammalian chromosomes were repeated rapidly in many cytogenetic laboratories around the world. Several other substances were soon found which resulted in fluorescent banding of chromosomes.

Another type of fluorescent banding produced R-bands. The R stands for "reverse" banding, since the light bands are dark and the dark bands are light when compared to Q-banding (Yunis 1977). It is also called T-banding since the telomeres are heavily banded. Acridine orange also has been used for reverse fluorescent banding, showing many variations in the color of chromosomes (Verma *et al.* 1977).

BANDING TECHNIQUES IN KARYOTYPING

G-Banding

Apparently the first to use an enzyme digestion of chromosomes as a pretreatment was Dutrillaux *et al.* (1971). They used pronase as the enzyme and obtained some banding of human chromosomes.

In 1971 Seabright published a procedure for banding chromosomes of humans by pretreatment of the slides with trypsin. This method, and its variations, has been the most frequently used for producing G-bands. The term G-bands was given because the slides, after treatment with trypsin, were stained with Giemsa stain.

Just prior to Seabright's (1971) report, Sumner *et al.* (1971) reported a banding technique which they called ASG (acetic/saline/Giemsa). Since the bands are apparently identical with those produced by trypsin treatment they generally are called G-bands. Sumner *et al.* noticed faint bands in some chromosomes which were being stained for centromeric heterochromatin, called C-banding. By modifying the technique they were able to produce well-defined bands consistently. They noted the great similarity, with some exceptions, to the Q-bands which had been recently reported. The advantage to their technique, compared with Q-banding, was its permanence. When slides are stained for fluorescence, the stain fades rapidly when the intense light of a microscope is passed through it. The banding pattern can only be preserved through photography. With G-banding the bands are permanent, even becoming more distinct over a storage period in the dark of a few weeks or months, so they can be observed, as well as photographed, over indefinite periods of time. The G-banding techniques are very consistent regardless of the type of pretreatment. Several other types of pretreatments yield G-banding patterns (Yunis 1977).

Following the discovery of the methods that induce G- and Q-banding in chromosomes, several people recognized that they had observed this characteristic of banding occasionally in chromosome preparations which had not been given one of these pretreatments, but they had thought that the banding was simply an artifact. Sanchez *et al.* (1973) found that faint bands could be produced by simply using a very dilute Giemsa stain (1:100).

New and improved techniques for producing G-bands continue to be discovered. In harvesting either lymphocyte or fibroblast cultures of cells, the goal is to have the largest possible number of cells in metaphase, since this is the stage of mitosis when the number and morphology of the chromosomes can be observed most accurately. The number of cells in metaphase can be increased by adding colchicine (or Colcemid) to the culture a few minutes, or up to four hours, immediately prior to harvest. Colchicine, an extract from the autumn crocus, disrupts the spindle fibers (microtubules) so that the cells do not proceed into anaphase, resulting in a larger number of cells in metaphase. This disruption of the spindle fibers also affects the formation of a normal equatorial plate, permitting a more random distribution of the chromosomes within the cell. Therefore, in cells harvested by the usual procedure (using colchicine), the orientation of the chromosomes is different from that found in cells in which colchicine has not been used. The harvesting procedures will be discussed more completely in Chapter 10.

Rønne *et al.* (1979B) found that adding Colcemid at 0.1 g/ml to 48-hour human lymphocyte cultures and then holding the cultures at 4° for 18 hours, or (1979A) adding Colcemid at 46 hours and then cooling to 0°C for 30 minutes at 47.5 hours resulted in more metaphase spreads of longer chromosomes than without the cooling. Storage of the cells overnight in the fixative (1 part glacial acetic acid to 3 parts methanol) after the third wash in fixative also enhanced the length of the chromosomes. Staining in dilute 1:20 Giemsa in a pH 8.0 buffer for only 10 minutes resulted in slight banding, but clarity was still enhanced by trypsin treatment.

Yunis *et al.* (1978) synchronized mitosis with special treatments and harvested the cells with a markedly different timing sequence to produce excellent banding in human chromosomes. Figure 6.4 illustrates G-banding in cattle.

C-Banding

One of the earliest banding techniques is called C-banding, the "C" referring to centromeric banding. This technique stains the het-

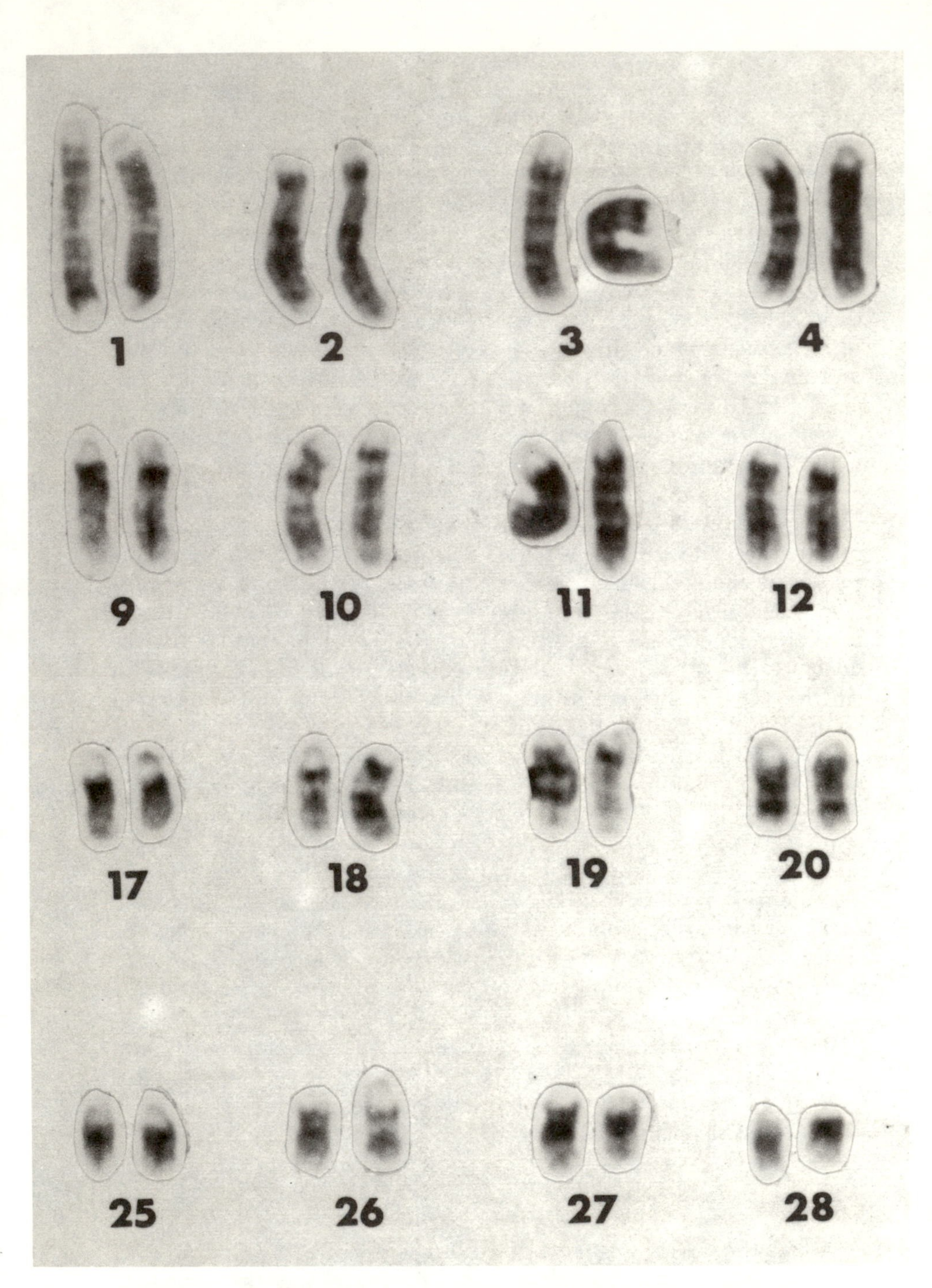

FIG. 6.4. Karyotype, G-banded, of a normal Brown Swiss bull.

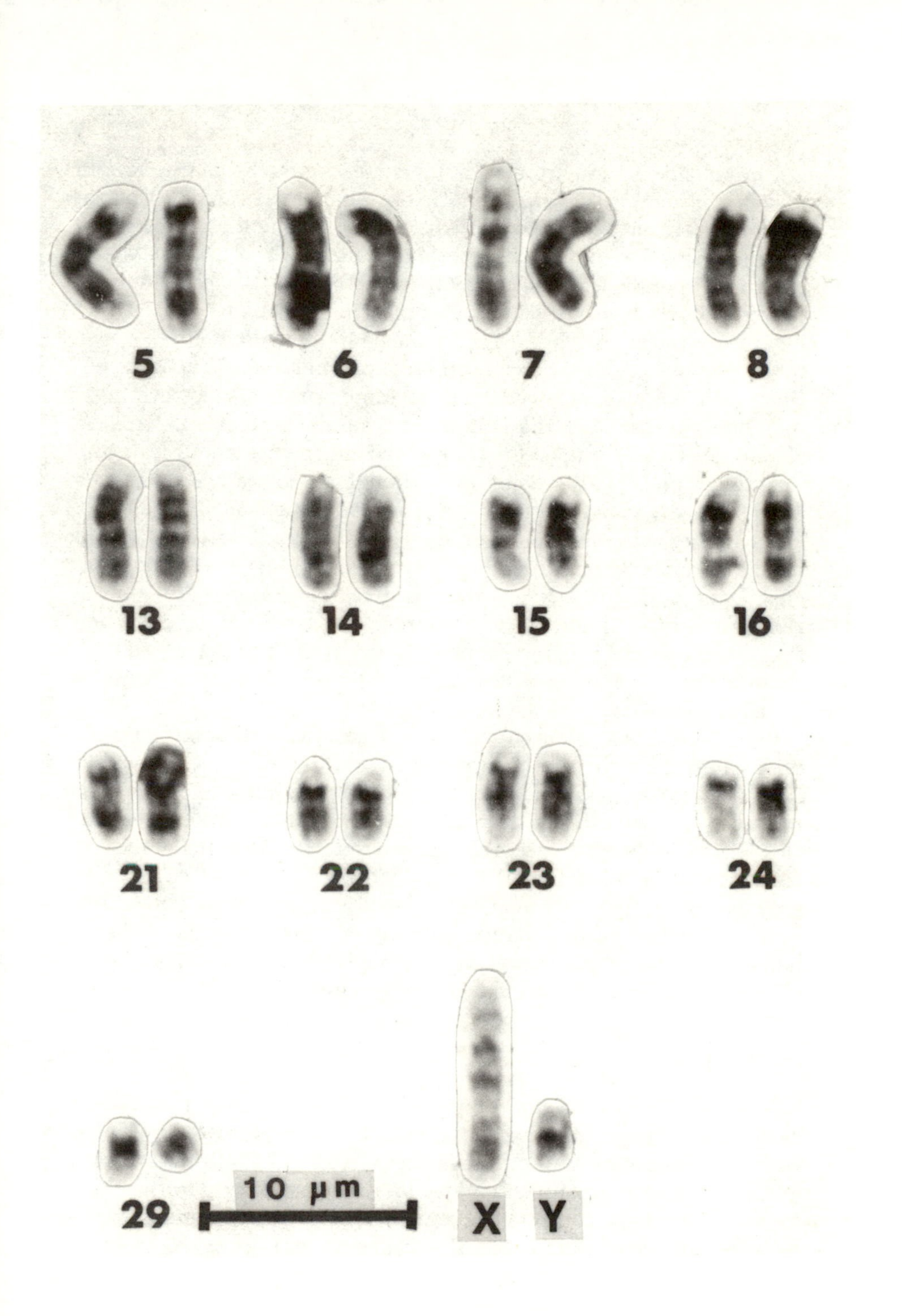
5
6
7
8
13
14
15
16
21
22
23
24
29
10 µm
X
Y

erochromatin of the chromosome and not the euchromatin, which makes up most of the chromatin in the chromosome. Heterochromatin stains more darkly during interphase and prophase than euchromatin, and is not as genetically active. It is located at the centromere, and occasionally in other parts of the chromosome, depending on the species. The Y chromosome is composed mostly of heterochromatin. Euchromatin makes up the genetically active, major portion of most chromosomes. Since heterochromatin is not limited to the centromeric region in many species, this technique may band areas other than the centromere. Therefore, C-banding should not be relied upon for characterizing the centromeric region in chromosomes of all species, even though all common livestock do have C-banding limited to the centromeric regions,

C-banding was first discovered by Pardue and Gall in 1970 in mouse chromosomes while they were hybridizing radioactive nucleic acids with the DNA of cytological preparations. After autoradiography they stained the preparations with Giemsa and found that the centromeric heterochromatin on chromosomes treated with sodium hydroxide (NaOH) stained more densely than the remainder of the chromosome.

In 1973 Hansen showed in cattle that all autosomes have C-bands, but the X chromosomes do not. The C-banding technique has been used to show some differences in Robertsonian translocations in cattle. Several different reports (presented in Chapter 11 on cattle), have identified dicentric chromosomes from Robertsonian translocations. Other Robertsonian translocations have resulted in only one centromere. Considerable variation, or polymorphism, has been demonstrated in C-bands of many species. The size of the C-band can vary greatly between chromosomes of homologous pairs within one animal. Figure 6.5 shows C-banding in cattle.

C_d-Banding

In a number of species C-banding identifies not only the heterochromatin in the centromeric area but also heterochromatin located in other parts of certain chromosomes. In 1974 Eiberg described a method that stains definitively two dots, corresponding to the chromatids, in only the centromeric region. This method is considered to be more definitive than C-banding for locating the centromeres. It has not been used extensively with livestock.

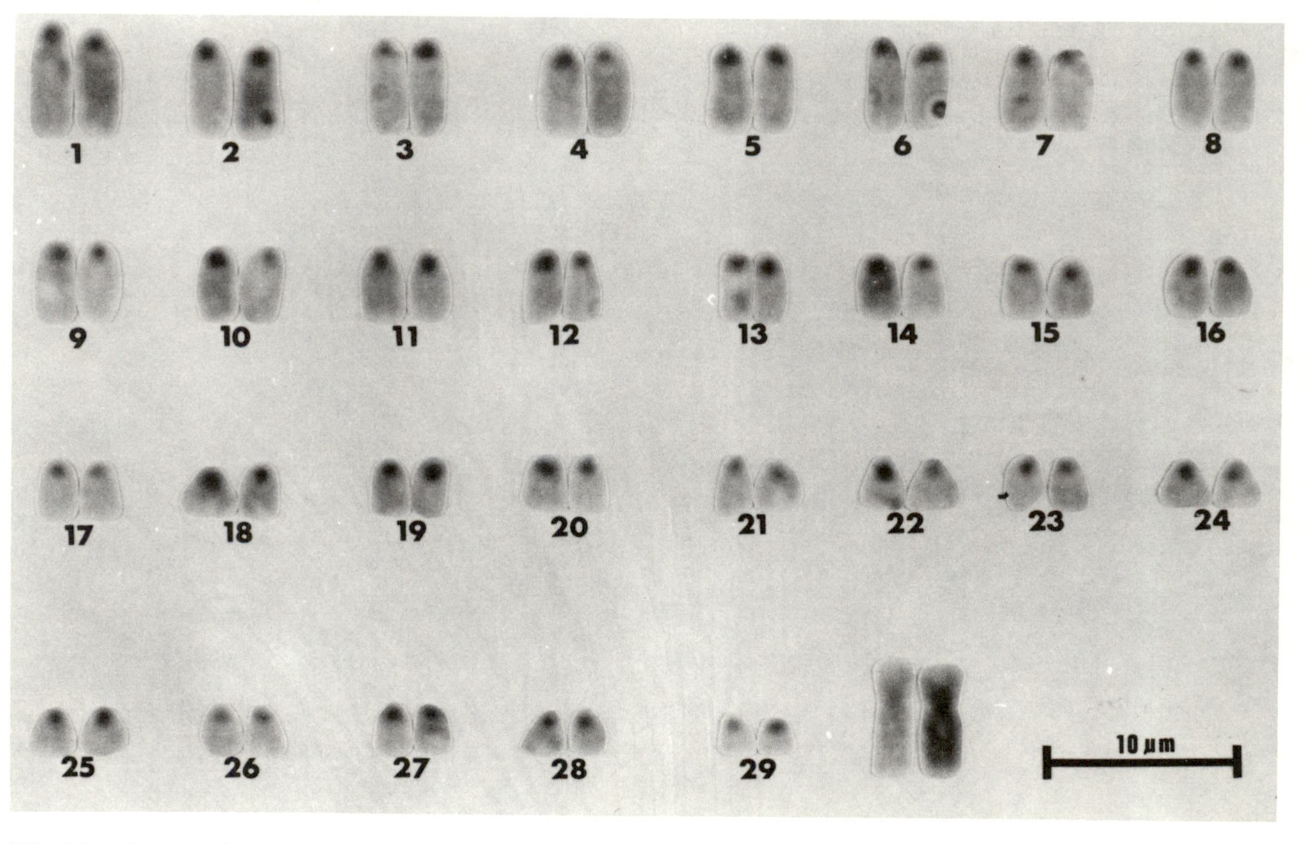

FIG. 6.5. C-banded bovine female karyotype. Each of the autosomes has a clearly defined C-band corresponding to the centromere, with no C-bands at other locations. The X chromosome has no C-band. The Y chromosome has an extensive C-banded area in the long arm.

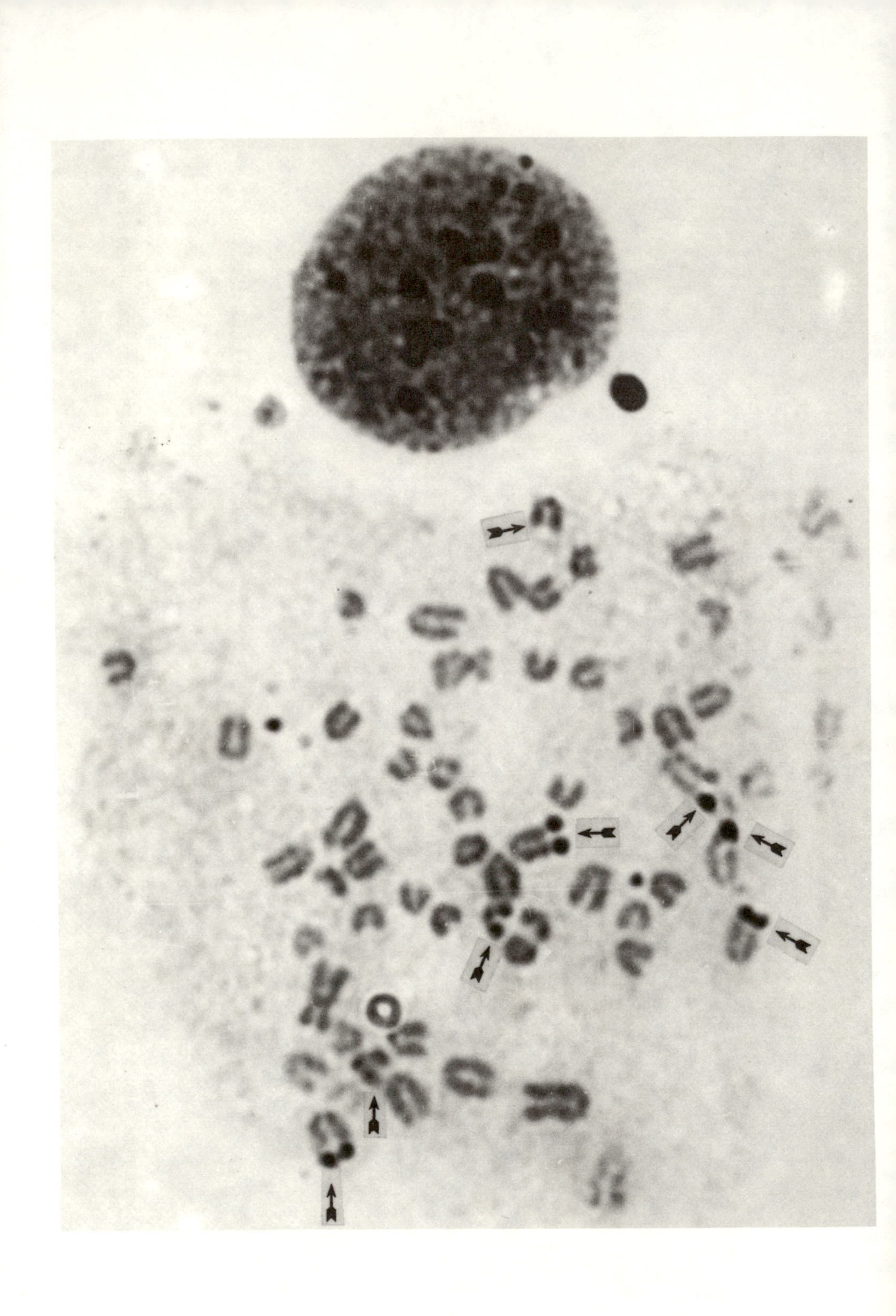

NOR-Banding

By another technique the nucleolar organizer regions (NOR) of the chromosomes in several species can be stained with ammoniacal silver. In human chromosomes the NORs have been found to be associated with the satellites, or secondary constrictions, of several chromosomes. These regions are also high in ribosomal DNA which has been found to stain with ammoniacal silver. DiBerardino *et al.* (1979) found 4 to 10 NOR regions per metaphase of Holstein cattle, all located at the telomeric ends of chromosomes 2, 3, 4, 11, and 29 [Reading Conference Standard (see Ford *et al.* 1980)]. The NORs could also be identified in meiosis of cattle at pachytene. Figure 6.6 shows the NORs in cattle.

REFERENCES

BRIDGES, C. B. 1935. Salivary chromosome maps. J. Hered. *26,* 60–64.

BROWN, W. V. 1972. Textbook of Cytogenetics. Mosby, St. Louis, MO.

CASPERSSON, T., FARBERS, S., FOLEY, G. E., KUDYNOWSKI, J., MODEST, E. J., SIMONSSON, E., WAGH, E., and ZECH, L. 1968. Chemical differentiation along metaphase chromosomes. Exp. Cell Res. *49,* 219–222.

CASPERSSON, T., ZECH, L., JOHANSSON, C., and MODEST, E. J. 1970. Identification of human chromosomes by DNA-binding fluorescent agents. Chromosoma *30,* 215–227.

DiBERARDINO, D., ARRIGHI, F. E., and KIEFFER, N. M. 1979. Nucleolus organizer regions in two species of Bovidae. J. Hered. *70,* 47–50.

DUTRILLAUX, B., deGROUCHY, J., FINAY, C., and LEJEUNE, J. 1971. Clarifying the fine structure of human chromosomes by enzymatic digestion (specifically pronase). C. R. Acad. Sci. Paris *273,* 587–588.

EIBERG, H. 1974. New selective Giemsa technique for human chromosomes, C_d staining. Nature (London) *248,* 55.

ELDRIDGE, F. E. 1975. A dicentric Robertsonian translocation in a Dexter cow. J. Hered. *65,* 353–355.

EVANS, H. J., BUCKLAND, R. A., and SUMNER, A. T. 1973. Chromosome homology and heterochromatin in goat, sheep and ox studied by banding techniques. Chromosoma *42,* 383–402.

FORD, C. E., POLLOCK, D. L., and GUSTAVSSON, I. 1980. Proceedings of the First International Conference for the Standardisation of Banded Karyotypes of Domestic Animals. Hereditas *92,* 145–162.

FIG. 6.6. Nucleolar organizer regions. Bovine NORs are found at the telomeric end of 5 pairs of chromosomes. Seldom are all 10 apparent in one cell. This cell has 8.

Photograph by Eldridge.

HANSEN, K. M. 1973. Heterochromatin (C bands) in bovine chromosomes. Hereditas *73,* 65–70.

KASTRITSIS, C. D., and CRUMPACKER, D. W. 1967. Gene arrangements in the third chromosome of *Drosophila pseudoobscura.* J. Hered. *58,* 113–129.

McKUSICK, V. A., and RUDDLE, F. H. 1977. The status of the gene map of the human chromosomes. Science *196,* 390–405.

PAINTER, T. S. 1934. A new method for the study of chromosome aberrations and the plotting of chromosome maps in *Drosophila melanogaster.* Genetics *19,* 175–188.

PARDUE, M. L., and GALL, J. G. 1970. Chromosomal localization of mouse satellite DNA. Science *168,* 1356–1358.

PARIS CONFERENCE 1975. Standardization in Human Cytogenetics, 1971. The National Foundation–March of Dimes, Washington, DC.

POPESCU, C. P. 1977. A new type of Robertsonian translocation in cattle. J. Hered. *68,* 138–141.

RØNNE, M., ANDERSEN, O., and ERLANDSEN, M, 1979A. Effect of Colcemid exposure and methanol acetic acid fixation on human metaphase chromosome structure. Hereditas *90,* 195–201.

RØNNE, M., NIELSEN, K. V., and ERLANDSEN, M. 1979B. Effect of controlled Colcemid exposure on human metaphase chromosome structure. Hereditas *91,* 49–52.

SANCHEZ, O., ESCOBAR, J. F., and YUNIS, J. J. 1973. A simple G-banding technique. Lancet *2,* 269.

SEABRIGHT, M. 1971. A rapidly banding technique for human chromosomes. Lancet *2,* 971–972.

SUMNER, A. T., EVANS, H. J., and BUCKLAND, R. A. 1971. New technique for distinguishing between human chromosomes. Nature New Biol. *232,* 31–32.

VERMA, R. S., DOSIK, H., and LUBS, H. A., JR. 1977. Demonstration of color and size polymorphisms in human acrocentric chromosomes by acridine orange reverse banding. J. Hered. *68,* 262–263.

YUNIS, J. J. 1977. New Chromosome Syndromes. Academic Press, New York.

YUNIS, J. J., SAWYER, J. R., and BALL, D. W. 1978. The characterization of high resolution G-banded chromosomes of man. Chromosoma *67,* 293–307.

ADDITIONAL REFERENCES

DOBZHANSKY, T. 1950. The chromosomes of *Drosophila willistoni.* J. Hered. *41,* 156–158.

GUSTAVSSON, I. 1969. Cytogenetics, distribution and phenotypic effects of a translocation in Swedish cattle. Hereditas *63,* 68–169.

METZ, C. W. 1935. Structure of the salivary gland chromosomes in *Sciara.* J. Hered. *26,* 176.

7

Fertilization, Parthenogenesis, and Sex Determination

FERTILIZATION

Fertilization of an ovum in mammals provides the normal stimulus for starting the second meiotic division. Ovulation of the ovum is under hormonal control, primarily the rapid increase of luteinizing hormone (LH), in balance with the other hormones affecting the reproductive system. Fertilization usually occurs in the upper third of the fallopian tube.

The ovum is surrounded by follicular cells (or granulosa cells) which are carried with the ovum at the time of ovulation (Balinsky 1976). The spermatozoa are attracted to the ovum in some species, apparently collide at random with the ovum in other species, and appear to have developed a stickiness that helps to keep the spermatozoa and ovum together. The mammalian spermatozoon produces an enzyme, hyaluronidase, which dissolves the material holding the follicular cells together. The zona pellucida, which is a protective shell for the ovum, provides another obstacle to penetration by the sperm. When the spermatozoon comes into contact with the zona, the acrosome (the cap on the spermatozoon) is lost, exposing the perforatorium of the spermatozoon. By some process not yet fully understood the zona and spermatozoon work together to permit entry of the sperm head. The last barrier is the vitellus, which appears to rupture and then, after the sperm head has entered the egg, fuses together again. This is the time at which the ovum nucleus begins to enter the second division of meiosis.

Apparently some change takes place in the zona, called the zona reaction, which makes it less penetrable after entry of one spermatozoon. This varies with species. In some species several spermatozoa may reach the perivitelline space, between the zona and vitellus. The reaction of the vitellus to penetration of a spermatozoon, which makes it less responsive to other spermatozoa, is called the vitelline block. A condition known as polyspermy occurs when more than one spermatozoon penetrates the vitelline. Polyspermy is infrequent in mammals and it is obviously disadvantageous. If two or more spermatozoa penetrate the vitellus and form pronuclei, polyploid embryos may be produced, which are not usually viable. It has also been suggested by Fechheimer and Harper (1980) that two spermatozoa penetrating one ovum might fertilize two maternal pronuclei, one of which might be a polar body, producing a chimeric individual. Several physiological or environmental factors, such as age of ovum at fertilization, time of insemination relative to ovulation, and genetic variations, may influence the probability of polyspermy. Digyny, which is the formation of a pronucleus from a polar body, may also be influenced by physiological and environmental factors. Superovulation in rabbits, which can be induced by injections of pregnant mare serum (Fujimoto *et al.* 1974), has been shown to result in an increase in triploid blastocysts over the number found without superovulation, so it is possible that polyspermy may be permitted through superovulation, or the female pronucleus may not undergo the normal process of division and extrusion of the second polar body. Superovulation results in the female releasing many more than the normal number of ova at one time.

After entry of the sperm head through the vitellus, the male pronucleus is formed. For a period of up to 15 hours the two pronuclei can be recognized as distinct bodies which, at the end of this period, migrate toward each other. The centrosome from the male pronucleus divides into its two centrioles and provides the spindle which identifies the metaphase plate. The membranes surrounding the two pronuclei disappear and the first mitotic division of chromosomes occurs. Only after this division does a nuclear membrane reform around each of the two new nuclei. The process of coalescence of the two pronuclei is called syngamy.

PARTHENOGENESIS

Parthenogenesis is the development of an embryo from an ovum that has been stimulated to develop by some means other than by a sperm. It is a natural form of reproduction in some species, such as the lizard *Lacerta saxicola armeniaca,* in which only parthenogenetic females live. Subjecting unfertilized ova to cold treatment, heat, a number of different chemicals, or pricking with a needle, or injecting material with a micropipette, can stimulate development in some amphibia. It is not unusual to find aged ova flushed from mammals which have started nuclear division, but these are usually defined as deteriorating ova. The animals resulting from parthenogenesis may be haploid or diploid. Parthenogenesis in chickens and turkeys is discussed in Chapter 15.

Two situations similar to parthenogenesis are androgenesis in which an ovum is stimulated to develop but the nucleus of the ovum is not involved in this development, and gynogenesis, in which the spermatozoon activates the egg but its nucleus is not involved in further development. These types of development are rarely, if ever, found naturally in mammals.

A common error made by many people is to assume that parthenogenetic progeny would be genetically identical to the dam since no spermatozoa are incorporated into the developing blastocyst. Why is this not true? The first meiotic division results in a random sample of chromosomes, with crossing-over having also occurred, so the resulting ovum nucleus and the first polar body are not identical. If the stimulus resulting in parthenogenesis permits a doubling of the chromosomes in the ovum nucleus, or if the second polar body should become one pronucleus and later fuse with the ovum pronucleus, an entirely homozygous progeny would result, but it would not be genetically identical with the dam. In androgenesis, where the sperm provides the chromosomes, the same result would occur, but the resulting embryo would not be genetically identical with the sire since each spermatozoon carries a random sample of chromosomes. Only if the first polar body, or one of its products from meiosis II, should unite with the ovum nucleus after extrusion of the second polar body would an individual result that would be genetically identical to the dam.

EARLY EMBRYOLOGY

For several days the fertilized embryo lives as a free body, first within the fallopian tube and later in the uterus of the female. It is nourished by uterine secretions at this time. The process of cell division within the fertilized ovum without increase in mass of the ovum is called cleavage. The cells within the fertilized ovum are called blastomeres. As the cells crowd into a ball within the zona, the morula stage begins. Fluid then begins to accumulate between the cells, and an inner cavity, or blastocoele, develops. The embryo is now called a blastocyst. All cells are identical at this stage, so any one of them could proceed to develop into a complete animal. No differentiation has occurred. It has been shown that, if the blastocyst is caused to divide into two parts, each may continue to develop into a complete organism. The twins resulting from such a division of the blastocyst will have identical genotypes.

In cattle, on about the 13th or 14th day the blastocyst will "hatch"; that is, it will break out of the zona and start elongating at an extremely rapid rate. Implantation begins soon after elongation. For several days after the 14th the embryo in cattle is extremely difficult to find because it is so long and delicate.

SEX DETERMINATION

In livestock, sex is determined by the sex chromosomes. In all species except poultry the male sex is determined by the presence of the Y chromosome. The female has two X chromosomes. In poultry the female is the heterogametic sex. having one Z and one W chromosome while the male has two Z chromosomes. Much research has been directed toward the determination of sex in animals and plants. The XX-XY (or ZW-ZZ) situation is most common, but other combinations also occur.

In 1891 Henking noticed one chromatin element in an insect which passed undivided to a daughter cell in meiosis, producing two types of spermatozoa. He labeled this element "X" as an unusual and little understood element. This terminology has persisted as the designation for one of the sex chromosomes.

In humans the absence of the Y chromosome, referred to as XO, is known as Turner's syndrome. Such individuals are females, but they are infertile and have infantile reproductive systems. XO in horses causes similar characteristics. In mice XO females also occur, but in this species they may be fertile (Welshons and Russell 1959). These observations have led to the conclusion that the Y chromosome is necessary to determine maleness in mammals. However, XO individuals in *Drosophila* are sterile males, indicating that in this species of insects the Y chromosome is not necessary for maleness but is necessary for fertility. In some species of insects, such as grasshopper, the male is XO and the female XX as the usual method for sex determination.

In addition to the Y chromosome, the gene for the H-Y antigen is necessary for the development of a normal male. In humans, this gene is one of the few normally carried on the Y chromosome. A gene on an autosome and one on the X chromosome have also been postulated to influence the expression of the H-Y antigen gene. The H-Y antigen is the histocompatibility Y antigen found when skin grafts from one sex are rejected by the other sex, even in a highly inbred strain of mice. It is widely conserved phylogenetically, so it is not species specific. The gene is necessary for early testis development from the bipotential embryonic gonad in mammals (Wachtel 1977). The absence of the H-Y antigen may explain some intersex cases.

In the muntjac deer the male has two Y chromosomes (Wurster and Benirschke 1970), so the diploid number differs by one chromosome between the sexes. Patterson and Petricianni (1973) found also in the muntjac deer that the X chromosome has become attached to one of the acrocentric autosomes. The Y chromosome is a very small metacentric chromosome. The female has six chromosomes and the male seven. Other species including the common shrew, *Sorex araneus,* are similar. Since sex is determined basically by a balance between the X and Y chromosomes in mammals, combinations of sex chromosomes which deviate from XY and XX usually result in sterility and/or other phenotypic modifications. The unusual pattern of sex chromosomes in the muntjac deer is an intriguing evolutionary development.

Klinefelter's syndrome in man results from a 47,XXY chromosome constitution. Such persons have male internal and external genitalia, may or may not have breast development, have small testes, and are sterile. A similar chromosomal constitution has been found in cattle, with small testicles and sterility as in man (see Chapter 11 on cattle chromosomes),

Female humans with 47,XXX, 48,XXXX, and 49,XXXXX have also been found. In these sex chromosome aberrations there is a tendency toward mental retardation, as in Klinefelter's syndrome.

Human males with 47,XYY are usually taller than average. Much controversy has centered on the aggressiveness of these individuals. Jacobs *et al.* (1965) first reported this association, but later publications have indicated that the sample was not representative.

Human males with 46,XX have been found, but the cause for this condition is obscure (Mittwoch 1973).

In the wood lemming (*Myopus schisticolor*) many more females are born than males. Fredga *et al.* (1976) found that females could be separated into two groups, one which is the normal XX and one which is XY producing only females. The XY females, however, had only XX cells in the oogonia, a result of apparent nondisjunction accompanied by loss of the Y chromosome. In 1978 Herbst *et al.* found through G-banding that the wood lemming had two types of X chromosomes, the mutated or X^* type being the one which resulted in females when combined with either AY or another X chromosome. Wachtel *et al.* (1976) also found that the X^*Y female was H-Y antigen negative and concluded that the X^* chromosome had a mutation that blocked the activity of the H-Y antigen locus. Ropers and Wiberg (1982) concluded that this locus was probably on the short arm of the X chromosome, which is not inactivated, and this corresponded to the location of the modification found by G-banding. If this locus could be transferred to other species, such as dairy cattle, then a modified sex ratio might be obtained, although the other species would also need to be able to accomplish the nondisjunction and perhaps some other modifications for it to be usable.

In 1962 Lyon found that one of the X chromosomes in vertebrates was inactive and appeared as a stained body in somatic interphase cells. This body persisted in interphase in contact with the nuclear envelope and was identified as the Barr body, or drumstick. By tritiated thymidine treatment one of the X chromosomes was found to be late-replicating. In 47,XXX human females two Barr bodies can be identified. This staining body is also known as sex chromatin and is facultative heterochromatin as distinguished from euchromatin. Facultative heterochromatin is euchromatin that has been heterochromatized; that is, it has become more heavily stained and is different from constitutive heterochromatin that is normally found in the centromeric region of chromosomes, which stains more heavily. The Barr body can be observed very easily from epithelial cells without culture.

An interesting illustration of sex determination and random inactivation of one of the X chromosomes is found in cats (Centerwall and Benirschke 1973). The gene for yellow color (sometimes appearing as orange or red) in cats and its allele, non-yellow, which may be gray, blue, brown, tabby, or black, are carried on the X chromosome. Most cats described as calico or tortoiseshell in color are females, since this color pattern is dependent upon two X chromosomes, one carrying yellow and one the non-yellow allele. Two colors appear as patches of yellow or non-yellow, and may be found also in combination with white areas which are under the control of a different set of alleles. Random inactivation of one X chromosome apparently occurs early in embryonic development, so that the cells descending from each of the two parent lines of cells become distributed throughout the body, including the skin. The skin cells in which the active X chromosome carries the yellow gene will have yellow-pigmented hair. The other skin cells which have descended from the cells with the active X chromosome carrying the non-yellow allele will have hair of the other color.

With such a mode of inheritance it can be seen that the frequency of male calico or tortoiseshell (T-C) cats would be expected to be zero, but rare cases of such animals have been reported. Usually such males are infertile. Many hypotheses have been presented for the occurrence of T-C males, but the most logical explanation had to wait for chromosomal analyses of such cats.

The presence of a Y chromosome determines the maleness of a cat, regardless of the number of X chromosomes. Two general types of karyotypes have been found in male T-C cats. One type is trisomic, 59,XXY, in which the two X chromsomes, probably resulting from nondisjunction in meiosis, each carried a different-color allele. As is expected, these animals are not fertile. The other type is caused by chimerism or mosaicism resulting in 38,XY/38,XY or 38,XX/38,XY or 38,XY/39,XXY, etc. The chimeras may develop from early fusion of two embryos, or fusion of placenta as in freemartin cattle. The mosaics also may result from a gene mutation at an early stage of embryonic development. Where chromosomal analysis shows a

38,XY karyotype the male has been assumed to be a chimera 38,XY/38,XY and may be fully fertile.

For study of X-chromosome inactivation the female mule or hinny has been particularly useful. The maternal X chromosome can be distinguished from the paternal X chromosome, regardless of whether the sire is a jack and the dam a mare or the sire a stallion and the dam a jennet. The gene for glucose-6-phosphate dehydrogenase is carried on the X chromosome. This enzyme is different in the two species as determined by starch-gel electrophoresis. Clones have been developed, biochemical measurements have been made, autoradiographs have identified late replication, and the relationships have been measured (Hamerton *et al.* 1971; Giannelli and Hamerton 1971). For further information, see Chapter 14 on chromosomes of horses, asses, and mules.

REFERENCES

BALINSKY, B. I. 1976. Embryology, 3rd Edition. Saunders, Philadelphia, PA.

CENTERWALL, W. R., and BENIRSCHKE, K. 1973. Male tortoiseshell and calico (T-C) cats. J. Hered. *64,* 272–278.

FECHHEIMER, N. S., and HARPER, R. L. 1980. Karyological examination of bovine fetuses collected at an abattoir. Proc. 4th Eur. Colloq. Cytogenet. Domest. Anim. 194–199.

FREDGA, K., GROPP, A., WINKING, H., and FRANK, F. 1976. Fertile XX- and XY-type females in the wood lemming *Myopus schisticolor.* Nature (London) *261,* 225–226.

FUJIMOTO, S., PAHLAVAN, N., and DUKELOW, W. R. 1974. Chromosome abnormalities in rabbit preimplantation blastocysts induced by a superovulation. J. Reprod. Fertil. *40,* 177–181.

GIANNELLI, F., and HAMERTON, J. L. 1971. Non-random late replication of X chromosomes in mules and hinnies. Nature (London) *232,* 315–319.

HAMERTON, J. L., RICHARDSON, B. J., and GEE, R. A. 1971. Non-random X-chromosome expression in female mules and hinnies. Nature (London) *232,* 312–315.

HENKING, H. 1891. Investigation of the first development in the ova of insects. II. Concerning spermatogenesis and its relation to development in *Pyrrhocoris apterus.* Z. Wiss. Zool. Abt. A *51,* 685–736.

HERBST, E. W., FREDGA, K., FRANK, F., WINKING, H., and GROPP, A. 1978. Cytological identification of two X-chromosome types in the wood lemming. Chromosoma *69,* 185–191.

JACOBS, P. A., BRENTON, M., MELVILLE, M. M., BRITTAIN, R. P., and McCLEMENT, W. F. 1965. Aggressive behaviour, mental subnormality and the XYY male. Nature (London) *208,* 1351–1353,

LYON, M. F. 1962. Sex chromatin and gene action in the mammalian X-chromosome. Am. J. Human Genet. *14,* 135–148.

MITTWOCH, U. 1973. Genetics of Sex Differentiation. Academic Press, New York.

PATTERSON, R. M., and PETRICIANNI, J. C. 1973. A comparison of prophase and metaphase G-bands in the muntjac. J. Hered. *64,* 80–82.

ROPERS, H. H., and WIBERG, U. 1982. Evidence for X-linkage and non-inactivation of steroid sulphatase locus in wood lemming. Nature (London) *296,* 766–768.

WACHTEL, S. 1977. H-Y antigen and the genetics of sex determination. Science *198,* 797–799.

WACHTEL, S., KOO, S. G. C., OHNO, S., GROPP, A., DEV, V. G., TANTRAVAHI, R., MILLER, D. A., and MILLER, O. J. 1976. H-Y antigen and the origin of XY female wood lemmings (*Myopus schisticolor*). Nature (London) *264,* 638–639.

WELSHONS, W. J., and RUSSELL, L. B. 1959. The Y-chromosome as the bearer of male determining factors in the mouse. Proc. Natl. Acad. Sci. U.S.A. *44,* 560–566.

WURSTER, D. H., and BENIRSCHKE, K. 1970. Indian muntjac, *Muntiacus muntjak:* A deer with a low diploid chromosome number. Science *168,* 1364–1366.

ADDITIONAL REFERENCES

BACCI, G. 1965. Sex Determination. Pergamon, New York.

BROWN, W. V. 1972. Textbook of Cytogenetics, Chapter 7. Mosby, St. Louis, MO.

McCARREY, J. R., and ABBOT, U. 1979. Mechanisms of genetic sex determination, gonadal sex differentiation and germ-cell development in animals. Adv. Genet. *20,* 217–290.

SOLARI, A. J. 1974. The behavior of the XY pair in mammals. Int. Rev. Cytol. *38,* 273–317.

8

Somatic Cell Hybridization

In 1960 Barski *et al.* cultured two different cell lines from mice, each of which had marker chromosomes. They cultured these two lines together and found some cells which had the total set of chromosomes from both lines in one cell. These cells were viable; that is, they could be maintained in culture without the original two cell lines. This was the first report of hybrid mammalian cells in culture.

Harris and Watkins (1965) reported the production of a true hybrid between cells from two species: human and mouse. They had used a technique originally used by Okada (1958) whereby an ultraviolet-inactivated virus from Japan—now widely known as Sendai virus—would cause fusion of cells. The original work by Barski (1960) had been done with *spontaneous* fusion, while the fusion using the Sendai virus as an induction agent increased the number of fused cells by a factor of 100.

Polyethylene glycol, PEG (Pontecorvo 1975), was found to induce fusion of mammalian cells in culture in a manner similar to Sendai virus. This chemical had been used earlier for fusing plant protoplasts. Techniques have been improved rapidly since the early discovery, and in many individual combinations the use of polyethylene glycol has resulted in frequencies of hybrid cells many times higher than those obtained by the use of Sendai virus (Wang *et al.* 1979). Most somatic cell hybridization has been done with human cells in combination with cell lines derived from laboratory animals such as mouse, rat, and hamster. Limited work has been reported using cattle, sheep, swine, or other livestock species. The polyethylene glycol techniques offer new opportunities to try these species in hybridization, although in some of these species Sendai virus might produce the best results. Polyethylene glycol has also produced more uniform results (Wang *et al.* 1979). Schneiderman *et al.* (1979) found that using Ca^{2+}-free media decreased the toxicity of PEG for some cultures of mammalian cells.

It has been observed that over several generations in a small number of hybrid cells some chromosomes were lost from one or the other of the parental cell lines. In the human × mouse cell hybrids, the cells maintained the full set of mouse chromosomes, but the human chromosomes were lost rapidly until only one to three human chromosomes were left in each cell. This loss of human chromosomes appeared to be random with respect to specific chromosomes. The selective loss of chromosomes from one of the parental lines varied from one combination of crosses to another. In early work with rat × mouse or hamster × mouse hybrids the loss seldom dropped below 10–20%, and the chromosomes lost were from either parental cell line. Irradiation of cells prior to hybridization results in the loss of chromosomes from the irradiated parental line.

When a line of hybrid cells has been cultured long enough to eliminate all but one or a very few chromosomes from one parental line, the individual clones can be cultured to produce larger volumes of cells, and biochemical tests can be used to identify the genes located on those chromosome(s). Starch-gel electrophoresis can be used to differentiate the ability of the cells to synthesize certain enzymes. These techniques can be used to construct chromosome maps.

Ringertz and Savage (1976) identified the following cell hybrids: man × mouse, man × Chinese hamster, man × Syrian hamster, man × mosquito, mouse × monkey, mouse × mule, mouse × rat, mouse × Chinese hamster, mouse × Syrian hamster, mouse × chick, Chinese hamster × kangaroo rat, and Chinese hamster × chick. Other cell hybrids that have been made include rat × Chinese hamster and chimpanzee × mouse. Other species probably have been used to produce somatic cell hybrids. A very clear review of the early history and development of somatic cell hybrids is found in Ephrussi (1972).

The ability to separate out individual chromosomes in cells which retain their viability—due to the complete set of chromosomes of the other cell line—offers an opportunity for study of characteristics of each chromosome. To some degree somatic cell hybridization is similar to working with species that are polyploid, such as wheat (which is hexaploid). Monosomic and nullisomic plants in polyploids retain their viability through the remainder of the genome.

REFERENCES

BARSKI, G., SORIEUL, S., and COMEFERT, F. 1960. Production dans des cultures *in vitro* de deux souches cellulaires en association, de cellules de caractere "hybride." C. R. Acad. Sci. Paris *251,* 1825–1827.

EPHRUSSI, B. 1972. Hybridization of Somatic Cells. Princeton University Press, Princeton, NJ.

HARRIS, H., and WATKINS, J. F. 1965. Hybrid cells derived from mouse and man: Artificial heterokaryons of mammalian cell from different species. Nature (London) *205,* 640–646.

OKADA, Y. 1958. The fusion of Ehrlich's tumor cells caused by HVJ virus in vitro. Biken's J. *1,* 103–110.

PONTECORVO, G. 1975. Production of mammalian somatic cell hybrids by means of polyethylene glycol treatment. Somatic Cell Genet. *1*(4), 397–400.

RINGERTZ, N. R., and SAVAGE, R. E. 1976. Cell Hybrids, 366 pp. Academic Press, New York.

SCHNEIDERMAN, S., FARBER, J. L., and BOSERGA, R. 1979. A simple method for decreasing the toxicity of polyethylene glycol in mammalian cell hybridization. Somatic Cell Genet. *5*(2), 263–269.

WANG, H. S., NIEWCZAS, V., NAXARETH, H. R. de S., and HAMERTON, J. H. 1979. Cytogenetic characteristics of 26 polyethylene glycol-induced human–hamster hybrid cell lines. Cytogenet. Cell Genet. *24*(4), 233–244.

9

Fertility As Affected by Chromosomes

INTRODUCTION

A high level of reproductive performance is vital to success in all types of livestock enterprises. Increasing the rate of reproduction (fertility) results in more offspring per female, which increases profitability in meat animal production. Keeping the calving interval at 12–13 months on the average in a dairy herd results in more milk produced in the lifetime of a cow than does longer calving intervals. A higher reproductive rate permits a higher selection intensity and a shorter generation interval, permitting faster genetic progress.

FACTORS AFFECTING FERTILITY

The level of fertility or reproductive performance is influenced by many factors. Before considering the known cytological causes of lowered fertility it is desirable to review briefly some of the other factors that influence fertility. The birth of a normal male or female animal is the result of a highly complex series of events, each of which is influenced by many independent as well as correlated factors. In studies of fertility, heritability, which is the portion of the total variance in a quantitative trait that is due to inheritance, has been found to be quite small, essentially zero (Everett *et al.* 1966). However, in a recent study on fertility of dairy *sires* where environmental variables could be partially adjusted, heritability was estimated at .20 to .35 (Gaunt *et al.* 1976). This higher level of heritability undoubtedly reflects a reduction of the effect of many management (or environmental) influences upon fertility. Low heritability of fertility also may be due to the "all-or-none" factor; that is, an offspring is born or it is not. Therefore, hereditary factors which influence livability from the time of conception until paturition cannot be measured on a continuum—they either cause death or permit life. Separation of "fertility" into smaller segments has resulted in higher heritability estimates of those fractions. For example, Zimmerman and Cunningham (1975) working with ovulation rate as one segment of fertility in swine found a heritability varying from .40 ± .08 to .52 ± .10.

Genes or chromosomal aberrations can affect fertility by the direct effect they may have on any of the hereditary factors which influence fertility, or indirectly by modifying the capability of the animal to react to variable management or environmental factors. Management factors obviously are not hereditary in themselves, but they affect measurements of fertility to such a great extent that they need to be identified and considered whenever livestock fertility is discussed.

Disease affects fertility. A general infection of the female reproductive tract may create conditions that impair the viability of sperm and/or the ability of the embryo to become implanted or be maintained during pregnancy. The female reproductive tract must be healthy from the time of fertilization to parturition in order to permit a successful pregnancy. General infection of the male reproductive organs may also impair the production of viable spermatozoa, and such infections may be transmitted to females. Such transmission of infection is more readily achieved through natural service but can also occur through artificial insemination. Specific venereal diseases, such as those caused by *Bacillus abortus, Trichomonas foetus, Vibrio foetus,* and other organisms, can result in an epidemic of abortions which greatly decreased herd fertility.

In dairy herds where a bull is not with the cows and heifers, heat detection has become a major management factor affecting fertility, especially in large herds. It is possible that some genetic factors influence this, since the tendency toward silent heat periods may have been unintentionally selected for as artificial insemination has become so widespread and selection for high production more intense. Heat detection is a management problem, but factors that influence the expression of estrus may have a genetic base.

Artificial insemination involves several other variables. The techniques of the inseminator influence fertility, as has been recognized by the variation achieved in conception rates by inseminators using the same sources of semen. Timing of insemination in relation to the onset of standing heat has been shown to have a significant effect upon success of breeding. Decreased fertility may be due, at least in part, to chromosomal aberrations related to the time of breeding in livestock. In rabbits (Shaver and Carr 1969) and hamsters (Adachi and Ingalls 1976; Yamamoto and Ingalls 1972), breeding females several hours after ovulation, induced by injection of chorionic gonadotropin, resulted in an increase in chromosomal anomalies and considerably lower fertility. The decrease in success when livestock are bred after the optimum time in the heat period is well known, but cytological studies have not been done with blastocysts from large animals in relationship to time of breeding.

Maurer *et al.* (1974) found that frozen semen apparently affected chromosome morphology in rabbits. Preimplantation losses of fertilized ova also increased with length of time frozen.

In dairy cattle very high production, over 16,000 pounds of milk, has been found to result in some decrease in fertility (Studer and Morrow 1978), although several studies, over many years, with cows producing at an average rate have indicated little relationship between fertility and milk or fat production. Apparently, as larger numbers of cows reach higher levels of production this stress begins to have an effect upon fertility. It has also been noted that, among reasons for culling cows, breeding problems have become more important in the last decade, at a time when production levels have been increasing rapidly. It is presumed at this time that the stress of milk production adversely affects reproduction. Modifications of economic producing ability in other types of livestock for characteristics other than milk production may also affect reproductive rates.

Fertility in dairy cattle can be measured in several ways. One standard method, widely used, is the 60- to 90-day non-return rate. By this measure all cows not returning to further service by 90 days following breeding are assumed to be pregnant. Another method is based on the number of females bred which produced a calf regardless of number of breedings. Also used is the number of services per conception, considering services at any one heat period as one, even though they may be bred more than once at one heat. Another is the length of calving interval, which is influenced by management decisions on the length of time after calving before a cow should be bred the first time.

Within each of the above measures several factors may be acting to reduce the number of successful pregnancies:

1. The egg may not be fertilized, due to characteristics either of the ovum or of the sperm.
2. The fertilized ovum may not continue development owing to a lack of viability, which may be caused by aberrations in chromosomes, genes, or the tubal or uterine environment.
3. Implantation may not occur, as the result of unsatisfactory uterine environment or because the zygote is deficient.
4. The implanted embryo may die, owing to either maternal causes or an embryonic deficiency.

Cytological aberrations could be operating in each of the above situations. Ova or sperm that are aneuploid, heteroploid, or carry a deletion or unbalanced translocation may produce zygotes which die early in embryonic development. Amann and Griel (1974) reported that 94% of the recovered ova from 51 cows bred with semen collected in the usual manner were fertilized, in females killed 26–96 hours after breeding. Bearden *et al.* (1956) found 96.6% of the ova fertilized when high-fertility bulls were used, and only 76.9% with low-fertility bulls when the females were slaughtered 3 days post-estrus. Kidder *et al.* (1954) found that 100% of the normal ova recovered from heifers bred to high-fertility bulls were fertilized, and only 71.9% fertilized from low-fertility bulls. These data all indicate that when high-fertility bulls are used, so that failure of the sperm is at a low level, nearly all ova are capable of being fertilized. The latter two authors reported a considerable decrease in number of embryos from both high- and low-fertility bulls by 33 days, when bred heifers were slaughtered, or by 60- to 90-day non-return rates. It would seem reasonable to assume that meiotic errors occurring in dairy females do not reduce the ability of an ovum to be fertilized, but may markedly affect subsequent viability. Koenig *et al.* (1983) have shown that 23% of the oocytes matured in vitro from slaughtered heifers had abnormal chromosome complements. The significance of this to reduced fertility in cattle still needs to be established, but it indicates that oocytes with abnormal chromosomes could be one cause for early embryonic loss.

Nutrition affects fertility although general malnutrition must become relatively severe before reproduction is affected. Flushing of ewes and sows is recognized as a method for improving fertility. Flushing is defined as the feeding of highly nutritious feeds in greater amounts to females shortly before breeding.

To conceive and maintain pregnancy females and males must possess reproductive tracts and organs that are normal anatomically. One of the earliest discoveries of anatomical disorders decreasing fertility in cattle was "white heifer disease" found in Shorthorns (Gilmore 1949). Apparently there was some pleiotropic effect of the gene causing white color, or a closely linked gene. A color gene in sheep has been associated with depressed fertility (Adalsteinsson 1975).

FREEMARTINISM

The condition known as freemartinism in cattle was recognized many years ago as a type of infertility associated with anatomical defects of the female reproductive tract in a heifer born twin to a bull (Marcum 1974). The reproductive tracts are usually underdeveloped or incomplete. It was first concluded in 1911, by Tandler and Keller, that choriovascular anastomosis of the placentas was the primary cause of the condition. It is still not known with certainty whether the development of the female is altered by hormonal effect of the earlier developing male reproductive organs or is related to the mixture of XX and XY cells in the developing fetus.

In heterosexual twins resulting in freemartins both the males and the females are chimeric. The term chimera is more precise than mosaic for this condition since it describes a zygote with cells from a different zygote, as compared with the term mosaic, which describes a zygote that has cells of different genetic composition but derived from the same zygote. The percentage of XX cells in males, or XY cells in females, varies greatly. At least three cases of females with 100% XY cells have been found within the total number observed (Marcum 1974). At the other extreme, typical freemartins have been found with fewer than 5% XY cells.

It seems logical to assume that if the vascular anastomosis occurred between some fairly large blood vessels the number of cells from the opposite sex would be large, and if the anastomosis was between minor blood vessels the number of cells from the opposite sex would be small. It also seems logical to assume that each embryo would provide at least half of its own type of lymphocytes, so that the maximum of the opposite type of cells, XY in females and XX in males, would be about 50%. Furthermore, it seems logical to assume that if the male of a pair of heterosexual twins had 25% XX cells the female would probably have 25% XY cells since the exchange of blood would be expected to be equal. However, observation has shown these assumptions to be false. Instead, Basrur and Kanagawa (1969) found that the percentages of cells of one type, or the ratio of male : female cells within twins, were usually parallel. If the male twin had 85% XY cells, the female twin usually had 85% XY cells. They concluded that there was some type of "dominance" of one cell type in twins with placental vascular anastomosis. The reason for this has not been determined.

Over a 15-month period several pairs of heterosexual twins were analyzed by Kanagawa *et al.* (1967), and the ratio of XX : XY cells in each individual was found to be relatively constant, indicating that the ratio for each individual was relatively constant for life. Marcum *et al.* (1972) also found a positive correlation between the members of heterosexual twins.

Many studies of bovine heterosexual twins (e.g., Marcum 1974) have indicated that about 90–91% are chimeric. Thus, approximately 9–10% of the females from heterosexual twins are fertile. Because of this low frequency, cattle breeders generally have assumed that any female born twin to a bull would have such a low probability for being fertile that such females are usually sent to slaughter. Until 1977, no chimeric females had been found that were fertile. Eldridge and Blazak (1977), Smith *et al.* (1977), and Miyake *et al.* (1980) each reported such a case. In the first paper the lymphocytes of 15 females which had produced a calf or had been diagnosed pregnant and had been born twin to a bull were examined. In one case a female was chimeric but still produced a calf. The other 14 animals were all non-chimeric, determined by examining at least 50 cells per animal for XY cells. In each of the other two reports, a chimeric female produced a calf. In the first and third reports the calves were born dead due to difficult delivery.

When a female is born twin to a male the breeder may wish to know whether it has a reasonable probability of being fertile. The first step in diagnosis is to examine the heifer for gross abnormalities of the external genitalia. In some cases the clitoris is enlarged and resembles a small penis, the vulva may be displaced downward, and extra hair may be present as on the prepuce of a male. These all indicate freemartinism. If the external genitalia appear normal, a 6-inch test tube can be lubricated and inserted into the vagina. A short, blind vagina will also indicate sterility. If the vagina appears normal in length the animal can be blood typed, or examined for gonosomal, sex chromosome, chimerism. Blood typing will reveal weak immune reactions to all blood types found also in the male, if chorionic anastomosis has occurred. If no anastomosis has occurred, the blood types of the male and female will usually differ in one or more blood antigens. Examination for gonosomal chimerism can be done by lymphocyte culture. If both XX and XY cells are found in the female, anastomosis of the placenta is indicated. Since one out of 15 fertile females born twin to bulls was found to be chimeric (Eldridge and Blazak 1977), the diagnosis can be considered to be greater than or equal to 90% accurate. It is desirable to examine the chromosomes of both members of the heterosexual twin pair, since examination of one corroborates the findings in the other. The chromosomal examination can be done as early as desired after the calf is born. This diagnosis is of special value to dairymen who normally would not keep the females that would be sterile. In beef herds where both females and bulls (or steers) may be kept for feeding out to slaughter the diagnosis may be delayed until the heifer exhibits, or does not exhibit estrus. She can then be bred if she comes in heat and if pregnancy results she is diagnosed to be fertile,

David *et al.* (1976) found five cases where Friesian dairy heifers were purchased from markets in England and from 10 to 28% were freemartins. Apparently dairymen had sold their females born twin to bulls because they knew the probability of fertility was very low. Either they did not inform the dealer, or the dealer sold the animals for dairy purposes without informing the buyers.

Marcum *et al.* (1972) found that there was no correlation between the percentage of XY cells in freemartins and the degree of virilization of the reproductive tracts. The basic mechanism for causing freemartinism is still not firmly established.

According to Makino *et al.* (1965), XY cells are also found in freemartins in tissues other than blood. From their studies, Makino *et al.* postulated that cells from the opposite twin affected tissues originating from mesoderm and endoderm but not from ectoderm.

Modification of the female reproductive tract similar to that observed in cattle has been found occasionally in sheep, goats, pigs, horses, and fowls. (Marcum 1974). A very few cases of blood-cell chimerism between heterosexual twin humans has also been found, but apparently in these cases fertility was not affected (Uchida *et al.* 1964).

Chimeric (XX/XY) male cattle have also been discovered among single-born animals (Fechheimer 1973; Makinen 1974; Dunn *et al.* 1979) and chimerism has been found in both male and female cattle (Wijeratne *et al.* 1977). Apparently twins may be conceived, their placentas anastomosed, and one twin die in utero. If the death of one twin occurs after anastomosis has been established, the chimeric condition continues in the remaining twin. Fechheimer and Harper (1980) examined the chromosomes of 224 fetuses collected from gravid uteruses obtained from a local slaughterhouse. They found 5 with sex chromosome chimerism and no evidence of twinning. From this they concluded that the chimerism was the result of fertilization of the ovum by two spermatozoa, with one fertilizing a polar body, or fusion of two zygotes. Swartz and Vogt (1983) found 1 single-born XX/XY heifer among the 71 animals which had not conceived during two successive breeding seasons. Some infertility in single-born bovine females, which seems unexplained, may be due to this type of chimerism.

Schroder (1975) has reviewed the evidence of transplacental transfer of blood, erythrocytes, and leukocytes in humans. A very low percentage of XY cells has been found in mothers who gave birth to males and in some cases where the mothers gave birth to females. In the latter case there was a question about the mother carrying such fetal cells from a previous male birth. So far no similar results have been found in cattle.

Multiple births of a number greater than two in cattle where the sexes are not alike also show XX/XY chimerism in all animals. This indicates anastomosis of the placentas for all animals from the pregnancy (Marcum 1974).

FERTILITY OF MALES WITH SEX CHROMOSOME CHIMERISM

Since bulls born twin to heifers are seldom sterile it has been concluded generally that fertility in the male bovine is not affected by sex chromosome chimerism. Few studies have been made, however, in which chimerism was specifically identified cytologically, and in which degree of fertility was measured in contrast to assignment of the bulls to just two groups, fertile or sterile. Dunn *et al.* (1979) identified 12 chimeric bulls selected for artificial insemination (AI) and compared them with 128 non-twin bulls, 48 of which were karyotyped and found to be non-chimeric. In a larger sample of 260 bulls from the same AI organization (including the 48 in the comparison sample) all were 60,XY. Therefore, the other 80 bulls in the comparison sample were also assumed to be non-chimeric. Seven (58.3%) of the 12 chimeric bulls were culled over a period of up to 10 years of age for reproductive reasons, compared to 7 (5.4%) of the 128 control bulls over the same age span. The difference was highly significant, $P < .001$. However, based upon 60- to 90-day non-return rates for 30,814 services to the 8 chimeric bulls used in AI, the deviation from breed average was not significant. Therefore, not all chimeric bulls are less fertile.

Gustavsson (1979) reported no difference in fertility between bulls born as dizygotic twins and single-born bulls. He studied 33 dizygotic twin bulls over a period of 20 years. Long (1979) reviewed the cases to date and, while she admitted some possibility of fertility problems, from a practical point of view, could see no reason to select against dizygotic twin bulls for that reason alone.

DeGiovanni *et al.* (1975) studied 39 bulls in an Italian AI center and found one chimeric XX/XY Holstein bull with a fertility of 57.8% from 365 first services in the first year of service, based upon 60–90 day non-return and 48.9% from 463 first services in the second year. This was compared to 62% for all bulls in the center. The bull sired 32 males and 116 females, a highly significant deviation from the 50/50 expected ratio. Further discussion on this bull will be found in Chapter 11.

INTERSEXES

Intersexes in livestock occur at a low frequency but occasionally are found in almost all, if not all, species. Dunn *et al.* (1970) found a true hermaphrodite in cattle, associated with a diploid–triploid chimerism evaluated cytologically. One of three intersex bovines studied by Kieffer and Sorenson (1971) was not chimeric, XX/XY. The reason for the intersex condition varied with cases, and was not completely determined, but freemartinism probably was the cause for two out of the three.

McFeely *et al.* (1967) studied 4 bovine intersexes and found 2 sex chromosome chimeras, one 60,XY and one case of a 60,X? in which only one normal X was found in a majority of the cells accompanied by an unidentified long acrocentric chromosome. In 7 out of 46 cells 2 apparently normal X's were found. The long acrocentric was postulated to be either a pericentric inversion in the X chromosome or a translocation of the short arm of the X chromosome to an autosome, and it was believed that this chromosome modification altered the formation of the genital tract.

In addition to bovines, McFeely (1967) also observed the chromosomes of four canines, two swine, three goats, and one cat which were intersexes. The chromosomal analyses varied, but both swine were cytologically 38,XX.

Sysa *et al.* (1974) reported one case obtained from a slaughterhouse of an intersex calf which had only XY sex chromosomes. No information was available on breeding, but the animal's cells were all XY. They also looked for but did not find Barr bodies, which would have indicated XX cells. They suggested that the intersex may have resulted from the absence of the "hypothetical X factor" (Jost), which has been postulated to have a lytic effect upon the Müllerian ducts.

In swine Makino *et al.* (1962) and Hard and Eisen (1965) each found one intersex which possessed the female XX chromosome constitution. Miyake (1973) found eight similar intersex cases in swine where the karyotypes were all XX. The gonads were of different characteristics:

2 cases both sides testis-like
2 cases both sides ovo-testis
2 cases one side ovary-like, the other side ovo-testis
1 case one side testis-like, the other side ovo-testis
1 case one side testis-like, the other side ovary-like

He gave a fairly complete review of a number of other studies on swine intersexes. These reports preceded the knowledge of banding, which might have revealed more detailed information. In one case (Hard and Eisen 1965), the animal most strongly resembled a male phenotypically. They reviewed several human cases of intersexes in which the karyotype was female.

One case of an intersex calf has been studied by Eldridge *et al.* (1984). The chromosomal analysis, from both lymphocytes and connective tissue cell culture, showed the animal to be a male, 60,XY. It was also found to be H-Y antigen negative, concurrently compared to similar cell cultures from a normal male and a normal female. The gene for the H-Y antigen, thought to be located on the Y chromosome, although possibly modified by one gene on the X chromosome and one on an autosome, is necessary for testicular development. This animal resembled a male in its forequarters, but urinated through a vulva displaced about 15 cm below the normal location, and had no penis or externally apparent testicles. The XY intersexes reported by McFeely *et al.* (1967) and Sysa *et al.* (1974) and the sterile XY female reported by Balakrishnan *et al.* (1979) also might have been the result of the H-Y antigen-negative condition.

TRISOMY

Several cases have been reported of sex chromosome trisomy in cattle. Norberg *et al.* (1976) found a Norwegian Red heifer, 18 months old, with a 61,XXX chromosome composition. She came in heat at 13 months but did not return to heat subsequently. The uterus was small, the ovaries underdeveloped, and there was one corpus luteum. Histologically there were few primary follicles but no degenerative changes. Forty-five cells were observed, all of which had 61,XXX chromosomes, with no chimerism. The dam and maternal half sibs were 60,XX.

Logue *et al.* (1977) studied a British Friesian bull with bilateral testicular hypoplasia and found it to be 61,XXY. The second X was late-replicating. Scott and Gregory (1965) also have reported a 61,XXY trisomic in an intersex in cattle. Both of these bulls were apparently sterile.

Lojda *et al.* (1975, 1976) reported a mosaic bull with chromosome composition 60,XY/61,XXY/59,XO which produced progeny, including 23% with gonadal hypoplasia. One of the affected male offspring was 60,XY/60,XX/59,XO.

GENERAL FERTILITY

Several reviews have been published (Foote 1970; Olds 1953; Pelissier 1972) which identify and discuss many causes for infertility in cattle, including genetic and cytogenetic causes (Foote 1970). These reviews emphasize the complex nature of fertility and the small genetic effects identifiable by statistical analyses of large amounts of data. Each of the components of fertility however, is probably subject to genetic and cytogenetic control.

Testis size and consistency have been shown to be correlated with number of spermatozoa produced and their morphology and motility (Foote *et al.* 1977). Hormones controlling these factors may be separately affected by different genes.

DIRECT CHROMOSOMAL EFFECTS

The most widely studied chromosomal aberration in cattle, the 1/29 Robertsonian translocation, apparently reduces fertility in females (Gustavsson 1971A,B; Blazak and Eldridge 1977; Refsdal 1976; Succi *et al.* 1979). When the 1/29 translocation males in use in artificial insemination in Sweden were compared with the bulls without the translocation no difference was found in fertility. A later study on unselected males (Dyrendahl and Gustavsson 1979) found an indication of nondisjunction in the second meiotic division of heterozygous 1/29 bulls to be about 5% higher than in normal bulls. It is currently postulated that the reduction in fertility of the daughters is due to some nondisjunction during meiosis (see Chapter 5 for illustration). The relationship between the number of adjacent and alternate segregations should determine the number of normal and aneuploid gametes produced. Alternate segregation permits balanced gametes, whereas adjacent segregation results in gametes being formed with a duplication or deficiency. In females only one functional gamete is usually produced at each ovulation. If it is aneuploid the chance for survival of the embryo is decreased. Refsdal (1976) showed that heterozygous 1/29 translocation animals had an increase in returns to service up to 90 days after breeding compared with normal cows. This indicates that the aneuploid embryos are lost before 90 days. No aneuploid calves at birth or later have been found from heterozygous 1/29 cows, indicating also that the aneuploid embryos are apparently inviable.

The male, on the other hand, produces many millions of spermatozoa in each ejaculate, as compared to one ovum usually being produced at each heat period in the female. Apparently, in heterozygous 1/29 males selected for high fertility the aneuploid spermatozoa are less likely to fertilize an egg, and the many spermatozoa with balanced sets of chromosomes may dominate the fertilization process. Thus, some males do not show a decrease in fertility.

It is interesting to note that, among the heterozygous translocation sires, a few had significantly more fertile daughters than was expected. Therefore, some unexplained factors may affect the frequency of adjacent segregation at meiosis. This also accounts for the variability in reports on the effect of the translocation on the fertility of females. It is also possible that in some heterozygous males there are some genetic factors that cause spermatozoa with duplications or deficiencies to be less viable and fertilize ova at a lower frequency.

Although more than 20 other Robertsonian translocations have been found in cattle, they have not usually been found in great numbers. Therefore, it has not been possible to determine the effect which these may have on fertility.

Another chromosomal anomaly that has affected fertility is the tandem fusion of chromosomes reported by Hansen (1970). This was found in the Red Danish milk breed. The translocation was transmitted to about 50% of the offspring as expected. The bulls had fertility reduced by about 10% and the females also showed some reduction in fertility.

Knudsen (1956) reported on bulls of low fertility in Sweden. He found two cases with an inversion and three with a translocation that affected fertility. He also found in meiosis some "sticky chromosomes" that resulted in chaotic meiotic events. A few normal sperm were produced, preventing complete sterility. The bulls also had bilateral testicular hypoplasia.

Bongso and Basrur (1976) investigated 58 Guernsey sires of low fertility in Canada. Five showed chromosomal peculiarities. One Guernsey bull carried the 1/29 translocation and was low in fertility. Another had a 27/29 translocation with two centromeres, one located near the end of the chromosome and one located medially. However, this bull was above average in fertility. Several hypotheses were presented to explain this. No other tissues were cultured to see if the aberration was limited to blood lymphocytes. Three other Guernsey bulls of low fertility had 10–15% of their cells with chromatid breaks, a higher frequency than found in other bulls.

In a study of 71 beef heifers which had not conceived through two successive breeding periods, Swartz and Vogt (1983) found 13 animals with chromosomal aberrations. Five animals (2 Red Polled, 2 Simmental, and 1 Marc II) were heterozygous for the 1/29 Robertsonian translocation. Two were 61,XXX, two were 59,XO/60,XX mosaics, two were mixoploid mosaics (one 59,XO/60,XX/61,XXX and one 59,XO/60,XX/61,XO), one was a tetraploid/diploid mosaic (120,XXXX/60,XX), and one was a sex chromosome chimera (60,XX/60,XY), which was recorded as a single birth, not a female born twin to a bull. Many population studies have been made of phenotypically normal cattle, and frequently these studies have found only chromosomally normal animals, or aberrations at an extremely low frequency. This study is another illustration of the value of studying animals which are not phenotypically normal in one or more characteristics. The 71 animals that Swartz and Vogt studied as controls were all free from chromosomal aberrations. The heritability of fertility, when studied quantitatively in all livestock species, is quite small, but there may be several chromosomal causes for low fertility that are identifiable, and which may occur randomly or may be highly heritable.

Bouters *et al.* (1974) found in swine some chromosomal abnormalities in embryos from boars of low fertility. Several causes were proposed for different results from each boar. Only 1 of 13 boars had a chromosomal translocation and 1 had no metaphase spreads, so 11 of 13 cases of lowered fertility did not have an apparent chromosomal problem. This illustrates how certain phenotypic variations, in this case lowered fertility, may be nearly identical with some due to chromosomal aberrations and some to other causes. Only by cytological analysis can the chromosomal aberrations be identified.

Alberman and Creasy (1977) reviewed several studies in humans which had found that 60% of the involuntary abortions in the first 3 months of pregnancy are associated with chromosomal abnormalities. Combining these data with information on the frequency of chromosomal errors in live births, stillborns, and late abortions, they concluded that in about 8% of all diagnosed pregnancies in humans the fetus is chromosomally abnormal.

REFERENCES

ADACHI, K., and INGALLS, T. H. 1976. Ovum aging and pH imbalance as a cause of chromosomal anomalies in the hamster. Science *194,* 946–948.

ADALSTEINSSON, S. 1975. Depressed fertility in *Icelandic* sheep caused by a single colour gene. Ann. Genet. Sel.Anim. *7,* 445–447.

ALBERMAN, E., and CREASY, M. R. 1977. Frequency of chromosomal abnormalities in miscarriages and perinatal deaths. J. Med. Genet. *14,* 313–315.

AMANN, R. P., and GRIEL, L. C., JR. 1974. Fertility of bovine spermatozoa from rete testis, cauda epididymis, and ejaculated semen. J. Dairy Sci. *57,* 212–219.

BALAKRISHNAN, C. R., YADAV, B. R., BHATTI, A. A., and NAIR, K. G. S. 1979. Unusual chromosome constitution of a bovine freemartin. Ind. J. Dairy Sci. *32,* 191–193.

BASRUR, P. K., and KANAGAWA, H. 1969. Parallelism in chimeric ratios in heterosexual cattle twins. Genetics, *63,* 419–425.

BEARDEN, H. J., HANSEL, W., and BRATTON, R. W. 1956. Fertilization and embryonic mortality rates of bulls with histories of either low or high fertility in artificial breeding. J. Dairy Sci. *39,* 312–318.

BLAZAK, W. R., and ELDRIDGE, F. E. 1977. A Robertsonian translocation and its effect upon fertility in Brown Swiss cattle. J. Dairy Sci. *60,* 1133–1142.

BONGSO, A., and BASRUR, P. K. 1976. Chromosome anomalies in Canadian Guernsey bulls. Cornell Vet. *66,* 476–488.

BOUTERS, R., BONTE, P., and VANDEPLASSCHE, M. 1974. Chromosomal abnormalities and embryonic death in pigs. 1st World Congr. Genet. Appl. Livestock Prod. *3:* 169–171.

DAVID, J. S. E., LONG, S. E., and EDDY, R. 1976. The incidence of freemartins in heifer calves purchased from markets. Vet. Rec. *98,* 417–418.

DEGIOVANNI, A., POPESCU, C. P., and SUCCI, G. 1975. The first cytogenetic study in an Italian artificial insemination center. Ann. Genet. Sel. Anim. *7,* 311–315.

DYRENDAHL, I., and GUSTAVSSON, I. 1979. Sexual functions, semen characteristics and fertility of bulls carrying the 1/29 chromosome translocation. Hereditas *90,* 281–289.

DUNN, H. O., MCENTEE, K., and HANSEL, W. 1970. Diploid-triploid chimerism in a bovine true hermaphrodite. Cytogenetics, *9,* 245–259.

DUNN, H. O., MCENTEE, K., HALL, C. E., JOHNSON, R. H., JR., and STONE, W. H. 1979. Cytogenetic and reproductive studies of bulls born co-twin with freemartins. J. Reprod. Fertil. *57,* 21–30.

ELDRIDGE, F. E., and BLAZAK, W. F. 1977. Chromosomal analysis of fertile female heterosexual twins in cattle. J. Dairy Sci. *60,* 458–463.

ELDRIDGE, F. E., KOENIG, J. L. F., and HARRIS, N. B. 1984. H-Y antigen

negative, 60,XY, intersex calf. Proc. 6th Eur. Colloq. Cytogenet. Domest. Anim. (In press.)
EVERETT, R. W., ARMSTRONG, D. V., and BOYD, L. J. 1966. Genetic relationship between production and breeding efficiency. J. Dairy Sci. *49,* 879–886.
FECHHEIMER, N. S. 1973. A cytogenetic survey of young bulls in the U.S.A. Vet. Rec. *93,* 535–536.
FECHHEIMER, N. S., and HARPER, R. L. 1980. Karyological examination of bovine fetuses collected at an abattoir. Proc. 4th Eur. Colloq. Cytogenet. Domest. Anim. 194–199.
FOOTE, R. H. 1970. Inheritance of fertility—facts, opinions, and speculation. J. Dairy Sci. *53,* 936–944.
FOOTE, R. H., SEIDEL, G. E., JR., HAHN, J., BERNDTSON, W. E., and COULTER, G. H. 1977. Seminal quality, spermatozoal output, and testicular changes in growing Holstein bulls. J. Dairy Sci. *60,* 85–88.
GAUNT, S. N., DAMON, R. A., JR., and BEAN, B. H. 1976. Heritability and repeatability of fertility of dairy sires. J. Dairy Sci. *59,* 1502–1504.
GILMORE, L. O. 1949. The inheritance of functional causes of reproductive inefficiency: A review. J. Dairy Sci. *32,* 71–91.
GUSTAVSSON, I. 1971A. Culling rates in daughters of sires with a translocation of centric fusion type. Hereditas *67,* 65–73.
GUSTAVSSON, I. 1971B. Chromosomes of repeat-breeder heifers. Hereditas *68,* 331–332.
GUSTAVSSON, I. 1979. Fertility of sires born as dizygotic twins and sex ratio in their progeny groups. Ann. Genet. Sel. Anim. *9,* 531.
HANSEN, K. M. 1970. Reduced fertility in a bull with chromosomal translocation. Bertn. XI Nord. Vet. Kongr. Bergen. (Anim. Breed. Abstr. *40,* 2961.)
HARD, W. L., and EISEN, J. D. 1965. A phenotypic male swine with a female karyotype. J. Hered. *56,* 254–258.
KANAGAWA, K., KAWATA, K., ISHIKAWA, T., and INOUE, T. 1967. Chromosome studies on heterosexual twins in cattle. IV. Long-term observations of sex-chromosome chimera ratio in cultured leukocytes. Jpn. J. Vet. Res. *15,* 31–36.
KIDDER, H. E., BLACK, W. G., WILTBANK, J. N., and ULBERG, L. C. 1954. Fertilization rates and embryonic death rates in cows bred to bulls of different levels of fertility. J. Dairy Sci. *37,* 691–697.
KIEFFER, N. M., and SORENSON, A. M., JR. 1971. Some cytogenetic aspects of intersexuality in the bovine. J. Anim. Sci. *32,* 1219–1228.
KNUDSEN, O. 1956. Chromosome investigations in bulls. Fortpflanzung, Zuchthyg. Haustierbesamung. *68,* 5–7.
KOENIG, J. L. F., ELDRIDGE, F. E., and HARRIS, N. 1983. A cytogenetic analysis of bovine oocytes cultured in vitro. J. Dairy Sci. *66* (Suppl. 1), 253.
LOGUE, D. N., and HARVEY, M. J. A. 1978. Meiosis and spermatogenesis in bulls heterozygous for a presumptive 1/29 Robertsonian translocation. J. Reprod. Fertil. *54,* 159–165.
LOGUE, D. N., HARVEY, M. J. A., ELDRIDGE, F. E., and POLLOCK, D. 1977. Identification of some chromosome anomalies in cattle. Personal communication.
LOJDA, L., and HAVRANKOVA, J. 1975. Hereditary testicular hypoplasia in

the progeny of a bull with chromosomal mosaicism (60,XY/61,XXY/59,X). 2nd Eur. Koll. Zytogenet. Vet. Tierzucht Säugetier 193–198.

LOJDA, L., RUBES, J., STAVIKOVA, M., and HAVRANKOVA, J. 1976. Chromosomal findings in some reproductive disorders in bulls. 7th Int. Congr. Anim. Reprod. Artif. Insem., Krakow 158.

LONG, S. E. 1979. The fertility of bulls born twin to freemartins: A review. Vet. Rec. *104,* 211–213.

MAKINEN, A. 1974. Chimerism in a bull calf. Hereditas *76,* 154–156.

MAKINO, S., SASAKI, M. S., SOFUNI, T., and ISHIKAWA, T. 1962. Chromosome condition of an intersex swine. Proc. Jpn. Acad. Sci. *38,* 686–689.

MAKINO, S., MURAMOTO, J., and ISHIKAWA, T. 1965. Notes on XX/XY mosaicism in cells of various tissues of heterosexual twins of cattle. Proc. Jpn. Acad. Sci. *41,* 414–418.

MARCUM, J. B. 1974. The freemartin syndrome. Anim. Breed. Abstr. *42,* 227–242.

MARCUM, J. B., LASLEY, J. F., and DAY, B. N. 1972. Variability of sex-chromosome chimerism in cattle from heterosexual multiple births. Cytogenetics *11,* 388–399.

MAURER, R., GELDERMAN, H., STRANZINGER, G., and PAUFLER, S. K. 1974. Invesitgation of chromosomal aberrations in rabbit blastocysts using frozen semen for insemination. Zuchthygiene *9,* 122–128.

McFEELY, R. A., HARE, W. C. D., and BIGGERS, J. D. 1967. Chromosome studies in 14 cases of intersex in domestic mammals. Cytogenetics *6,* 242–253.

MIYAKE, Y. 1973. Cytogenetical studies on swine intersexes. Jpn. J. Vet. Res. *21,* 41–49.

MIYAKE, Y., ISHIKAWA, T., ABE, T., KOMATSU, M., and KODAMA, Y. 1980. A fertile case of a bovine heterosexual twin female with sex-chromosomal chimerism. Zuchthygiene *15,* 103–106.

NORBERG, H. S., REFSDAL, A. O., GARM, O. N., and NES, N. 1976. A case report on X-trisomy in cattle. Hereditas *82,* 69–72.

OLDS, D. 1953. Infertility in cattle—A review. J. Am. Vet. Med. Assoc. *122,* 276–287.

PELISSIER, C. L. 1972. Herd breeding problems and their consequences. J. Dairy Sci. *55,* 385–391.

REFSDAL, A. O. 1976. Low fertility in daughters of bulls with 1/29 translocation. Acta Vet. Scand. *17,* 190–195.

SCHRODER, J. 1975. Transplacental passage of blood cells. J. Med. Genet. *12,* 230–242.

SCOTT, C. O., and GREGORY, P. W. 1965. An XXY trisomic in an intersex of *Bos taurus.* Genetics *53,* 473.

SHAVER, E. L., and CARR, D. A. 1969. The chromosome complement of rabbit blastocysts in relation to the time of mating and ovulation. Can. J. Genet. Cytol. *11,* 287–293.

SMITH, G. S., VAN CAMP, S. D., and BASRUR, P. K. 1977. A fertile female cotwin to a male calf. Can. Vet. J. *18,* 287–289.

STUDER, E., and MORROW, D. A. 1978. Postpartum evaluation of bovine reproductive potential: Comparison of findings from genital tract examination per rectum, uterine culture, and endometrial biopsy. J. Am. Vet. Med. Assoc. *172,* 489–494.

SUCCI, G., MOLTENI, L., and DEGIOVANNI, A.M. 1979. Preliminary observations on the fertility of daughters of bulls carrying the 1/29 translocation. Atti Soc. Ital. Sci. Vet. Avic. *33,* 215.

SWARTZ, H. A., and VOGT, D. W. 1983. Chromosome abnormalities as a cause of reproductive inefficiency in heifers. J. Hered. *74,* 320–324.

SYSA, P., BERNACKI, Z., and KUNSKA, A. 1974. Intersexuality in cattle—A case of male pseudohermaphrodismus with a 60,XY karyotype. Vet. Rec. *94,* 30–31.

TANDLER, J., and KELLER, K. 1911. The chorion in differently sexed twin pregnancy of the cow and the morphology of the female animals which result. Dtsch. Tieraerzt. Wochenschr. *19,* 148.

UCHIDA, I. A., WANG, H. C., and RAY, M. 1964. Dizygotic twins with XX/XY chimerism. Nature (London) *204,* 191.

WIJERATNE, W. V. S., MUNRO, I. V., and WILKES, P. R. 1977. Heifer sterility associated with single-birth freemartinism. Vet. Rec. *100,* 333–336.

YAMAMOTO, M., and INGALLS, T. H. 1972. Delayed fertilization and chromosome anomalies in the hamster embryo. Science *176,* 518–521.

ZIMMERMAN, D. R., and CUNNINGHAM, P. J. 1975. Selection for ovulation rate in swine: Population, procedures and ovulation response. J. Anim. Sci. *40,* 61–69.

10

Laboratory Procedures for Chromosome Studies

INTRODUCTION

A fundamental principle of biological research is to work as closely as possible with the living organism in its natural environment. In spite of this, it is frequently necessary, even imperative, to modify the natural environment in order to perform certain experiments. It is often desirable to control stringently certain variables, or to remove certain substances for analysis, or to simulate in vivo actions with in vitro methods.

It would be very desirable to study chromosomes in living tissue in vivo, but obviously this is not practicable because magnification is necessary to view chromosomes. Light microscopy and electron microscopy both require very thin sections, or layers, of material. In animal tissue this means removing material from the body, except for some special cases like the membrane between the toes of frogs. Microorganisms can be studied live, and some living plant tissues like root tips can be observed while attached to the living organisms. For ease of handling, however, even plant tissues are removed from the living plant and observed in a medium that will permit continuing growth for a period of time.

Early chromosome studies were usually made by sectioning the plant or animal material with microtome, using rapidly dividing tissues such as root tips from plants or testicular tissue from animals. These tissues were placed in a fluid that would kill and fix the material rapidly. Then the pieces were mounted in paraffin blocks or other media and sectioned, mounted on slides, and stained. With small numbers of chromosomes, this method was reasonably useful, but with larger numbers of chromosomes, typical of livestock, this method did not provide clearly distinct and separated chromosomes. More recently, tissue culture has permitted the growth of cells in media, which improves ease of handling, gives wider separation of chromosomes, and allows more accurate observation. Tissue culture has an interesting history.

The first reported in vitro tissue culture was done by Wilhelm Roux in 1885. He took the medullary plate from an embryo developing in an avian egg and grew these cells in a saline solution for a few days. The next report of tissue culture was made by J. Arnold, a German, in 1887, who put some alder pith into a frog where it was invaded by blood leukocytes. After the pith had been in the frog for a few days, he removed it and placed it in a warm saline solution where he noticed that the leukocytes migrated out of the pith into the saline solution. In 1903 Jolly maintained salamander leukocytes in hanging drops for up to a month by placing one drop of fluid on a coverslip and inverting it over a depression in a glass slide. In 1907 Harrison reported growth in a culture of nerve cells and showed that the cells maintained their integrity. Carrel, about 1912, was able to develop further refinements in tissue culture and, working with a group of people, developed several new concepts of tissue culture. Carrel started one series of mammalian cells which he kept alive for 40 years by a very elaborate and highly controlled procedure. Even though considerable tissue culture work was done over several decades, tissue culture was not used for chromosome studies until recent years (Paul 1975; Hamerton 1971).

DEVELOPMENT OF SYNTHETIC MEDIA

Prior to 1930–1940 natural media was used in tissue culture research. For example, some cells placed into a blood clot would grow and divide. Although the cells were observed growing in the blood clot, there was always the question whether the cells were from the tissue being cultured or originated from the blood clot. It was necessary to develop a synthetic medium in order to be sure that the cells growing in the culture were from the tissue being studied. Synthetic media also permitted more precise control of growth since the ingredients were known and each could be varied quantitatively and qualitatively. Natural media are very complex mixtures with minute quantities of numerous unknowns, and also vary from one source to another.

The search for synthetic media was started around 1911 or 1912, but it was in the 1930s before an effective synthetic medium was formulated. The development of synthetic media for tissue culture work proceeded very slowly until about the 1940s and 1950s, when there was a tremendous expansion. Today there are many different culture media reported in the literature, including TC-199, McCoy's 5A, Ham's F-10, MEM, RPMI, and many more. Cancer research gave a big impetus to tissue culture, since sizable financial support was available and the characteristics of cells in the growth process had been thought to be the key to understanding malignant growths. Most of the earlier tissue culture work was related to histology, embryology, and endocrinology.

Although growth can be induced and maintained in synthetic tissue culture media, better results usually are obtained when the medium also has 10–25% of natural media such as blood serum, plasma, or embryo extract. Fetal calf serum seems to be the best, but other sources of serum can be used, and good cultures of lymphocytes can be obtained by using the plasma from the collected blood samples that provide the lymphocytes. Some substances, not yet identified, found in blood serum or plasma increase growth of cells in culture. The perfect synthetic medium is still to be formulated. Biochemistry has provided much knowledge to scientists working on tissue culture.

CONTROL OF CONTAMINATION

A universal problem encountered in tissue culture is contamination, which has probably caused more failures than any other single reason. Since a medium provides all necessary material for tissue cell growth, it also permits contaminating cells to grow very rapidly. If bacteria or mold spores get into the culture, they will grow as rapidly or more rapidly than the lymphocytes or other tissue cells being studied. Therefore, all steps in tissue culture must be done under sterile conditions. The tissue, glassware, culture medium, and air surrounding all activities concerned with making and transferring cultures must be kept as free from contamination as possible. Even with the best precautions contamination may occur, so antibiotics such as penicillin, streptomycin, and a fungicide are used routinely to suppress the growth of contaminants. A look at the literature will reveal that tissue culture developed slowly until the discovery and exploitation of antibiotics in the late 1930s and early 1940s.

TISSUE CULTURE

"Tissue culture" is a broad term which can be broken down into three subdivisions: tissue culture, organ culture, and cell culture. Organ culture is an attempt to maintain a piece of tissue from one organ intact so that its biochemistry, endocrinology, and histology can be studied. Tissue culture in its narrower sense is a study of tissues which may form monolayers of cells that adhere to glass or plastic surfaces and which maintain a type of structure, even though different from the same tissue in vivo. Cell culture is a type of tissue culture in which the cells are dispersed or suspended in the medium. An example of cell culture is the growth of lymphocytes. Other sources of animal cells can be induced to grow as cell cultures. Cultures of fibroblasts from connective tissue can be easily grown between two glass coverslips immersed in culture medium. Fibroblasts and several other types of tissues need to adhere to a surface to grow. Placing small pieces of tissue between coverslips in small test tubes, such as Leighton tubes, is time consuming and cumbersome. The use of Falcon (plastic) flasks is simpler. The tissue can be minced up with fine pointed scissors in a small amount of culture medium and streaked across one flat side of the flask with a minimum amount of fluid. This can be done with a sterile Pasteur pipette and a small rubber bulb. Extra fluid can then be removed with the pipette. The flasks are placed on their sides in the incubator for 2 hours, which permits the pieces to adhere to the surface. Then additional medium and serum (20%) can be carefully added so as to not wash the particles free. Initial growth usually takes about 4–8 days. The medium should be changed about every 5 days. Further details for transferring, harvesting, and slide preparation can be found in many recent journal articles, or see Paul (1975). Roller tubes, shaker flasks, Leighton tubes, petri dishes, and other equipment have also been adapted to tissue culture.

TABLE 10.1. Normal Body Temperatures of Different Species of Animals

Animal	Average temperature (°C)
Horse	37.7
Cattle	38.5
Sheep	39.1
Goat	39.7
Pig	39.2
Dog	38.9
Cat	38.6
Rabbit	39.5
Chicken	41.7

Incubator temperature influences the rate of growth of cells in culture. In the literature most references are to an incubation temperature of 37.5°C, which reflects the large amount of work done with human cells. The best incubation temperature for cells is the normal body temperature of the species whose cells are being grown. For humans, that is 37.5°C (98.6°F). For cattle normal body temperature is 38.5°C (100°F), so that is the best incubation temperature. Table 10.1 gives the normal body temperatures for other species.

Cells in culture are susceptible to temperature fluctuations, particularly temperatures above the optimum. Lower temperatures, down to 4°C, will slow growth but usually will not kill the cells. Since incubators vary in quality and their state of repair, thermometers should be placed in several locations in the incubator and checked every 30 minutes for a day or two before cultures are started. Fluctuations should not exceed ±.5°C for repeatable results. Water-jacketed incubators are most desirable because all sides but the door can be kept at a relatively constant temperature. A low level of air circulation in the incubator is also desirable but not necessary.

The pH of the medium is critical. It should be maintained at about 7.1 to 7.2. The pH for Ham's F-10 culture medium is usually adjusted by addition of a 10% solution of $NaHCO_3$. If the cultures are in tightly stoppered containers which are about one sixth to one fourth full, the metabolism of the cells and the $NaHCO_3$ in the medium will maintain a correct CO_2 atmosphere to keep the pH at the correct level. If the cultures are in petri dishes or bottles with loose cotton plugs, then a CO_2 atmosphere must be maintained in the incubator or the pH will drop too low (below 6.8).

The culture medium should be made with triple-distilled water, the last distillation over glass to remove metal ions. The source of water can be a problem. Two cases illustrate this. In one laboratory, river water was the primary source of tap water for the city. Upstream, detergents had been added that had about the same vapor pressure as water and were carried over into the distillate. Cells could not be grown until the laboratory changed its primary water source to a well. In another case the original source of water that was being put through the first distillation was the recovery water from a steam plant. A rust inhibitor being used by the steam plant had the same vapor pressure as water and was toxic to cell cultures.

Many sources of tissue will provide cells that will grow actively without any specific agents to induce mitosis. These include connective tissue from directly under the skin, or lung, spleen, liver, etc. Until 1960, blood leukocytes, specifically lymphocytes, were thought never to divide after entry into the bloodstream either in vivo or in vitro, except in cases of leukemia. Therefore, lymphocytes had not been useful for chromosome studies. In 1960 Peter Nowell accidentally discovered that phytohemagglutinin was a mitogenic agent; that is, it would induce mitosis in lymphocytes. *Phyto* means plants, *hem* refers to blood, and *agglutinin* means to agglutinate or cause to clump. Phytohemagglutinin (PHA) is a plant substance, a saline extract of navy beans (*Phaseolus vulgaris*), which agglutinates the red cells of blood. It has been used for many years to separate white blood cells (leukocytes) from red blood cells (erythrocytes). Nowell accidentally used a solution that was much more dilute than intended. When he observed the cells, he noticed that some were undergoing mitosis. He was a perceptive scientist. Further tests indicated the most appropriate dilution for this purpose. This discovery marked the beginning of a new era in tissue culture studies for research on chromosomes.

Blood is an easy tissue to obtain. Other tissues, such as skin or testicular tissue, can be obtained by biopsy, but this is more difficult. Livestock owners are less willing for such samples to be obtained. Obtaining tissue samples from internal organs presents even more difficult problems. Therefore, blood lymphocytes have become one of the most widely used tissues for chromosome studies, so a detailed outline of procedures is desirable.

LYMPHOCYTE CULTURES

Blood is obtained from the external jugular vein or the middle coccygeal vein in 10-ml Vacutainer tubes with 143 units of heparin per tube. The skin area covering the vein is thoroughly cleaned with 70% ethyl alcohol to reduce contamination. The blood is immediately cooled, either in a refrigerator if handy or in a Styrofoam chest with a container of about 3–4 pounds of crushed ice located at one end. The blood samples are placed in a rack at the other end of the chest so that the collection tubes are not in direct contact with the ice. The tubes are gently inverted several times immediately after collection to permit even distribution of the heparin throughout the sample to prevent clotting. Vigorous shaking breaks many cells and decreases the quality of the sample. The samples are then returned to the laboratory and stored in a refrigerator for up to 24 hours.

The culture medium is prepared as follows. Best results are obtained with freshly prepared medium, although it can be stored for a few days in a refrigerator in airtight bottles.

For 100 ml of Ham's F-10 medium

1.00 g of dried Ham's F-10
97.7 ml of glass-distilled (triple-distilled) water
1.0 ml of antibiotic-antimycotic solution (10,000 IU of penicillin, 10,000 μg streptomycin and 25 μg Fungizone®)
1.00 ml of phytohemagglutinin (M form), no specific units identified

The concentration of the antibiotic–antimycotic solution becomes 100 IU of penicillin, 100 μg of streptomycin and 25 μg Fungizone per milliliter of culture medium. Phytohemagglutinin potency seems to vary, so that each new batch purchased should be tested to determine the optimum amount for maximum mitotic stimulation. These materials are placed in a 250-ml beaker or Erlenmeyer flask with a plastic-coated magnetic stirrer and stirred until fully dissolved. Then 10% (w/v) sodium bicarbonate solution is added until the solution reaches a pH of approximately 6.9–7.0. In the process of putting the medium through a Millipore filter, under vacuum, the pH in the medium will increase approximately .2 to .3, resulting in medium with a pH of 7.2, which is optimum. The 10% sodium bicarbonate solution can be kept in the refrigerator for approximately 30 days, but after that time chemical changes have occurred which reduce its buffering capacity.

All the glassware and laboratory utensils must be clean. Cleaning is accomplished by washing with agents manufactured for cleaning tissue culture material, such as Hemosol, Solution 7X-Linbro Scientific, and Alconox. After washing, the glassware is rinsed three times in tap water and five times in distilled water, dried either in air or in an oven at about 100°C overnight, and then sterilized in an autoclave at 123°C for 23 to 30 minutes.

The culture medium is put through a Millipore filter, size .22 μm. The flasks and culture bottles or storage bottles for collecting and/or storing the media must have been sterilized, usually in an autoclave. The filter removes any microorganisms. If a large batch is to be made it is desirable to test its sterility by putting .5 to 1.0 ml into a tube of 5.0 ml of thioglycollate broth and incubating at 37°–38.5°C. Use of fresh media, which is desirable, reduces but does not eliminate the value of this step.

The medium is transferred to sterile culture bottles (medicine bottles of 25-ml capacity are quite satisfactory), 4 ml per bottle, in a sterile room, box, or laminar flow hood. The hood should be clean, and the table top and sides sprayed with 70% ethyl alcohol. All the materials placed in the hood, including the blood collection tubes, should be clean and, where necessary, sprayed with alcohol solution. The operator can most efficiently work with arms clear of clothing to the elbow, and should wash thoroughly and spray hands and under fingernails with the alcohol solution just prior to preparation of the cultures.

To each culture bottle .4 ml of whole blood is added. A 3-ml disposable syringe and an 18- or 20-gauge 1½-inch needle is convenient for transferring the blood from the collection tube. The green stopper on the collection tube should be thoroughly cleaned with 70% ethyl alcohol before inserting the needle. Since some vacuum may still be in each tube, the plunger should be pulled out to allow about 2 ml of air to be injected into the tube before withdrawing the sample. The collection tube should again be gently inverted several times to mix the cells prior to withdrawal of the sample.

To each culture bottle .5 ml of fetal calf serum is added either before or after the blood is added. Sterile, silicon rubber stoppers are then tightly inserted, the cultures are gently swirled to mix all ingredients, and the bottles are placed in an incubator.

Because of the cost of fetal calf serum, and occasionally its lack of availability, an alternate procedure has been found to be very satisfactory in our laboratory for all species. Add 6 ml of medium to a culture bottle and 1 ml instead of .4 ml of blood. The extra blood provides additional autologous plasma which replaces the need for an external source of calf serum. Centrifugation of the blood collection tubes and the use of supernatant plasma are no more effective than use of the additional whole blood, and more lymphocytes are harvested. Harvest is also started at 45 ± 1 hours instead of the standard 3-day, or 66- to 68-hour, harvest. Since the preparation of duplicate blood cultures for each blood sample is standard procedure, the second culture can be harvested the third day if the second-day culture is not good.

Since incubators may not have exactly the same temperature at all places, duplicate samples should not be placed near each other but distributed on the shelves. Uniform culture temperatures are also enhanced by leaving some space for air circulation between bottles.

Cultures of cattle, swine, sheep, or goat blood do not need any further attention until time to harvest, except to note the incubator temperature periodically. Horse blood cultures, however, need the addition of 12 IU of heparin per milliliter of culture and should be agitated gently an hour or two after being placed in the incubator, since horse blood has a greater tendency to clot than does cattle or swine blood. This added amount of heparin approximately doubles the amount of heparin obtained from the collection tubes.

Harvesting of lymphocytes by centrifugation of the cultures, to separate the cells from the supernatant solution, should start about 66 ± 2 hours (or 45 ± 1 hours by the alternative method) after initiation of incubation. Two to three hours before the first centrifugation, .04 ml of a 1 mg/ml solution of colchicine diluted 1:19 with sterile, glass-distilled water is added to each culture bottle (.05 ml in the alternate method). The final concentration of colchicine should be approximately 1 μg/ml of culture. If the chromosomes are to be banded, so that individual chromosomes can be identified, they should be as long as possible without too many overlaps. The concentration of the colchicine and the length of time cultured after addition of colchicine can be varied to achieve these results. A technique by Rønne *et al.* (1979A) results in longer, slightly banded chromosomes, which can be banded more clearly by the trypsin techniques. In this method the colchicine (or Colcemid) can be added the night before harvesting, about 18 hours, and the culture placed in the refrigerator. It is then placed back into the incubator for approximately 10 minutes before harvesting begins. If the major purpose is to count the chromosomes per cell, then shorter chromosomes can be more easily counted.

Colchicine is a spindle fiber poison. It disrupts the microtubules, or spindle fibers, and prevents the cells in metaphase of mitosis from proceeding into anaphase. The result from adding colchicine to the culture 1–4 hours before harvesting is the accumulation of much larger numbers of cells in metaphase. Since the lymphocytes are not synchronized, nuclei will be found in all stages from early prophase through late metaphase, but no anaphase of telophase stages will be found.

Colcemid is a more highly refined derivative of colchicine. With reduction in concentration to about one tenth it can be used to replace colchicine.

After the incubation period with colchicine, each culture bottle is agitated gently to get all the cells into suspension, and the contents are then poured into a 12-ml, siliconized, graduated centrifuge tube. A graduated tube is preferred because additions of subsequent solutions can be more easily measured. Siliconizing the tubes will prevent the cells from adhering to the sides, but it is not mandatory.

The tubes are then centrifuged at 1200 rpm, or 300 *g*, for 10 minutes and the supernatant is aspirated down to 1.2 ml.

A hypotonic solution of .067 *M* KCl is then added to a volume of 8 ml and the packed cells are put into a uniform suspension with a Pasteur pipette and rubber bulb alternately removing the solution and forcing it gently back into the centrifuge tube. The hypotonic solution will cause the erythrocytes to lyse, and in some cases swell up and burst, and will permit swelling of the leukocytes to a stage short of bursting.

Hungerford, in 1965, developed the technique that utilizes the hypotonic swelling of cells. It has been known for many decades that osmotic pressure causes cells to swell. Hungerford took this simple concept of adding a hypotonic potassium chloride solution, sodium citrate solution, or hypotonic solution prepared from one of several other ingredients and caused the cells and nuclei to swell so that the chromosomes were separated more widely. When the cell flattens out on the slide the chromosomes are still separated. The level of hypotonicity must be controlled. If a hypotonic solution is too dilute the cells will burst and the chromosomes will be lost. Too little hypotonicity will not produce enough spread. Two things are very important: the right strength of solution and the length of time in solution. Twenty-two minutes in the hypotonic solution at room temperature and another 10 minutes in the centrifuge is satisfactory for lymphocyte cultures. The cells are in the hypotonic solution about 35 minutes, which permits the maximum amount of swelling without bursting of the cells. A shorter time can be used if the cells in the hypotonic solution are placed in an incubator at 38.5°C.

After 22 minutes in the hypotonic solution, 1 ml of fixative (1 part glacial acetic acid and 3 parts methyl alcohol) is then added and mixed thoroughly with a Pasteur pipette, using a separate pipette for each culture. Absolute methyl alcohol of reagent grade is needed, since even a small amount of water through condensation or otherwise can markedly affect the quality of the harvest. The solution, upon mixing, will turn from red to a dark brown or almost black color. The tubes are immediately placed in the centrifuge and spun down again for 10 minutes at 300 *g*.

The fixative must be freshly prepared and kept cold. It is convenient to prepare the fixative in a plastic squeeze bottle and keep it in a beaker of crushed ice and water. If allowed to stay at room temperature the mixture of glacial acetic acid and methyl alcohol will form crystals. If such a mixture is added to the cells in the tube the cells will be destroyed by the crystals.

The 1 ml of fixative added at the end of the hypotonic treatment appears to toughen the cell membranes. The swelled cells are very fragile at this time and without the addition of fixative many are apparently broken and lost during centrifugation. Because of the fragility of the cells at this stage of harvest the agitation with the pipette should be reasonably gentle.

After centrifugation the supernatant is aspirated down to a point slightly above the packed cells. The packed cells, debris, and supernatant will be very dark in color and the line separating the packed cells from the supernatant may be difficult to discern. Fixative is added to a volume of 7 ml and thoroughly but gently mixed. The resulting solution will vary in color from light amber to light brown. It is spun down again for 10 minutes at 300 *g*. Rønne *et al.* (1979A,B) have indicated that holding the cells for 30 minutes at room temperature before this centrifugation, and holding the suspended mixture before the final centrifugation overnight at 4°C, will enhance the G-banding of the chromosomes and will result in a larger number of long chromosomes suitable for further banding treatment. Extended time in the fixative, however, will decrease the level of success in C-banding. G- and C-banding have been discussed in Chapter 6.

Following this centriguation a small amount, or pellet, of white cells can be seen at the bottom of the tube. The solution still will be quite dark. Aspirate down to approximately 1 cm above this pellet. Fixative is then added up to a volume of 5 ml, and the cells are thoroughly but gently dispersed and again centrifuged. After the fixative has been added this time the further steps can proceed immediately or the material can be stored in a refrigerator at 4°C for an hour or two, or even overnight before centrifuging and continuing. The solution will be nearly clear or light amber in color.

The supernatant is again aspirated and the cells are washed one more time in fixative up to a volume of 3 ml. After the next centrifugation and removal of supernatant the cells are dispersed by flipping the bottom of the tube with the index finger. Enough fixative is added to bring the suspension to a desired level of turbidity. This varies with different samples. Experience will dictate the desired level of turbidity. If the cells are too thick on the slide, reduce the turbidity. If they are so widely separated that too much scanning of the slide under the microscope is necessary, increase the turbidity. If necessary the tube can be centrifuged again and the desired concentration of cells in suspension can be obtained.

The glass microscope slides used for the cells must be very clean. Regardless of the source, the slides need to be stored for 24 hours or longer in methyl alcohol with a few added drops of 1 *N* hydrochloric acid. Each slide is then rinsed in distilled water and polished clean with an absorbent, lintfree towel. The culture number should be placed on the slide before the cells are placed on it. Slides with one end etched are convenient for marking with a lead pencil and the markings will not be removed by later staining procedures. If slides are not etched then a diamond stylus can be used for identification.

Three drops of the suspension of lymphocytes can be dropped on each slide from about 20–35 cm above the slide. The slides should be placed on a level table for preparation and can be permitted to air-dry. Air-dried slides are best for G-banding. The slides can be observed under phase contrast to check quality. It is desirable to make only one slide per culture until quality has been checked. If the chromosomes are not well spread, another slide should be prepared and flame-dried. This is done by holding the slide with tweezers and touching the edge to the flame of a Bunsen burner. Do not leave the slide in the flame. The alcohol will ignite leaving the slide quite warm but not hot. The slide can then be grasped firmly between thumb and forefinger and the remaining drops of glacial acetic acid can be removed by waving through the air in a vigorous and rapid arc. Be careful that the acetic acid is not splashed on clothes or other material, which it will damage. An alternative method for removing the acetic acid is to rinse the slide in fixative, stand on end, and allow to drain and air-dry.

As soon as the slide is dry it can be stained. A simple procedure is to place the slides on a pair of parallel glass rods or tubes, connected by rubber or plastic tubing at each end, which are placed over a sink. A stock Giemsa stain solution is diluted 1:20 with distilled water buffered to a pH of 6.8. Buffering can be done simply and conveniently with Gurr buffer tablets. The stain is flooded over each slide until the slide is covered. The diluted stain needs to be freshly prepared. The stain is left on for 20 minutes, then rinsed under a slow flow of tap water followed by distilled water. The slides can then be carefully blotted to remove excess water and allowed to air-dry. Examination of quality of slides is easier after staining than under phase contrast. An alternative staining procedure can be followed by diluting the stock Giemsa solution 1 : 50 and leaving for 90 minutes, or 1 : 15 for 10 minutes.

When the slides are dry they can be permanently mounted, using such material as Mounting Medium (Technicon). The size of the coverslips should be either 24 × 50 mm or 24 × 60 mm. Permanent mounting improves the clarity of the chromosomes under high-dry objectives. If the slides are thoroughly dry when permanently mounted they will stay in excellent condition for years. Oil placed on the slide for viewing with oil immersion objectives can be easily removed from the coverslip with Kleenex or similar tissues. Oil placed on an unmounted slide will need to be removed with xylene.

SOMATIC CELL GENETICS

The tissue culture methods described so far are called primary cultures. This term describes the cultures of cells obtained from organisms and their initial growth. As the cells are transferred to new or fresh medium, a process of "passaging," the cell cultures which have been placed in suspension for transfer are called primary cell lines. A majority of primary cell lines eventually enter a period of senescence, with many chromosome modifications, and the culture dies. If, however, the primary cell lines survive many passages, and few will do this spontaneously, they are called established cell lines. Different tissue sources and different species respond differently. In some cases a "transformation" seems to occur, and even though the resulting cells form an established cell line, the chromosome numbers and morphology may vary from the original primary cell line.

In livestock cytogenetics nearly all of the work being done has as its objective the characterization of the animal providing the primary cell culture. Modification of chromosome complement introduces variables which are of little value in meeting this primary objective. Therefore, primary cell cultures from very early passages are the most accurate and reliable for this purpose. However, the study of cells in culture may provide some additional insight into the effects of chromosome modifications. The study of cells in culture from the standpoint of genetics of the cell lines is called somatic cell genetics (see Chapter 8).

Cell cultures can also be used to determine biochemical reactions. The field of biochemical genetics grew out of the study of microorganisms in culture. Knowledge of the biochemical genetics of large animals will be enhanced by tissue culture work.

Another application of cell culture to cytogenetics has been the production of allophenic mice. Mintz and Slemmer (1969) removed the zona pellucida of mouse blastocysts and induced them to fuse. These mosaic embryos were then transferred to recipient females. The resulting animals were mixtures of the two types or strains of mice with two separate sources of diploid cells. Would such an animal exhibit hybrid vigor?

Markert and Petters (1977) reported the removal of one of the pronuclei from the fertilized ovum of a highly inbred mouse, induction of doubling of the number of chromosomes to restore the diploid number, and successful transplanation of the ovum into a recipient female. The resulting mice were completely homozygous. This achievement raises many possibilities concerning the manipulation of early embryos.

CHROMOSOMES OF BLASTOCYSTS

Chromosomes can also be studied from early blastocysts. Rabbit blastocysts at about 6 days after breeding can be obtained from the uterus by flushing. The female may be sacrificed or flushing can be done by laparotomy in an anesthetized animal. With the uterus removed from the animal a blunt needle can be inserted through the cervix and the uterus can be cut just before the tubal-uterine junction. The flushing medium is about 10 ml of Ham's F-10 without serum. Blastocysts can be easily identified without magnification, but manipulation can best be done under a dissecting microscope with a variable magnification of 10× to 70×.

The blastocysts are transferred with a Pasteur pipette in which the diameter of the opening is just large enough for the blastocyst. For 6-day rabbit blastocysts the tip may need to be removed just above the point at which it gets larger, and fire-polished. The pipette should be siliconized. The blastocysts are transferred into 20 ml of Ham's F-10 culture medium with streptomycin and penicillin added. The streptomycin and pencillin are put in at a concentration to provide 50 to 100 IU of penicillin and 50 to 100 μg of streptomycin per milliliter of medium. The blastocysts are incubated at 38.5°C for 2 hours, after .4 ml of Colcemid in Hank's solution (concentration, 10 μg/ml) has been added.

Each blastocyst is then transferred to a siliconized, 3-ml centrifuge tube, in which 1 ml of culture medium plus 25% fetal calf serum has been placed. These are kept in the incubator. Three drops of .25% trypsin are added to one tube and it is placed back in the incubator for 15 minutes. Then under the dissection microscope, with a separate siliconized Pasteur pipette for each blastocyst, and a rubber bulb, the zona pellucida is broken and the contents are freed. With care the zona and a minimum of medium can be removed with the pipette. The cellular material is then broken up by gently agitating it in the medium. The tube can be set aside at room temperature until all blastocysts have been treated similarly. When all are completed each tube can again be agitated with the pipette to disperse the cells further in the medium with trypsin.

The tubes are centrifuged for 10 minutes at 200 *g* and the supernatant is carefully removed. Two milliliters of hypotonic solution .07 M KCl is added to each tube and, using the same pipettes, mixed thoroughly but gently. After 15 minutes a small amount of cold, fresh fixative (1 part glacial acetic acid and 3 parts methyl alcohol) is added, and the suspension is mixed. Centrifugation and fixation proceed as with lymphocyte cultures. The amount of cellular material is much smaller, so care is needed to preserve it at all steps. It is difficult to see the pellet of cells without magnification.

After centriguation following the third fixative, leave only 3–4 drops in the tube. Flip the tube with the forefinger to mix, then with the pipette drop all of the material on a *clean,* labeled slide. Allow to air-dry. Stain and observe as with slides from lymphocyte cultures.

Cattle blastocysts are flushed from bred females using about 200–300 ml of Dulbecco's solution and a three-way Foley catheter. For heifers a 5-cm^3 balloon on the catheter is acceptable. Flushing is recommended at 9–13 days after onset of estrus, which is counted as Day 0. The female is given an epidural anesthetic of 5.0 ml of procaine hydrochloride, $2\frac{1}{2}$. The flushing medium (which has been placed in the uterine horn), about 10–20 ml, is first put into three or four egg collection cups, 80-ml capacity, then the remainder is collected in Kelly infusion jars. The cups of material are kept warm on a warming plate and the Kelly jars are kept in an incubator. The solution is then searched at 10× magnification under a dissection microscope for blastocysts. They are treated as described for rabbit blastocysts, but since cattle chromosomes appear to be more reactive to Colcemid or colchicine, the concentration should be considerably lower.

Studies of chromosomes from lymphocytes characterize the animals from which the blood is obtained. The chromosome studies of blastocysts permit the characterization of all the early zygotes, some of which may not be viable beyond 12–30 days. Zygotes with abnormal chromosome complements may be one cause of lowered fertility in livestock.

MEIOTIC STUDIES

The study of chromosomes in meiosis in livestock can most easily be done with testicular material (Evans *et al.* 1964; Meredith 1969). This material can be obtained either by biopsy or by castration or at slaughter from sexually mature males. The tubules can be teased out in a 2.2% citrate solution. The tubule contents can be milked out into the solution. The material is treated with a hypotonic solution,

.07 *M* KCl, for 30–45 minutes and then fixed in 3:1 methyl alcohol:glacial acetic acid. After dispersing in 60% acetic acid the solution is put on a clean slide by dropping about 3 drops; it is then sucked back into the pipette. This is repeated three or four times. The cells adhere to the glass slide. The cells can then be stained and observed (Meredith 1969).

Meiotic cells may also be studied in the oocytes from the female. Cattle ovaries can be obtained from a slaughterhouse from freshly killed cattle. Collection of the oocytes should proceed as rapidly as possible (Jagiello *et al.* 1974; Koenig 1982).

In the laboratory the remaining mesentery is removed from the ovaries which are held in McCoy's 5A modified medium with 800 IU of penicillin and 800 μg of streptomycin per milliliter at 38.5°C. The ovaries are split so the halves lie flat in a petri dish; the follicles are opened with fine pointed scissors as pressure is exerted with a pair of curved forceps. The contents of each follicle, including the oocyte, are expelled. After all follicles have been ruptured, the ovary is rinsed into the petri dish and the oocytes are transferred with a fine-tipped Pasteur pipette to a small petri dish containing 1 ml of medium with 100 IU of penicillin and 100 μg of streptomycin per milliliter. The oocytes are again transferred to another small petri dish, 1½ inches in diameter, with 2 ml of the same medium which has been kept equilibrated with 5% CO_2 in the air, and put into the incubator at 38.5°C in a 5% CO_2 in air atmosphere. The double transfer permits removal of follicular fluid which tends to agglutinate. Oocytes with cumulus cells tightly surrounding them, and off white or light tan in color, have the highest probability of successful culture through both the first and second meiotic divisions. Naked or dark-colored oocytes, or those that have a wrinkled exterior, have a much lower probability of successful culture.

The peak number of oocytes in the first meiotic division will be found after about 14 hours in culture. After 24 hours a majority of the oocytes will be at metaphase II.

The oocytes are transferred with a pipette to a slide with a depression, into one drop of .1% pronase (protease) in Earle's balanced salt solution, adjusted to pH 7.2. The pronase dissolves the material which holds the cumulus cells together. In about 30 seconds the pronase is neutralized with several drops of medium with 50% serum and the oocytes are transferred to a second depression slide into medium with 50% serum. They are then transferred to a hypotonic solution of .7 *M* sodium citrate in another depression slide and are cleaned of cumulus cells with mechanical agitation. Several transfers in the hypotonic solution will aid in cleaning the oocytes. The total time in hypotonic is about 20 to 25 minutes, to allow for spreading of the chromosomes. Several drops of distilled water are added to the last transfer for about 5 more minutes.

Each oocyte is then transferred to a microscope slide, removing as much liquid from the slide with the transferring pipette as possible. One drop of fresh fixative at room temperature is dropped on the oocyte. Two more drops of fixative are put on after the oocyte can no longer be seen under the dissecting microscope.

The slides are air-dried and temporarily stained with 2% toluidine blue in resin and a coverslip is put on, avoiding air bubbles. They can then be observed and photographed, and the coverslip and the stain can be removed. For permanent slides, stain with 5% Giemsa in 6.8 pH buffer for 20 minutes and permanently mount.

Another method for recovery of oocytes is by trypsin digestion (Strickland *et al.* 1976). The whole ovary is sectioned and the outer layer of tissue is diced up into .5 mm^3 fractions, which are then treated for 1 hour with trypsin. Large numbers of oocytes can be recovered.

SQUASH TECHNIQUE

Another method for observing chromosomes, which was developed mostly by the plant cytogeneticists, is called the squash technique. This is not a very sophisticated name. It simply means to place the cells or small pieces of tissue on a slide, put a coverslip over them, and with your thumb put considerable weight on the coverslip to "squash" the tissue to a layer one cell thick. This spreads the cells so the chromosomes can be seen. This technique was first reported in the late 1930s. It first was applied to humans and other mammals starting in 1944 (LaCour), and later with improved techniques by Ford and Hamerton (1956). The air- and flame-drying techniques for cells dispersed in fixative have replaced the squash method for many tissues. However, the squash technique still has some uses.

Although lymphocytes, other cells, oocytes, blastocysts, spermatozoa, etc., are very small, the techniques of micromanipulation permit a limited amount of physical manipulation. Micromanipula-

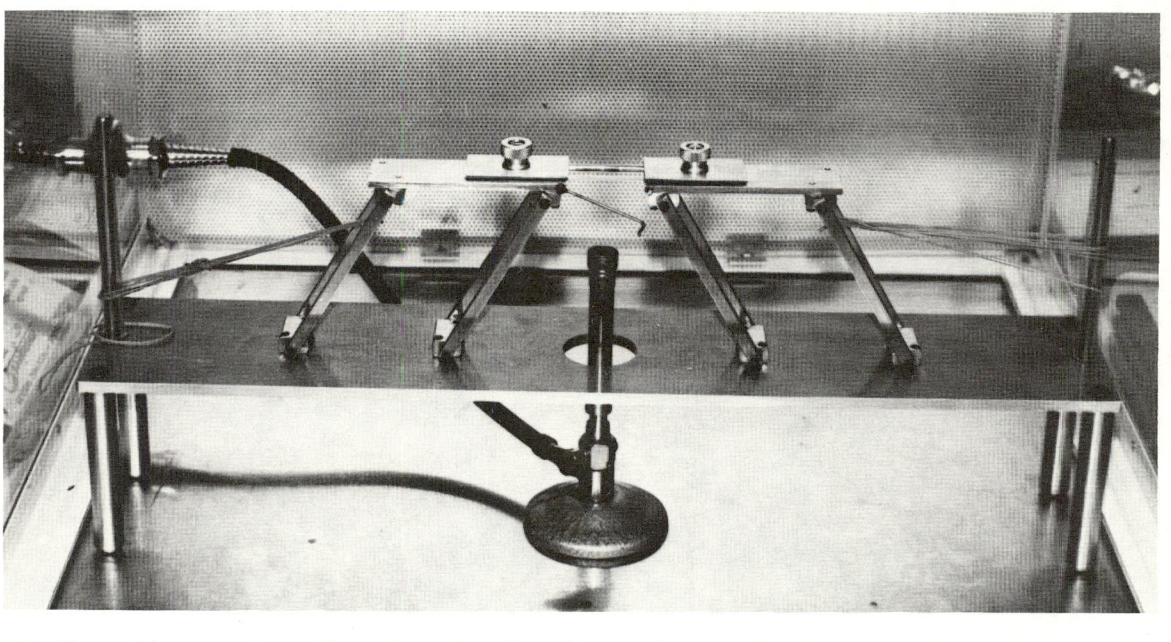

FIG. 10.1. A homemade capillary tube puller. Size of the resulting capillary tube tip can be varied by the force (number of rubber bands) and the distance from the flame. The thumbscrew plates have a rubber liner and the base an appropriate-size groove.

FIG. 10.2. Micromanipulators, two different makes. A matched pair would be better. They need three-way adjustments.

tion apparatus and capillary tube pullers are available. The pullers permit drawing out glass needles with points fine enough to be inserted into cells and cell nuclei. Capillary tubes can be made with openings small enough that specific parts of cells such as pronuclei can be withdrawn. A puller and a pair of micromanipulators are shown in Figures 10.1 and 10.2.

Microscopes can be equipped with cameras for microphotography. Prints of metaphases of chromosomes are used for making karyotypes and are useful for some detailed examinations of chromosomes. Photographic prints are essential for studying banded chromosomes.

REFERENCES

EVANS, E. P., BRECKON, G., and FORD, C. E. 1964. An air-drying method for meiotic preparations from mammalian testes. Cytogenet. Cell Genet. *3*, 289–294.

FORD, C. E., and HAMERTON, J. L. 1956. The chromosomes of man. Nature (London) *178*, 1010–1023.

HAMERTON, J. L. 1971. Human Cytogenetics, Vol. 1. Academic Press. New York.
HUNGERFORD, D. A. 1965. Leucocytes cultured from small inocula of whole blood and the preparation of metaphase chromosomes by treatment with hypotonic KCl. Stain Technol. *40,* 333–338.
JAGIELLO, G. M., MILLER, W. A., DUCAYAN, M. B., and LIN, J. S. 1974. Chiasma frequency and disjunctional behavior of ewe and cow oocytes matured in vitro. Biol. Reprod. *10,* 354–363.
KOENIG, J. L. F. 1982. A cytogenetic analysis of bovine oocytes cultures in vivo. M. S. Thesis, Univ. of Nebraska.
LaCOUR, L. F. 1944. Mitosis and cell differentiation in the blood. Proc. Roy. Soc. Edinburgh *B62,* 73–85.
MARKERT, C. L., and PETTERS, R. M. 1977. Homozygous mouse embryos produced by microsurgery. J. Exp. Zool. *201,* 295–302.
MEREDITH, R. 1969. A simple method for preparing meiotic chromosomes from mammalian testis. Chromosoma *26,* 254–258.
MINTZ, B., and SLEMMER, G. 1969. Gene control of neoplasia. I. Genotypic mosaicism in normal and preneoplastic mammary glands of allophenic mice. J. Nat. Cancer Inst. *43,* 87–94.
PAUL, J. 1975. Cell and Tissue Culture. E. & S. Livingstone Ltd., Great Britain.
RØNNE, M., NIELSEN, K. V., and ERLANDSEN, M. 1979A. Effect of controlled Colcemid exposure on human metaphase chromosome structure. Hereditas *91,* 49–52.
RØNNE, M., ANDERSEN, O., and ERLANDSEN, M. 1979B. Effect of Colcemid exposure and methanol acetic acid fixation on human metaphase chromosome structure. Hereditas *90,* 195–201.
STRICKLAND, J. D., DORGAN, W. J., and MOODY, E. L. 1976. Recovery of bovine oocytes from whole ovaries. J. Anim. Sci. *42,* 1565.

11

Cattle Chromosomes

Cattle, both *Bos taurus* and *Bos indicus,* have 60 chromosomes. The only distinguishable difference between these two species is in the Y chromosome, which is the smallest acrocentric in *Bos indicus* and is a small submetacentric or metacentric chromosome in *Bos taurus*. As microscopic and other techniques for observing chromosomes have improved, the accepted number has changed from quite small, $2n = 16$, to the well-established number of $2n = 60$. The accuracy of morphological description also has improved.

HISTORICAL BACKGROUND

According to Makino (1956) the earliest published report on the chromosome numbers in cattle was by von Bardeleben in 1892, who indicated from spermatogonial cells that the $2n$ number was 16. In 1902 Schoenfeld gave the $2n$ number as a range of from 20 to 25 based on spermatogonial cells, and an n number of 12 in the male from primary and secondary spermatocytes. Von Hoof in 1913 published the number as 20–24, from spermatocytes. Then in 1919 Masui indicated a $2n$ number of 33 in spermatocytes and a haploid number of 17 for the primary spermatocyte and either 16 or 17 for secondary spermatocytes. He concluded that the sex chromosome composition in cattle was XO.

Wodsedalek (1920) counted 37 chromosomes in spermatogonial cells and 38 in oogonial cells, and also concluded that the male was XO and the female XX. He reported 19 chromosomes in primary spermatocytes and 18 or 19 in secondary spermatocytes. Then

Krallinger (1931) from Germany reported 60 chromosomes in spermatogonial cells, with 30 in both primary and secondary spermatocytes. He also stated that the sex chromosome composition was XY in male cattle. Although several people expressed doubt about Krallinger's reports, later publications verified the accuracy of his work. In 1943 and 1944 Makino (1956) verified the number of 60, and the sex chromosomes as XX–XY. He also indicated that there were no racial differences in number or sex mechanism between *Bos taurus* and *Bos indicus*. He had not observed the morphological difference in the Y chromosomes of the two species that was reported by Kieffer and Cartwright (1968) and later confirmed by others.

Melander and Knudsen (1953), using the squash technique, were able to spread the chromosomes from testicular tissue, reconfirming the number and sex chromosome composition of Swedish Red and White bulls.

Sasaki and Makino (1962) reviewed the development of knowledge on the chromosome number of cattle, and by skin and kidney tissue cultures confirmed the established numbers. Also, Crossley and Clarke (1962) stated: "Until 1960 all the chromosome counts performed on cells of this species [*Bos taurus*] were done on histologically prepared specimens of testicular tissue." They credit Chiarelli, deCarli, and Nuzzo (1960) with the first report from other tissue. From their own work with blood and muscle cells in culture they found 60,XX and 60,XY chromosomes from both sources.

The discoveries of Peter Nowell (1960), Hsu and Pomerat (1953), and Moorhead *et al.* (1960) gave great impetus to the study of mammalian chromosomes. Nowell (1960) discovered that phytohemagglutinin (PHA) was a mitogenic agent, causing blood leukocytes to transform into blast cells and enter a cycle of cell division. This was the first time that leukocytes from normal subjects had been found to divide with sufficient frequency for study. Hsu and Pomerat (1953), using the principle of osmosis that had been recognized for many decades, showed that by adding a hypotonic solution the tissue culture cells could be caused to swell, which would permit the individual chromosomes to be more widely dispersed on a glass microscope slide. This principle had been applied by Makino and Nishimura (1952) to cells prior to use of the squash technique but had not been used previously on cultured cells. Such an example of a well-known technique being used in a new way causes surprise that it had not been done earlier. Moorhead *et al.* (1960) combined these

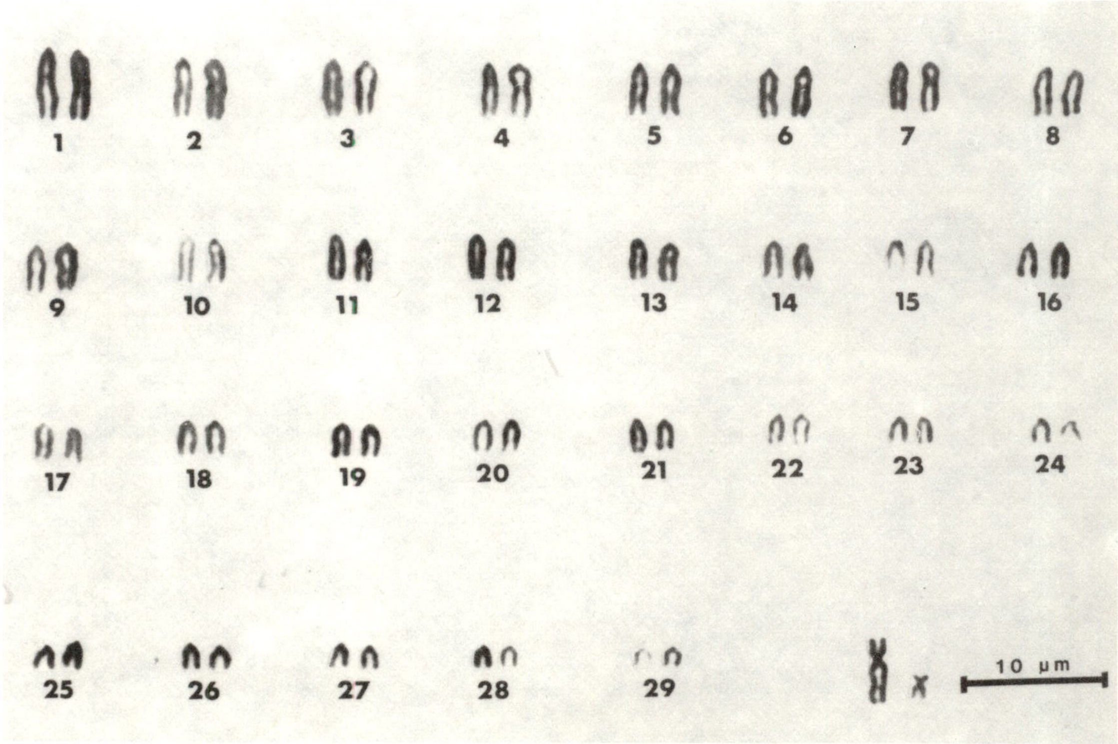

FIG. 11.1. Karyotype of a normal Brown Swiss bull.

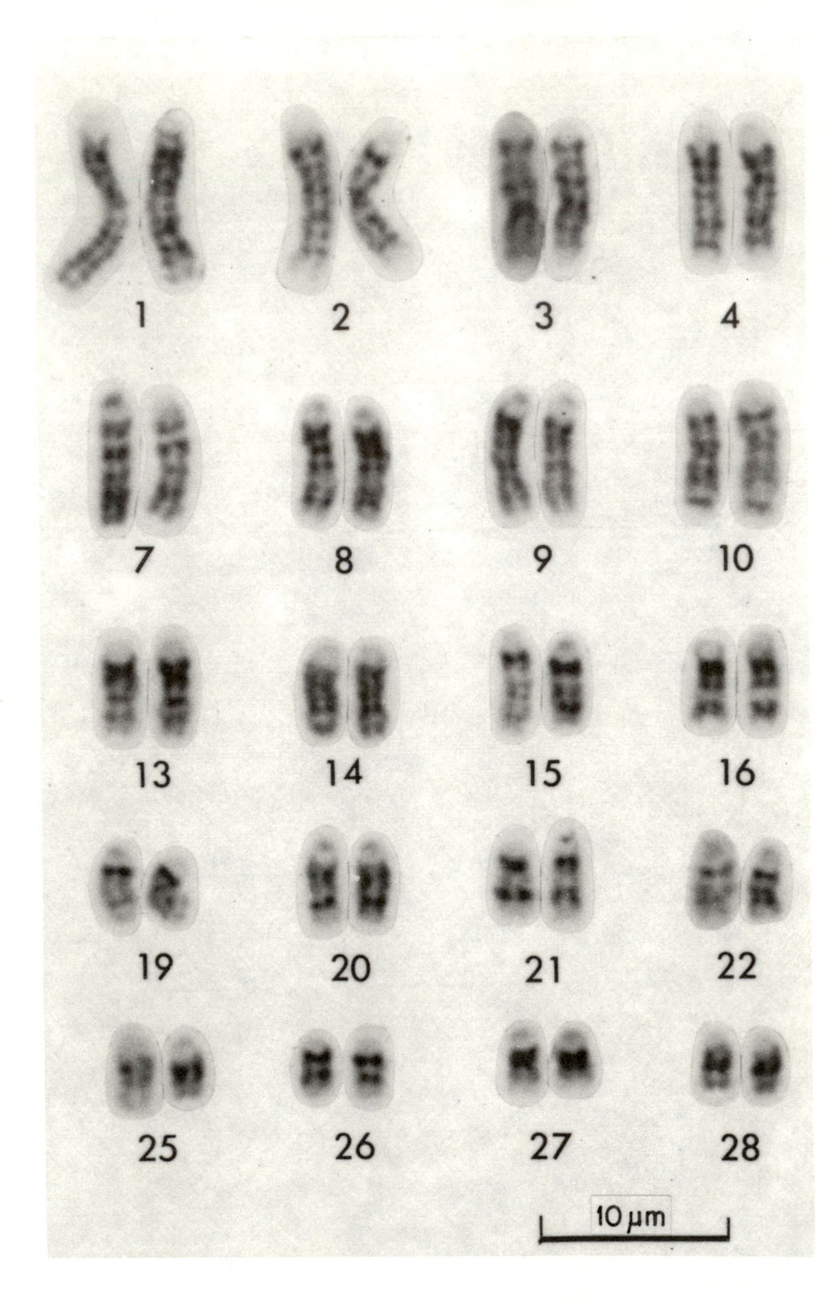
1
2
3
4
7
8
9
10
13
14
15
16
19
20
21
22
25
26
27
28
10 µm

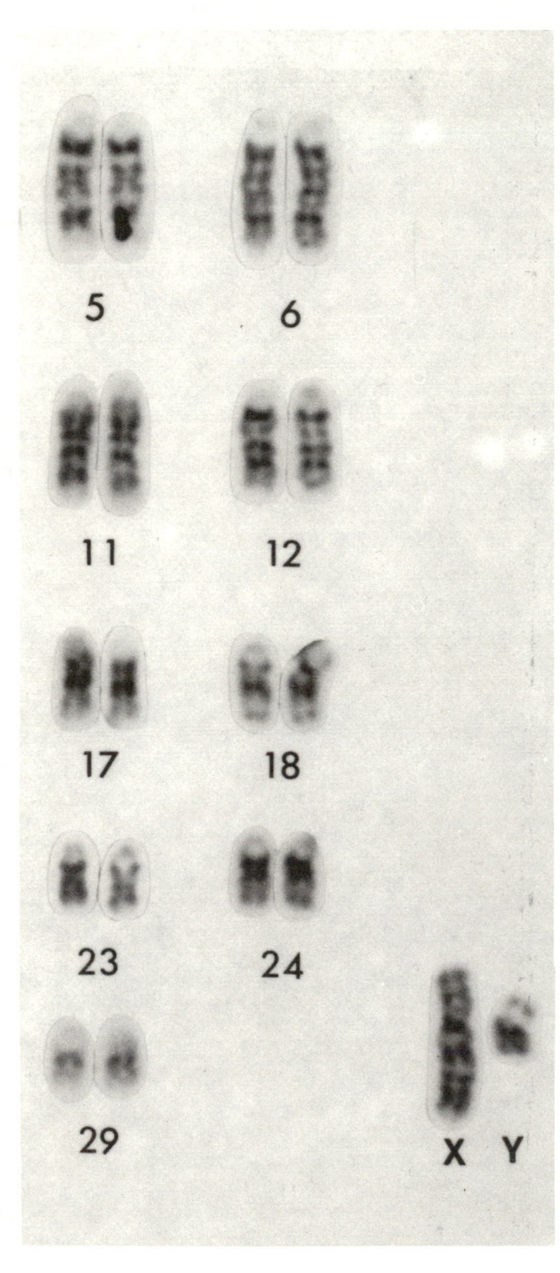

FIG. 11.2. Karyotype of banded bovine chromosomes and diagrams of the banding patterns. The chromosomes are rearranged to correspond with the Reading Conference of 1976.

From Lin et al. (1977).

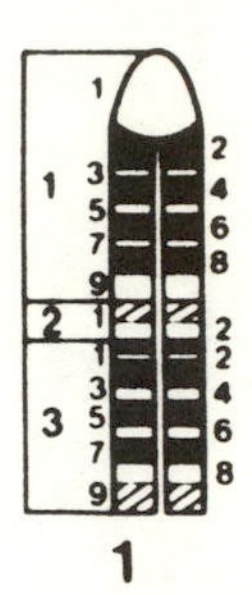

1

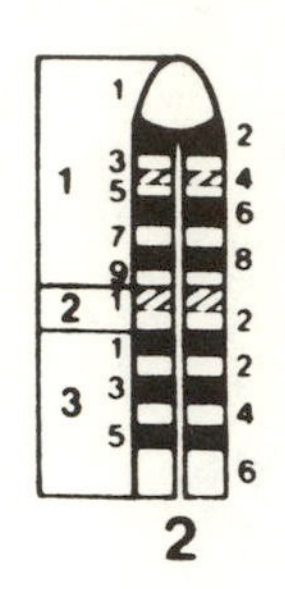

2

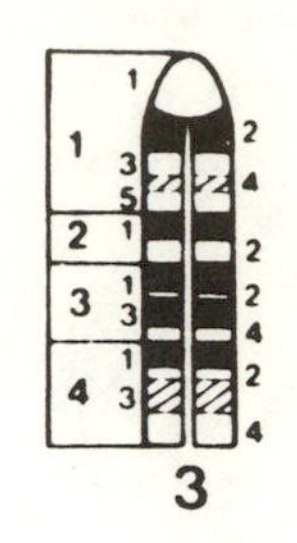

3

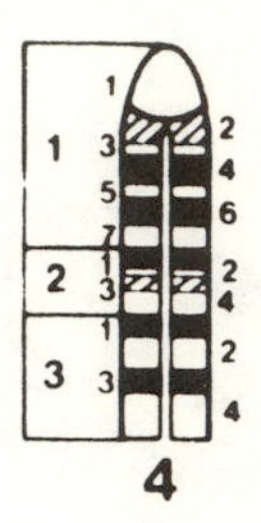

4

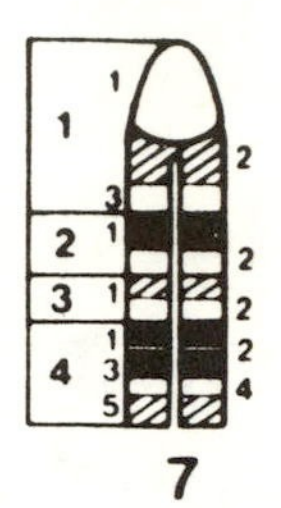

7

8

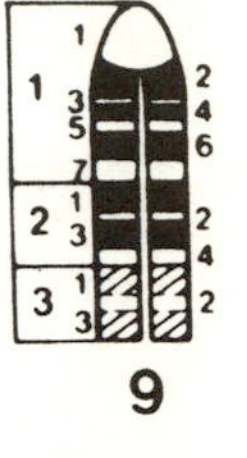

9

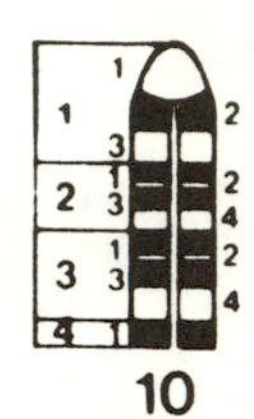

10

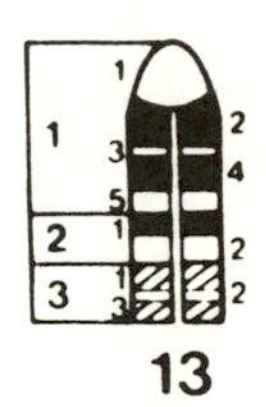

13

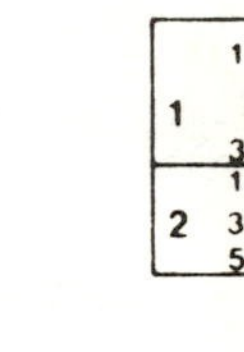

14

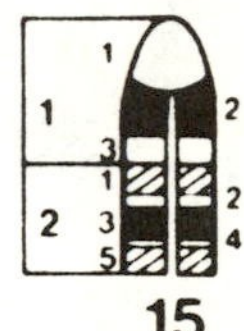

15

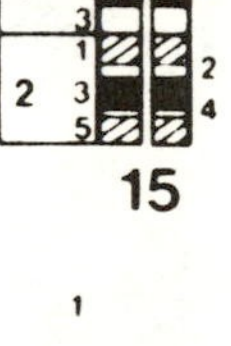

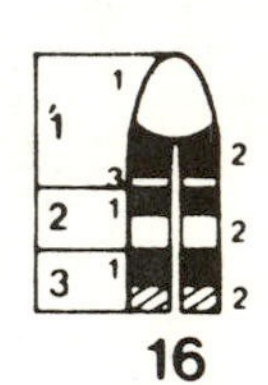

16

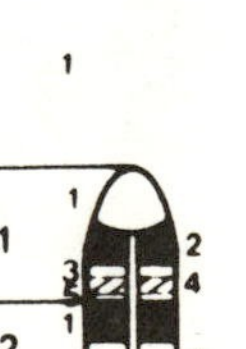

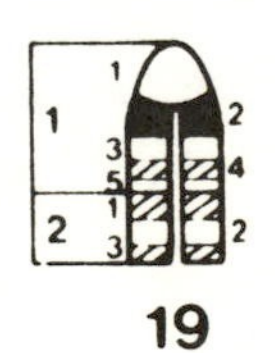

19

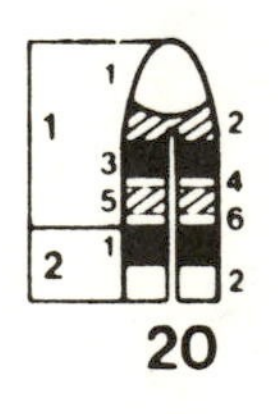

20

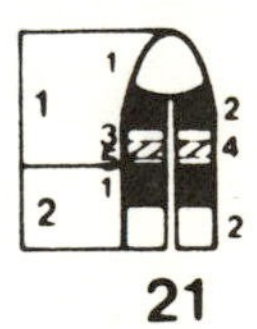

21

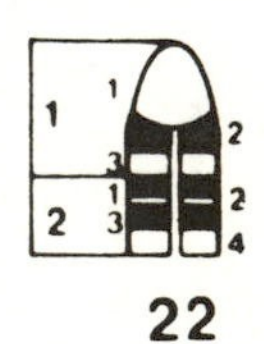

22

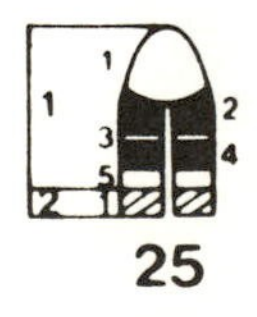

25

26

27

28

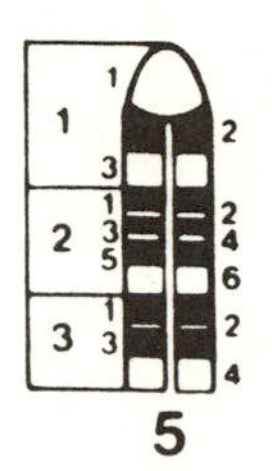

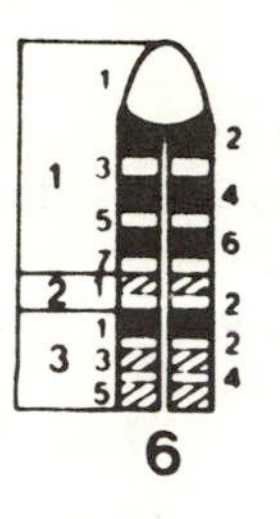

POSITIVE (Dark)

PALE (Lighter)

NEGATIVE

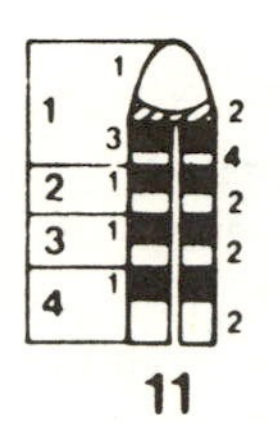

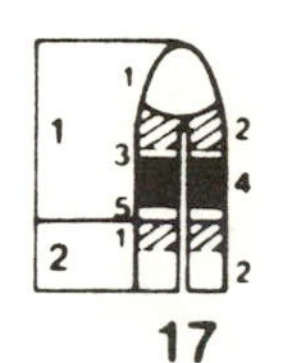

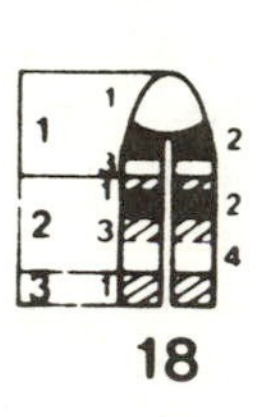

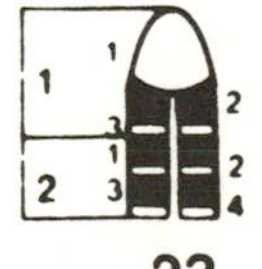

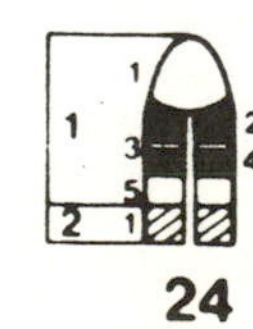

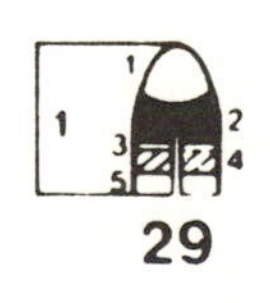

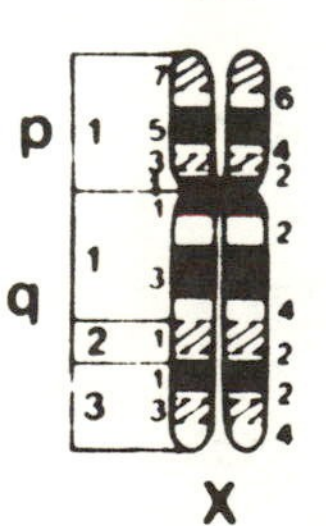

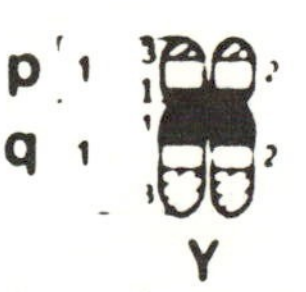

FIG. 11.3. Diagrammatic representation of G-banded bovine chromosomes. Rearranged to correspond with the standardized karyotype of the 1976 Reading Conference.

From Lin et al. (1977).

new techniques into a system for studying chromosomes using blood leukocytes. This technique, modified in several details by numerous scientists, is now in use in laboratories worldwide.

With the application of these new techniques, 1960 can be considered a turning point in the chromosome studies on mammals. Several papers published on cattle chromosomes soon after this date demonstrated the clear, well-separated chromosomes that could be seen and photographed using these new techniques (Sasaki and Makino 1962; Crossley and Clarke 1962; Herschler *et al.* 1962; Ulbrich *et al.* 1963; Nichols *et al.* 1962; Biggers and McFeely 1963; Chiarelli *et al.* 1960; Melander 1959; Yosida and Lamontain 1964). For cattle (*Bos taurus*), the normal number, morphology, and sex-determination pattern of the chromosomes were finally established with repetition by a sufficient number of researchers to be considered completely reliable. With this firmly established background on the normal chromosome complement of cattle it was then possible to search for variations. Figure 11.1 shows the established karyotype for *Bos taurus* cattle.

Banding of chromosomes has permitted more detailed study and identification of cattle chromosomes. The longest pair of autosomes usually can be identified with considerable certainty with Giemsa stain in cells where the chromosomes are fairly long, but the other 56 autosomes show such small differences in length when arranged from the longest to the smallest that identifying homologous chromosome pairs is nearly impossible (Cribiu and Popescu 1974; Gustavsson 1969). Lin *et al.* (1977) published one of the clearest sets of photographs and diagrams illustrating the G-banded bovine (*Bos taurus*) chromosomes, following the standard agreed upon at the 1976 Reading Karyotype Conference (Ford *et al.* 1980) (Figs. 11.2 and 11.3). For Q-banding see Hansen (1972), for R-banding see Gustavsson and Hageltorn (1976), and for C-banding see Figure 9.5 of Hansen (1973).

ROBERTSONIAN TRANSLOCATIONS

The 1/29 Robertsonian Translocation

In 1964 Gustavsson and Rockborn, while doing research in Sweden on lymphatic leukemia in cattle, found the first clearly observable chromosomal aberration in cattle, a centric fusion, that additional research identified as a 1/29 Robertsonian translocation. Knudsen (1956) stated that his cytomorphological studies of low-fertility bulls "seemed" to indicate inversions and translocations, but his illustrations and further description did not verify the actual observation of such structural changes. The English summary states the conclusion more positively than he apparently intended. Gustavsson (1969) pursued the original discovery and published a classic paper on this chromosomal anomaly. In 1966 Herschler and Fechheimer found the same type of Robertsonian translocation in a set of bovine triplets while studying the cytogenetics of freemartins. Since the publication of Gustavsson's (1969) extensive study of the 1/29 Robertsonian translocation in Swedish Red and White cattle, many more reports on this widely distributed chromosomal aberration have appeared. A review by Gustavsson (1979) cites most of them. It has been found in at least 38 breeds, distributed throughout the world, where cytological studies have been made (see Table 11.1).

The frequency of occurrence of this translocation varies widely among breeds. In Holstein–Friesian in the United States and other countries the frequency has been found to be zero. The translocation has not been found in several other breeds extensively sampled: Jersey, Angus, Hereford, Shorthorn, and Normandy. In other breeds

TABLE 11.1. Breeds of Cattle in Which the 1/29 Robertsonian Translocation Has Been Found

Austrian Simmental	Limousin
Baoule'	Marche
Blonde Aquitaine	Modica
British White	Montbeliard
Brown Atlas	Norwegian Red
Brown Mountain (W. Germany)	Oberinntal (Austrian) Grey
Brown Swiss	Pisa
Charolais	Podolian
Chianina	Polish Red
Czechoslovakian Red Pied	Red Poll
Gascony	Romagna
German Red Pied	Russian Black Pied
German Simmental	Santa Gertrudis
Guernsey	Siamese
Hungarian Gray	Simmental
Hungarian Simmental	Swedish Red and White
Japanese Black	Swiss Brown
Kuban (Zebu)	Thai
Kuri	Vosges

it occurs in frequencies as high as 32%, and in one herd of British White cattle in Scotland (Eldridge 1975) a breed with relatively small numbers of animals, the frequency was found to be 66%. Further study of additional herds of this breed by Wilkes (1979) led to a frequency estimate of 40% for the breed. Fischer (1971) found the 1/29 translocation in Thai cattle.

Since the 1/29 translocation chromosome can be identified cytologically with relative ease, and is transmitted from heterozygous parents in a standard Mendelian pattern as a dominant, there was an early interest in the possibility that it would be associated with certain phenotypic characteristics. Only one characteristic, fertility, has been found to have correlation (for discussion, see chapter on fertility). Other characters studied by Gustavsson (1969, 1971A,C) for which no clear correlation could be established, include (a) milk production, butterfat production, and fat percentage; (b) body conformation including shape of head and neck, chest, shoulders, withers, back, pelvic region, thighs, flanks, legs, general appearance, and ease of movement; (c) color markings; (d) udder shape and attachment and teat shape and placement; (e) nine different loci for blood antigens; (f) transferrin; and (g) carbonic anhydrase. Other investigations have indicated some positive effects of the 1/29 translocation on milk production and meat production, which apparently has resulted in the selection of 1/29 carrier bulls for artificial insemination (Darre *et al.* 1972).

No significant differences in quality, freezability, or fertilizing ability were found by Moustafa *et al.* (1983) between 15 Simmental bulls with Robertsonian translocation (13 with 1/29, 1 with 5/18, and 1 with 14/21) and 384 Simmental bulls without translocations. However, it has been found that among bulls producing semen of sufficiently high quality for use in artificial insemination the correlation between these measurements of semen quality and fertility level is not high.

Two studies (Gustavsson 1971C; Swartz and Vogt 1983) of repeat breeder heifers found the 1/29 translocation at a higher rate than expected among the females which did not conceive. Although fertility is not decreased markedly by this Robertsonian translocation, it is found more frequently among problem breeders.

In Sweden the elimination of 1/29 heterozygotes and homozygotes from artificial insemination by screening all selected males has apparently reversed a trend toward lowered fertility in the Swedish

Red and White breed (Gustavsson 1975). Australia has passed a law requiring that all imported bulls and imported semen be free from the 1/29 translocation.

The decrease in fertility has been shown (Refsdal 1976) to be associated with return to estrus before 90 days. It was assumed that nondisjunction may have occurred in some cases during meiosis. The resulting germ cells had either a deficiency or a duplication. Zygotes resulting from such germ cells died as early embryos. Since many sires used widely in artificial insemination had fertility records which were quite normal, it has been further assumed that defective spermatozoa in some males had much lower probabilities of fertilizing an ovum than did the normal spermatozoa. More recent studies by Dyrendahl and Gustavsson (1979) have shown that lower fertility also occurs in 1/29 translocation males, when these males have not been selected previously for high fertility. In females when an ovum with a deficiency or duplication is produced it could be fertilized but would die prior to 90 days into the gestation period.

Logue and Harvey (1978) in meiotic studies of 1/29 translocation bulls compared to bulls without the translocation found that the increase of hypermodal cells (for example, in cattle with $n = 30$, cells with 31 or more chromosomes) was from 2.8% in the normals to 6.4% in the heterozygous translocation bulls. Some other studies have indicated a higher number of cells in the second division of meiosis deviating from the expected, but these results correspond fairly closely to the reduction in fertility of heterozygotes. The fertility percentages of bulls and of their progeny vary greatly between individuals, so it might be expected that the amount of nondisjunction, and consequent unbalanced gametes, may also vary among individuals as the basic genetic constitution of the animals vary (Fig. 11.4).

Since unbalanced gametes resulting from nondisjunction may be selected against, the study of early blastocysts from cattle with one heterozygous 1/29 parent is also useful. Popescu (1980) found 2 of 52 embryos lacking the number 1 chromosome. King *et al.* (1980) found 2 embryos out of 18, one at 3 days and one at 7 days after breeding, which had 60 chromosomes including the 1/29 translocation, and apparently 2 normal number 1 chromosomes. These two reports would indicate that the problem of fertility is definitely the result of gametes with a deficiency or a duplication participating in the fertilization process. No live monosomic or trisomic animals have been

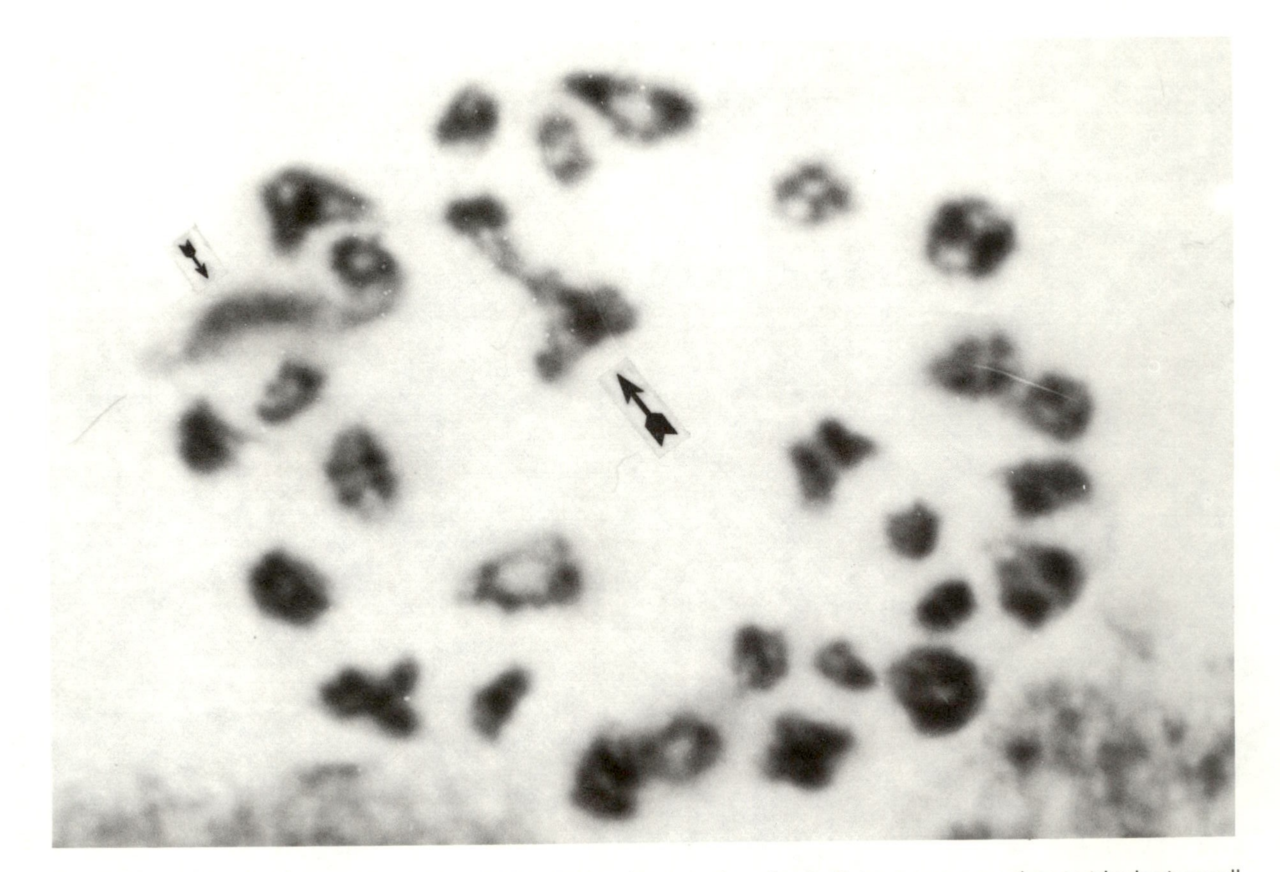

FIG. 11.4. Meiosis of a heterozygous 1/29 Robertsonian translocation bull. Large arrow points to trivalent; small arrow points to more lightly stained XY bivalent with a nearby bivalent more darkly stained.

Photograph by Eldridge.

found among progeny of 1/29 parents, so it has been concluded that the embryos die sometime soon after fertilization. Further observation by Linares *et al.* (1980) confirmed that a higher frequency of degenerated embryos, or embryos in the process of degeneration, was found from three heterozygous 1/29 carrier bulls bred to superovulated cows than from one control bull.

Many questions have been raised concerning the origin and widespread occurrence of the 1/29 Robertsonian translocation. Its occurrence in so many breeds of cattle of European origin has led to the hypothesis that the first translocation(s) may have occurred centuries ago. With no major deleterious effects associated with it, the increase in frequency may have been just random drift. Since no proven cases of de novo occurrence have been reported, the hypothesis for an ancient origin is strengthened. The single C-band in the 1/29 translocation chromosome (Fig. 11.5) has been accepted by some as evidence for an old, well-established translocation. Robertsonian translocations of a more recent origin seem to have larger, or dicentric, centromeres as shown by C-banding (Evans *et al.* 1973; Eldridge and Balakrishnan 1977; Eldridge 1974; Niehbuhr 1972; Popescu 1977; DiBerardino *et al.* 1979; Masuda *et al.* 1978). There is no evidence to date that would indicate a higher rate of fusion of chromosomes 1 and 29 than any other pair of chromosomes.

G-, Q-, and R-banding of the chromosomes has shown beyond question that the number 1 chromosome, the longest, is the chromosome that provides the long arms of the 1/29 translocation. The evidence from banding is not quite so certain with respect to the short arm being the number 29, the shortest chromosome. It is still difficult to establish with complete certainty the three or four shortest chromosomes by banding. However, the evidence is very strongly supportive as to the short arm being the number 29 chromosome.

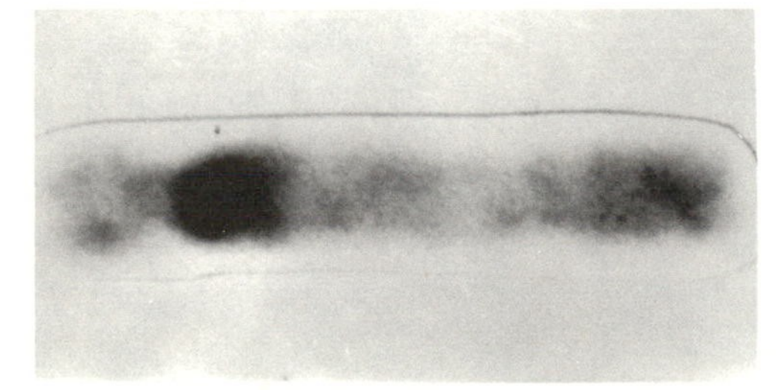

FIG. 11.5. C-banded 1/29 Robertsonian translocation. The monocentric C-band is located on the long-arm side of the centromere.

Photograph from the thesis of Master's W. F. Blazak, University of Nebraska.

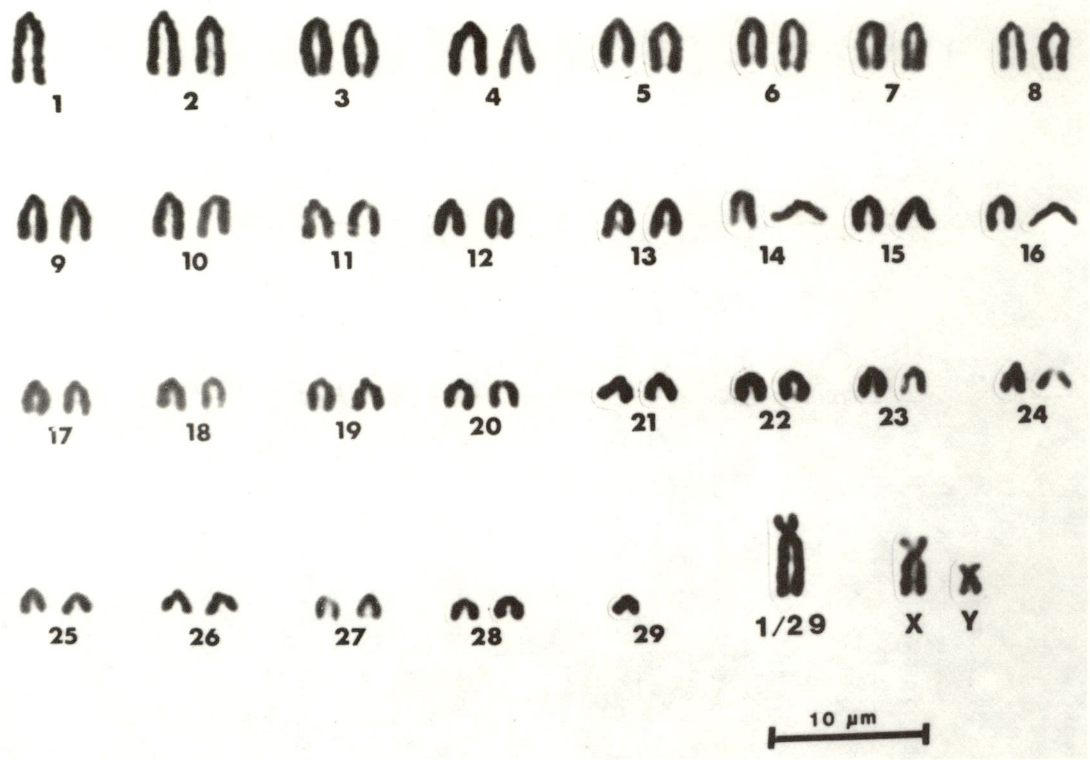

FIG. 11.6. Giemsa-stained karyotype of a Brown Swiss bull with a 1/29 Robertsonian translocation. The chromosomes are paired primarily upon length. Accurate pairing of chromosomes can be accomplished only with G-banded preparations.

Photograph by Eldridge.

For example, even though the first publication on the 1/29 translocation in British White cattle reported it to be 1/27, it is now conceded that it is probably 1/29 (Eldridge 1975).

Figure 11.6 illustrates the 1/29 Robertsonian translocation in cattle. Figure 11.7 shows quite clearly the bands by G-banding. Four cells in Figure 11.8 illustrate by G-banding the formation of the 1/29 Robertsonian translocation.

Figures 5.1 and 5.2 (see pp. 33 and 34) show the three different mechanisms by which Robertsonian translocations can be formed. Since the 1/29 translocation appears to be monocentric it is assumed that either mechanism (a) or (c) must have occurred.

Other Robertsonian Translocations

The 1/29 Robertsonian translocation is the most widely known and widely studied Robertsonian translocation. In the limited amount of research on chromosomes of cattle, 25 other Robertsonian translocations have been found (Table 11.2). In a number of these reports either the researchers did not band the chromosomes or, if they tried to band them, they encountered too much difficulty in identifying them to be certain of the specific chromosomes involved. In addition to the 25 given in Table 11.2, there are at least four other reports of Robertsonian translocations in which there was no attempt to identify the specific chromosomes involved in the translocation. Since the results from the Reading Conference were not available when many of these studies were made, there could be some banded translocations which have used numbers for the chromosomes that are different from the standardized karyotypes. Therefore, the list in Table 11.2 and the descriptions that follow may not be precisely the same as the results that will be found in the future if the same translocation is studied further.

It would seem reasonable to anticipate that breaks would occur more frequently in some chromosomes than in others. In Table 11.2 there are more translocations involving chromosomes 1, 5, 11, 14, 21, and 28 than any other chromosomes, but the pattern is not clearly a deviation from randomly occurring breaks. As more data are accumulated a pattern may develop. So far, four chromosomes, 12, 17, 19, and 26, have not been reported in Robertsonian translocations in cattle.

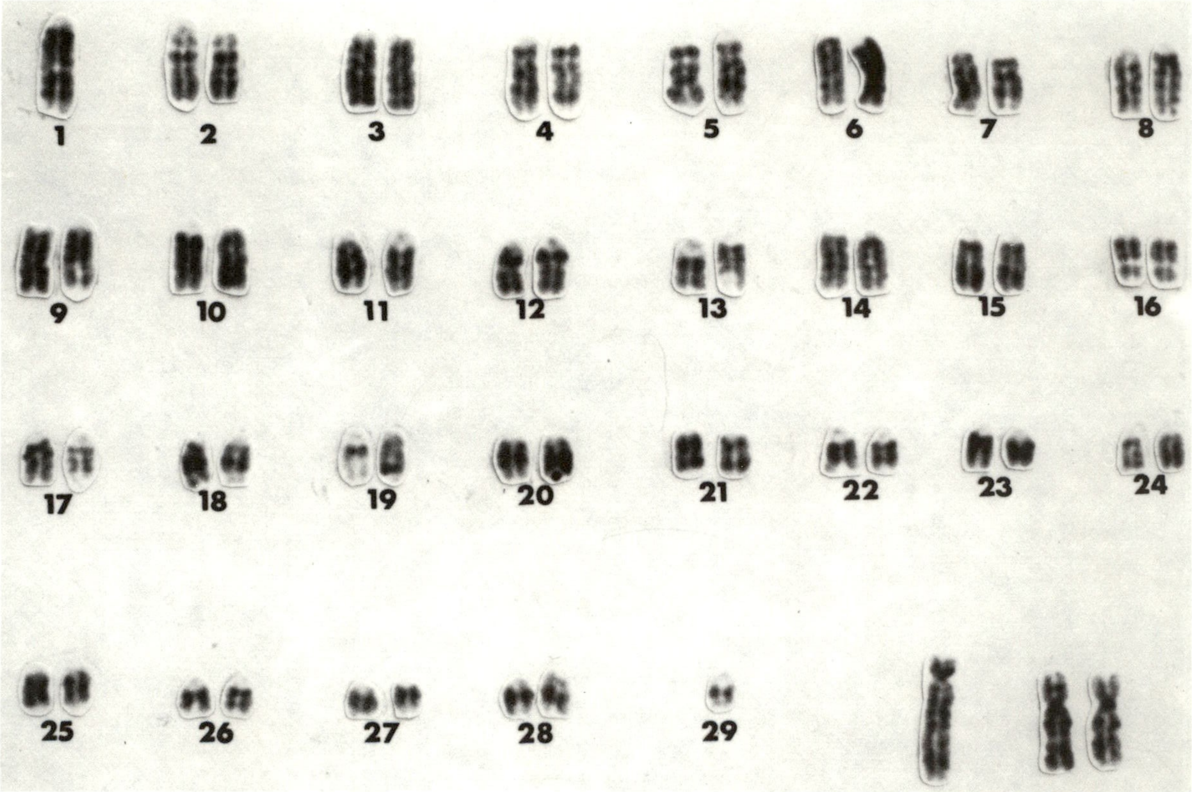

FIG. 11.7. G-banded karyotype of a bovine female with a 1/29 Robertsonian translocation.
Photograph by Eldridge.

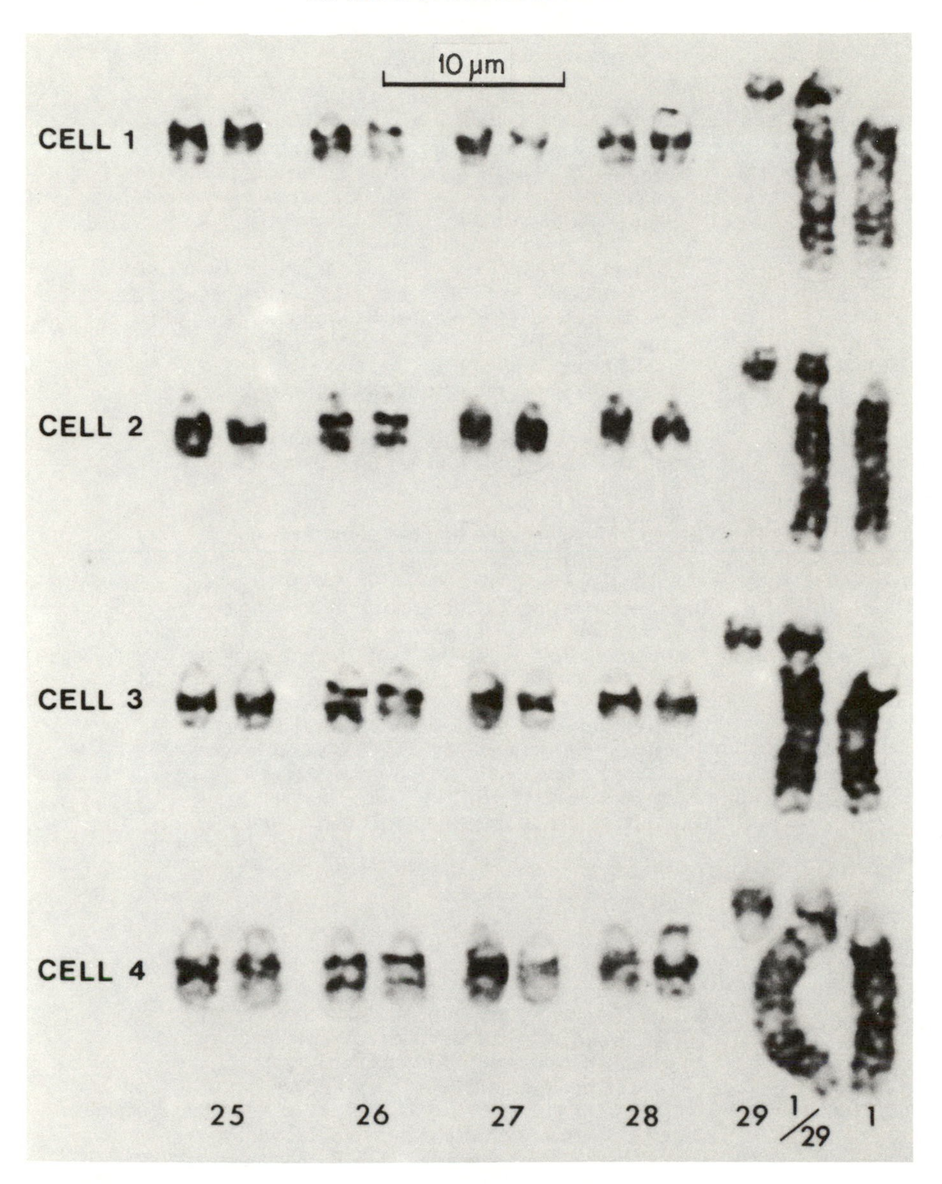

FIG. 11.8. Four G-banded cells showing the formation of the 1/29 Robertsonian translocation in cattle.
Courtesy of C. C. Lin and with the permission of the Genetics Society of Canada.

TABLE 11.2. Robertsonian Translocations in Cattle

Chromosomes	Reference	Breed
	Group A. Involving the number 1 chromosome	
1/4	Lojda *et al.* (1976)	Czechoslovakian cattle
1/23	Lojda *et al.* (1976)	Czechoslovakian cattle
1/25[a]	Stranzinger and Forster (1976)	Simmental
1/28	Lojda *et al.* (1976)	Czechoslovakian cattle
1/29	Many authors (1964–1978)	38 different breeds
	Group B. Involving the number 3 chromosome	
3/4	Popescu (1977)	Limousin
3/27	Samarineanu *et al.* (1976)	Friesian
	Group C. Involving the number 5 chromosome	
5/18	Papp and Kovacs (1980)	Simmental
5/23	Samarineanu *et al.* (1976)	Romanian Brown
	Group D. Involving the number 6 chromosome	
6/16[b]	Eldridge (1974)	Dexter
6/28	Lojda *et al.* (1976)	Czechoslovakian cattle
	Group E. Involving the number 8 chromosome	
2/8	Pollock (1974)	British Friesian
8/9	Tschudi *et al.* (1977)	Swiss
	Group F. Involving the number 11 chromosome	
11/16[b]	Kovacs and Papp (1977)	Hungarian Simmental
11–12/15–16	Bruere and Chapman (1973)	Blonde Aquitaine × Simmental
11/21[c]	Samarineanu *et al.* (1976)	Romanian Brown cattle
11/22	Lojda *et al.* (1976)	Czechoslovakian cattle
	Group G. Involving the number 14 chromosome	
14/20	Harvey and Logue (1975)	Swiss Simmental
14/24	DiBerardino *et al.* (1979)	Podolian
14/28	Ellsworth *et al.* (1979)	Holstein
	Group H. Involving the number 21 chromosome	
13/21[c]	Kovacs and Papp (1977)	Holstein–Friesian
7/21 (originally reported as 5/21)[c]	Hanada *et al.* (1981)	Japanese Black
5/21	Masuda *et al.* (1978); Okamoto *et al.* (1981)	Japanese Black
	Group J. Involving the number 27 chromosome	
25/27[a]	DeGiovanni *et al.* (1979)	Alpine Grey
	Group K. Involving other chromosomes	
10/15	Pinheiro and Ferrari (1980)	Petangueiras
7–11/20–25	Darre *et al.* (1975)	Blonde Aquitaine × Limousin
	Group L. Involving chromosomes not identified	
Froget *et al.* (1972); Knudsen (1956); Kovacs *et al.* (1973); Soldatovic *et al.* (1977)		
Chromosomes not yet reported to be involved in translocations are 12, 17, 19, and 26.		

[a] Both involve chromosome 25.
[b] Both involve chromosome 16.
[c] Three involve chromosome 21.

Involving the Number 1 Chromosome. Five reported translocations involve the number 1 chromosome. Lojda *et al.* (1976) reported Robertsonian translocations 1/4, 1/23, and 1/28 in Czechoslovakian cattle. Stranzinger and Forster (1976) found a 1/25 Robertsonian translocation in Piebald cattle in Germany. Eldridge (1975) identified a translocation in British White cattle in Scotland as a 1/27, but slides from these animals banded by Gustavsson and further study following the Reading Conference confirmed it as the 1/29.

Involving the Number 11 Chromosome. The next most frequent chromosome to be involved in translocations is the number 11. One 11/16 was reported by Kovacs (1976) in Hungarian Simmentals. Samarineanu *et al.* (1976) reported an 11/21 in a Brown cow in Romania. Bruere and Chapman (1973) found a young Simmental bull with a translocation between either 11 or 12 and 15 or 16 chromosomes. This case could have been the same (11/16) as reported by Kovacs (1976). The animal was too young for any measure of fertility but appeared phenotypically normal.

Lojda *et al.* (1976) found an 11/22 Robertsonian translocation in Czechoslovakian cattle.

Darre *et al.* (1975) reported another translocation in a Blonde Aquitaine × Limousin crossbred male which they called "F," which was only generally identified. They indicated that the q arm was from chromosomes 7–11, and the p arm from chromosomes 20–25. This translocation could have been the same as the 11/21 (Samarineanu *et al.* (1976) or 11/22 (Lojda *et al.* 1976), or all three translocations could have been the same. This case (Darre *et al.* 1975) is of great interest, however, because the bull in which this translocation was found also had a 1/29 translocation. One of his sons received the 1/29 translocation without the "F" and one of his daughters carried the "F" without the 1/29.

Involving the Number 3 Chromosome. Two distinctly different Robertsonian translocations have been reported. One came from Popescu (1977) on another dicentric Robertsonian translocation in a Limousin cow in France between the number 3 and 4 chromosomes. Identification was positive using 5-bromodeoxyuridine (BUdR) and acridine orange staining. No breeding results were reported, so no estimate of its effect upon fertility was made. The dicentric condition was established by C-banding. It was assumed that the translocation was relatively new because of the dicentric condition.

Samarineanu *et al.* (1976) discovered a 3/27 translocation in a Holstein freemartin. Apparently the animal was a mosaic with only part of its XX cells carrying the translocation chromosome.

Involving the Number 5 Chromosome. Four reports have been published on Robertsonian translocations involving the number 5 autosome. Samarineanu *et al.* (1976) found a cow of the Romanian Brown breed that had osteoarticular problems, as well as a 5/23 translocation. The phenotypic variation may have been only coincidental.

Papp and Kovacs (1980), by C-banding, found a 5/10 translocation in a Simmental bull to be dicentric. By use of G-banding they concluded that the number 5 chromosome was identical with the number 5 in the report by Masuda *et al.* (1978), but that the short chromosome involved was definitely different. They also concluded that the number 5 chromosome was different from the one reported by Eldridge (1975A) as a 5–6/15–16, and thus further confirmed the report by Logue *et al.* (1977) that this was the number 6 chromosome in the Dexter cow. Further examination of offspring will be made since the bull was to be used for breeding. Publication of good-quality photographs permits this type of comparison and analysis.

Study of the 5/21 translocation has been continued by Masuda *et al.* (1980). Among 20 cattle, 13 had the 5/21 translocation, 5 had 1/29, and 2 had both (Fig. 11.9). No detectable decrease in fertility has been found. The 5/21 translocation was dicentric.

Okamoto *et al.* (1981) also found the dicentric 5/21 in Japanese Black cattle, independently from Masuda *et al.* (1978) (Fig. 11.10). It is possible that the frequency of this translocation in Japanese Black cattle may be rather high and widespread.

Involving the Number 14 Chromosome. Three cases have been reported involving the number 14 chromosome. Harvey and Logue (1975) found a translocation in three Swiss Simmental cattle, which, by length of arms from Giemsa-stained slides, was identified generally as a fusion of chromosomes from the 11–15 group and the 19–23 group. By G-banding they identified more positively and accurately the translocation as 13/21. Following the Reading Conference in which the karyotype of cattle was standarized, the nomenclature for this translocation was finally identified as 14/20 (Logue *et al.* 1977). C-banding and G-banding showed "two blocks of

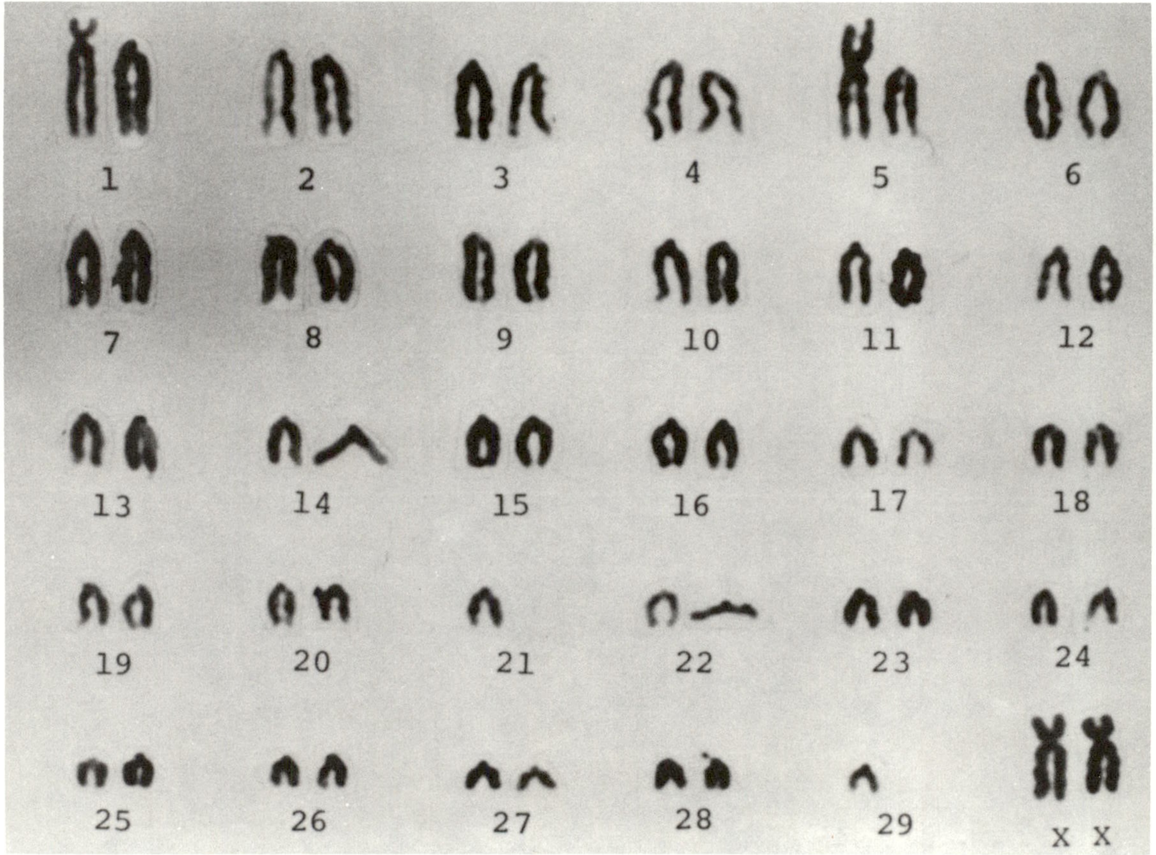

FIG. 11.9. Karyotype of a bovine female with two different Robertsonian translocations, 1/29 and 5/21. *Photograph by H. Masuda. From Masuda et al. (1980).*

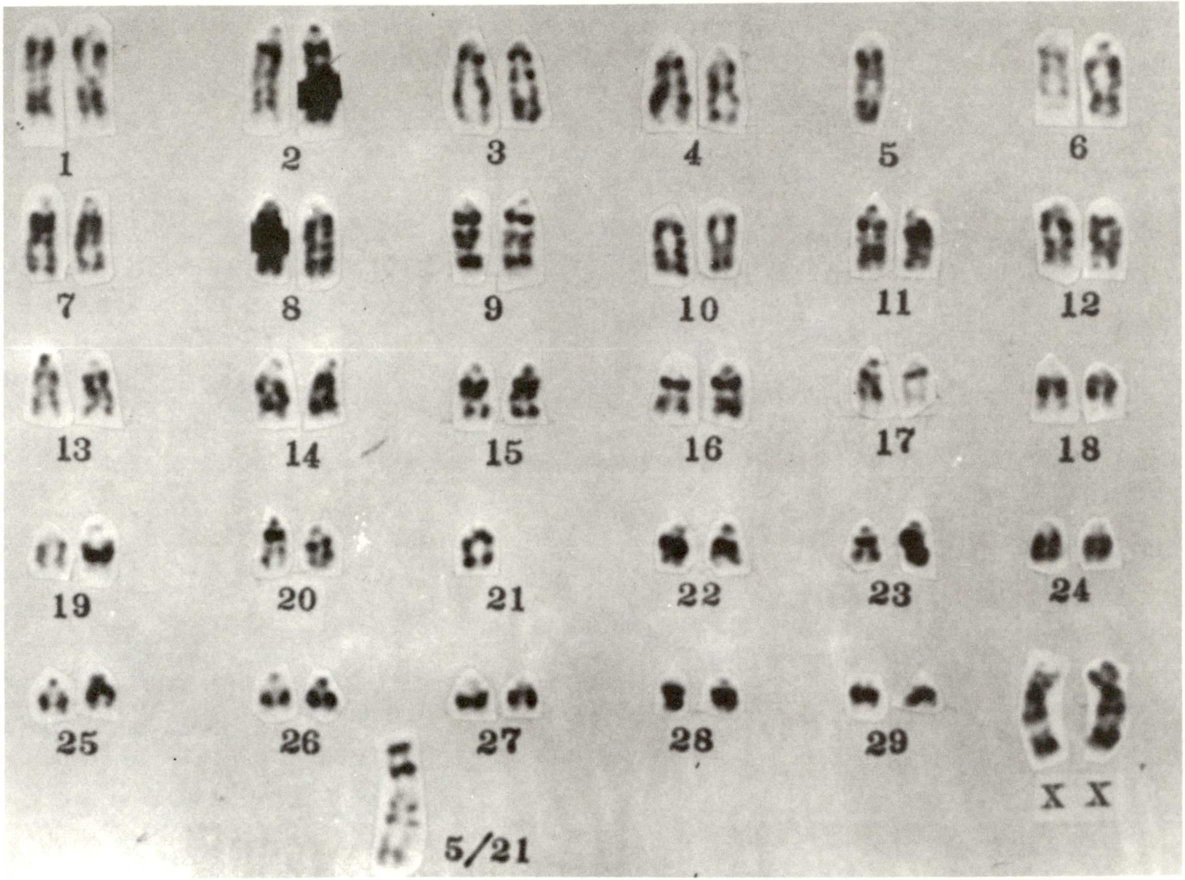

FIG. 11.10. G-banded karyotype of a bovine female with the 5/21 Robertsonian translocation.
Photograph by H. Masuda. From Masuda et al. (1978).

heterochromatin to be present at the primary constriction of the translocation metacentric chromosome" (Figs. 11.11 and 11.12).

Some other Swiss Simmental cattle in Switzerland, related to the ones studied by Harvey and Logue (1975), appear to have the same translocation. Since the animal studied in Bruere and Chapman's

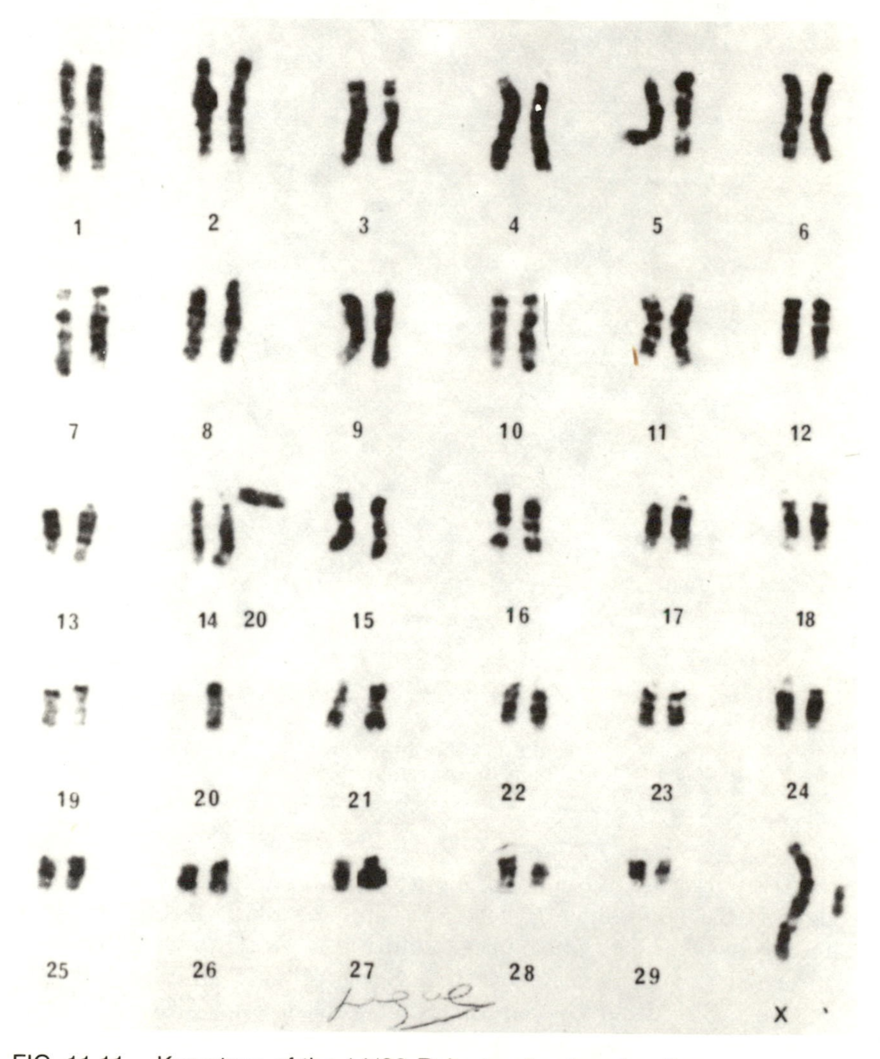

FIG. 11.11. Karyotype of the 14/20 Robertsonian translocation.
Photograph by D. Logue. From Logue et al. (1977).

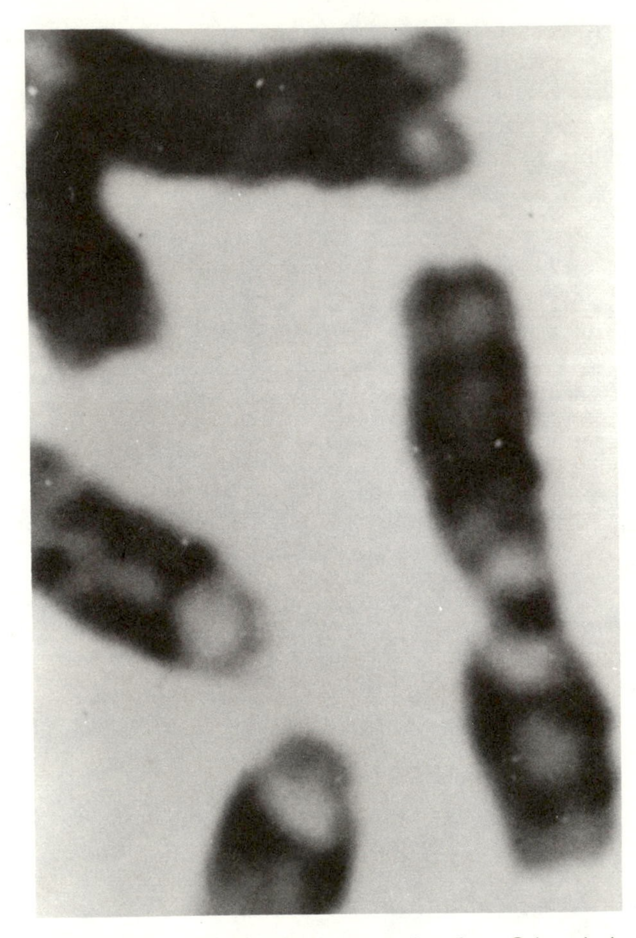

FIG. 11.12. Enlarged photograph of a G-banded 14/20 Robertsonian translocation.
Phograph by D. Logue. From Logue et al. (1977).

(1973) report was a descendant of one animal studied by Harvey and Logue (1976), and since Bruere and Chapman (1973) did not band the chromosomes, it can be reasonably assumed that both reports concerned the same translocation. Some other translocations not positively identified by banding, but reported as different, may eventually turn out to be identical with each other.

The 14/20 translocation (Harvey and Logue 1975) apparently segregated in a 1 : 1 ratio, with no variation between sexes, similar to other translocations. The meiotic studies revealed a trivalent at diakinesis as expected.

DiBerardino *et al.* (1979) reported a 14/24 Robertsonian translocation in a Podolian-type heifer in Italy. The translocation was identified by acridine orange R-banding. It was also found to be dicentric by C-banding. In addition to the new translocation the animal was also found to have a 1/29 Robertsonian translocation.

Ellsworth *et al.* (1979) found a Holstein Friesian cow with a 14/28 translocation. She had an unusually wide head at eye level and muzzle, and a very masculine body. She was among the top 20% of the herd in production and had 6 calves by 8 years of age. Two daughters examined were chromosomally normal, as were her maternal and paternal half-sibs, so it was suggested that the translocation originated with her, and may not have been present in the ovaries.

Involving the Number 8 Chromosome. Pollock (1974) reported a chromosome abnormality in British Friesian cattle which he identified at that time as a 2/4. In a much more complete report (Pollock and Bowman 1974) the chromosomes involved were described: "The banded karyotype showed that, according to the idiogram of the authors, the second and fourth autosomes were probably involved, as previously suggested (Pollock 1972). This idiogram may, however, have been different from those prepared by other workers and it would be advantageous if standard banded karyotypes for cattle, and possibly other agriculturally important animals, were compounded." Following the Reading Conference in 1976 the designation was changed to correspond with the standardized karyotype (Logue *et al.* 1977). It is now recognized as a 2/8 (Fig. 11.13).

When the last report (Pollock and Bowman 1974) was published the 89 calves from mating of the heterozygous sire had been classified as follows: 31–60,XY; 18–59,XYt; 18–60,XX; and 22–59,XXt (the term 59,XYt means 59 chromosomes including XY and 1 translocation). The ratio of 40 translocation animals to 49 animals without the translocation did not differ significantly from the usual 1 : 1 ratio, nor did the distribution of the translocation differ significantly between the sexes, 18 : 22. The offspring were not old enough when the paper was written to obtain any measure of the effect of the translocation on fertility. No phenotypic effects were noted to be associated with the translocation.

The other translocation involving chromosome 8 was only provi-

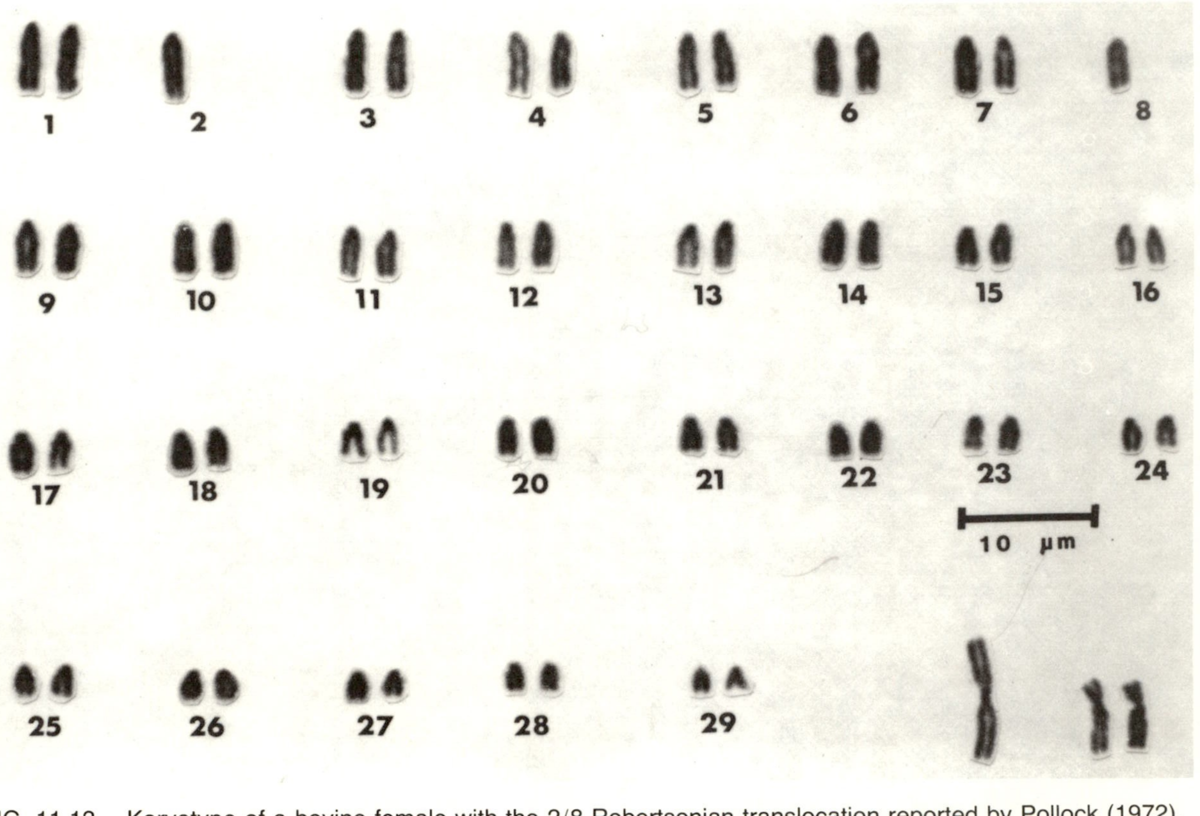

FIG. 11.13. Karyotype of a bovine female with the 2/8 Robertsonian translocation reported by Pollock (1972). *Photography by Eldridge.*

sionally identified as 8/9 since no banding was done. Tschudi *et al.* (1977) found this translocation in a Braunvieh or Brown Swiss bull among 1141 animals studied. Since the translocation bulls in this study were eliminated from service, no information was given on the transmission. Apparently the bull with this translocation was phenotypically normal.

Involving the Number 6 Chromosome. Two reports have been published of Robertsonian translocations in cattle involving the number 6 chromosome. Eldridge (1974) found a translocation in a Dexter cow in the Animal Breeding Research Organisation herd in Edinburgh, Scotland, which was identified by banding as a centric fusion between the 5th or 6th and 15th or 16th chromosomes. Following the Reading Conference which established the standardized karyotype, and with further work done by Logue *et al.* (1977), the translocation was specifically identified as 6/16 (Fig. 11.14). The cow produced three calves, none of which possessed the translocation chromosome. An interesting deviation in this case was the fact that 39 out of 486 cells observed, or 8.0%, had 60 chromosomes with no translocation. She was not a twin, and blood typing gave no indication that there might have been a female twin lost early in the pregnancy. Nothing certain could be given to explain the mosaicism, although several hypotheses could be advanced. The female was sold to the Veterinary College in Glasgow where further study was conducted. Although the first pregnancy took eight inseminations, it was not implied that the translocation caused the lowered fertility since the entire herd at that time was having fertility problems. C-banding identified the translocation as dicentric, but no C_d-banding was done.

Lojda *et al.* (1976) found a 6/28 translocation in Czechoslovakian cattle.

Involving the Number 21 Chromosome. Samarineanu *et al.* (1976) found an 11/21 translocation in Romanian Brown cattle, although it could have been 11/22, and Masuda *et al.* (1978) clearly identified the 5/21, both discussed earlier under the 11 and 5 groups, respectively. In addition, Kovacs and Papp (1977) reported a 13/21 in a mosaic Holstein–Friesian.

Involving the Number 27 Chromosome. Through R-banding and microdensitometric profiles, DeGiovanni *et al.* (1979) quite clearly

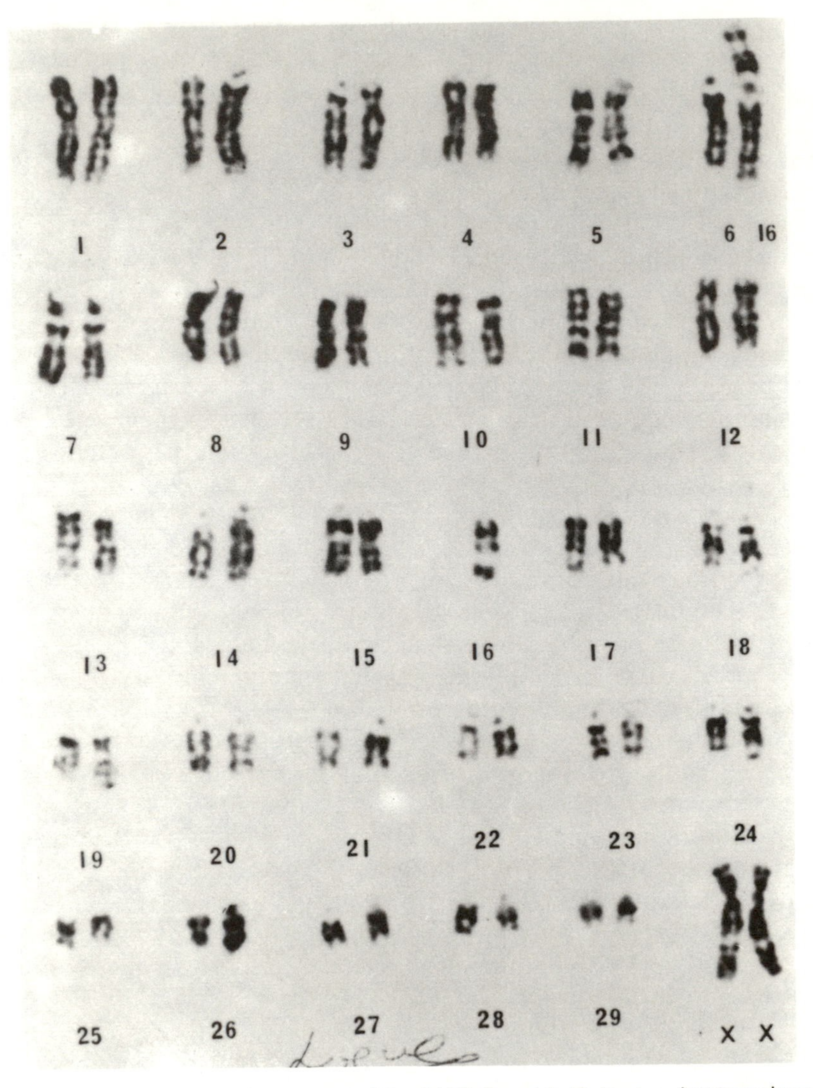

FIG. 11.14. G-banded karyotype of the 6/16 dicentric Robertsonian translocation. This was first reported by Eldridge as (5 or 6)/(15 or 16), later confirmed by Logue as 6/16.

Photograph by D. Logue. From Logue et al. (1977).

showed that a Robertsonian translocation chromosome was 25/27 in Alpine Grey cattle in Italy. By photoelectric scanning, variations in density of photographs of chromosomes can be converted into graphs which show the relative density of the banding pattern along the length of the chromosome. Five out of 20 bulls sampled carried this translocation, which did not appear to be dicentric. Four of these were related through a common ancestor in the third generation, and the fifth had no common close ancestors. Both semen quality and libido were poor in most of the five bulls. Further studies (DeGiovanni *et al.* 1980) of meioses from one bull indicated a rate of nondisjunction of 46.7%, which seemed high, and they suggested that additional bulls need to be studied. The frequency in the breed has been estimated at 9.4% by Succi *et al.* (1980).

Another translocation, not Robertsonian but involving chromosome 27, was reported by Bongso and Basrur (1976). This is considered in greater detail under Dicentric Translocations.

Involving Other Chromosomes. Pinheiro and Ferrari (1980) reported that one Petangueirias bull had a 10/15 Robertsonian translocation. The sire was normal and the dam unavailable. Eldridge (1974) found a dicentric Robertsonian translocation involving chromosome 16. This is discussed under number 6 (p. 141) and under Dicentric Translocations.

Darre *et al.* (1975) reported a translocation only broadly defined as 7–11/20–25. This might be the same as some others reported.

Involving Chromosomes Not Identified. Four other authors (Froget *et al.* 1972; Knudsen 1956; Kovacs *et al.* 1973; Soldatovic *et al.* 1977) reported one or more translocations without specifying the chromosomes involved. Some of these may duplicate ones described previously, or they may be different translocations.

Dicentric Translocations

The presence of two blocks of heterochromatin, by C-banding, at the centromeric region of Robertsonian translocation chromosomes has been accepted by several persons (Eldridge 1974; Popescu 1977; DiBerardino *et al.* 1979) as evidence for the dicentric condition. Arguments against this have been presented by Pathak and Wurster-Hill (1977), since heterochromatin has been found in many species

TABLE 11.3. Dicentric Translocations

Chromosomes	Breed	Reference
6/16	Dexter	Eldridge (1974)
3/4	Limousin	Popescu (1977)
14/24	Podolian	DiBerardino *et al.* (1979)
14/28	Holstein	Ellsworth *et al.* (1979)
5/21	Japanese Black	Masuda *et al.* (1978); Okamoto *et al.* (1981)
5/18	Simmental	Papp and Kovacs (1980)

to be distributed in various parts of the chromosomes, and not limited to the centromeric region. However, in cattle where C-banding identifies heterochromatin in only the centromeric region, and in Robertsonian translocation chromosomes where the C-bands are apparently two bodies, it seems reasonable to consider such chromosomes to be dicentric. Six different dicentric Robertsonian translocations have been reported (Table 11.3).

The first such dicentric Robertsonian translocation, 6/16, was reported by Eldridge (1974) (Fig. 11.15). Popescu (1977) reported a 3/4 translocation in the Limousin breed which was dicentric. In 1979 DiBerardino *et al.* reported on a 14/24 dicentric Robertsonian translocation in a heifer of the Podolian type in Italy, which was accompanied by a 1/29 Robertsonian translocation in the same animal. The 1/29 had only one block of centromeric heterochromatin, similar to other reports of this translocation. The 14/24, however, showed a very extended area of heterochromatin completely across the centromeric region. Masuda *et al.* (1980) reported that 13 Japanese Black cattle out of 55 examined had a 5/21 Robertsonian translocation. C-banding again revealed that the 5/21 translocation was dicentric, with two blocks of heterochromatin in the centromeric region. In this group of 55 cattle 5 were heterozygous for the 1/29 translocation, and 2 heifers had both the 1/29 and the 5/21. Okamoto *et al.* (1981) confirmed the dicentric nature of the 5/21 translocation in Japanese Black cattle.

Ellsworth *et al.* (1979) found in a Holstein cow a dicentric 14/28 translocation and Papp and Kovacs (1980) described another clear case of C-banding extending across the centromere and appearing to be dicentric.

The simultaneous occurrence of two different Robertsonian translocations, 14/24 and 5/21, accompanied by 1/29 translocations in the same animals offered opportunities to compare C-banding in the

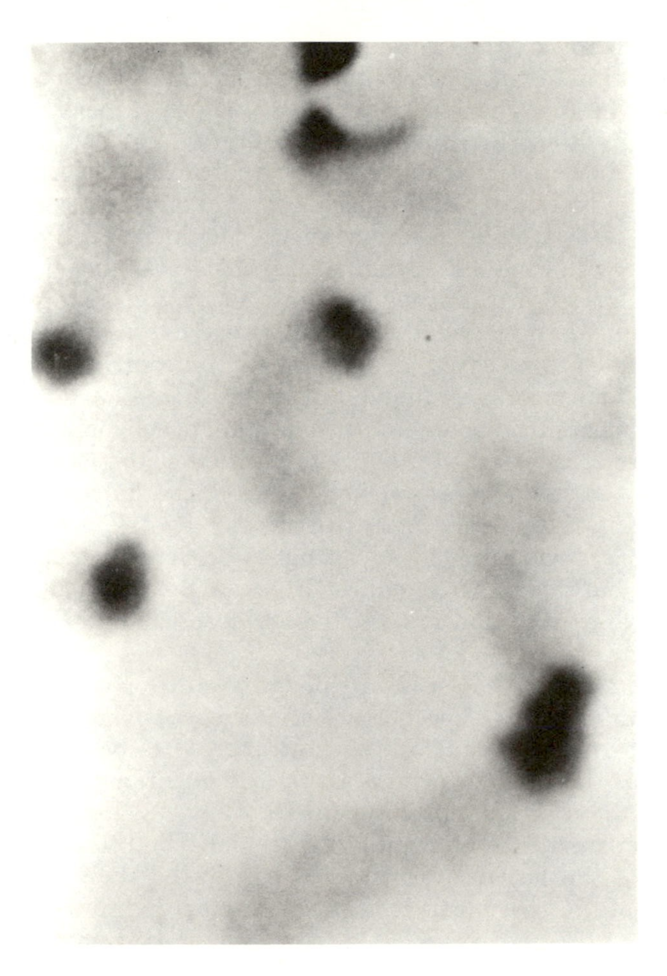

FIG. 11.15. Dicentric Robertsonian translocation, 6/16, as illustrated by C-banding.
Photograph by D. Logue.

same cells. The widespread occurrence of 1/29 appears to indicate an older origin of this translocation, and the dicentric nature of the other translocations tends to support the concept that newer, less frequently found Robertsonian translocations are dicentric. It must be recognized, however, that other explanations may also be made for these differences. For example, when the number 1 chromosome fuses with number 29 there may be a tendency for one of the chromosomes to break either through the centromere or on the long-arm side, resulting in one block of heterochromatin, while a number of other chromosomes have a tendency to break on the short-arm side, resulting in dicentric chromosomes.

Not all Robertsonian translocations in cattle have been studied by C-banding, and none has reported C_d-banding, so comparisons across all cases are difficult to make. Seldom has it been possible to determine with confidence whether a translocation has occurred de novo or has been transmitted for many generations. Therefore, it is difficult to determine whether the dicentric nature of a translocation is related to its age. Since a number of dicentric Robertsonian translocations have been found, but several more have been found that are apparently monocentric, it would appear more likely that the nature of the translocation is a function of the location of the breaks at the time of centric fusion rather than a function of the age of the translocation. The 1/29 translocation with its monocentric appearance is widespread, indicating its occurrence many generations ago, but the 5/21 in Japanese Black cattle, which has not been studied extensively, appears also to be widespread in that breed, but is dicentric.

The 27/29 translocation in a Guernsey bull reported by Bongso and Basrur (1976) is a special case of a dicentric translocation. It is not a Robertsonian translocation but appears to be fusion after a break near the telomeric end of the q arm of chromosome 27 and a break in the p arm of chromosome 29. They suggest that the two centromeres may not be equally functional. The relatively normal fertility of the bull was considered also to be the result of less viability of the sperm carrying the unbalanced chromosomes. The daughters were not investigated to determine transmission of this translocation.

The tandem translocation in cattle studied by Hansen (1979) could also be classified as dicentric. Apparently when the number 18 chromosome became fused to the telomeric end of chromosome 1, not all of the centromeric heterochromatin was lost. C-banding revealed some heterochromatin at that position. No measure has been made concerning the degree to which this heterochromatin region was involved in spindle fiber attachment. Other species have heterochromatin located at numerous regions along the length of the chromosomes without any apparent effect on mitosis at anaphase.

X-AUTOSOME TRANSLOCATIONS

In 1968 Gustavsson *et al.* found a Swedish Red and White heifer with one X chromosome longer than the other, the additional length being in the short arm. It was presumed that a piece of an autosome from an unidentified chromosome had been translocated to the short arm of one X chromosome. The sire had the 1/29 translocation and the dam was a normal 60,XX. The heifer, in addition to the X-autosome translocation, also had the 1/29 translocation. The heifer was bred 12 times using various combinations of natural service and artificial insemination with both fresh and frozen semen. She conceived on the 12th insemination, which was done artificially with fresh semen. The calf was a full-term, stillborn male without any apparent abnormalities. Skin fibroblast cultures of the calf showed it to have 59 chromosomes, including the 1/29 autosome translocation, and apparently normal sex chromosomes.

Lymphocyte cells cultured from the heifer were labeled with tritiated thymidine, and in 50 cells the normal X always was found to be late-replicating. In a later report (Gustavsson 1971B) 600 more cells were studied with the same results. These results are in general agreement with studies of X-chromosome rearrangements in other species. The random inactivation of one X chromosome, when both are normal, is disturbed in cases where one X chromosome has become fused with an autosome. According to the Lyon (1961) hypothesis, inactivation of one X chromosome is at random, and approximately equal numbers of the X chromosomes in the somatic cells will be of paternal or maternal origin. It seems possible that the translocated piece of the autosome may have initiated synthesis at the same time as the other autosomes and progressed throughout the translocation chromosome, and therefore the normal X chromosome was late-replicating. The animal never conceived a second time.

Eldridge (1980) reported an X-autosome translocation in a 3/4 Holstein × 1/4 Brown Swiss heifer (Fig. 11.16). From Giemsa-stained slides the translocation appeared to be identical to the case published by Gustavsson (1971B). By banding, the translocation was found to be between the number 13 chromosome and the short arm of one X chromosome (Fig. 11.17).

The telomeric portion of one of the number 13 chromosomes, in-

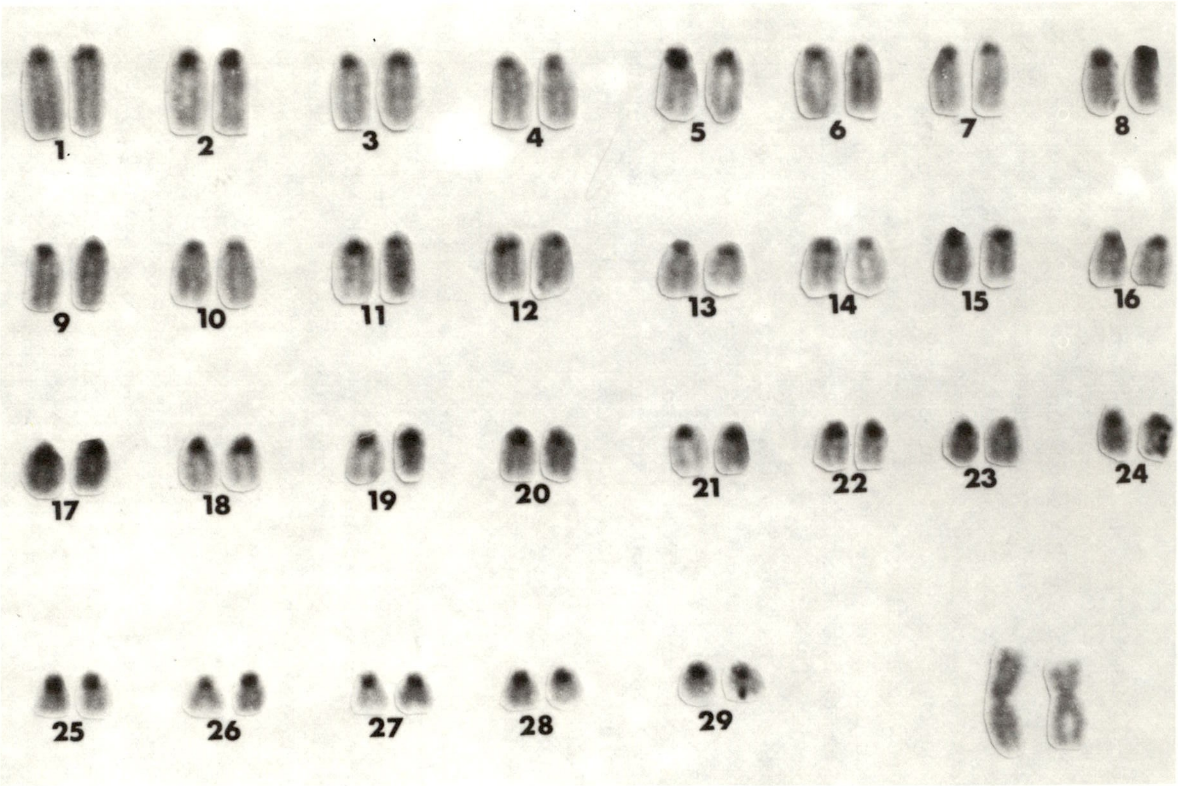

FIG. 11.16. A C-banded karyotype illustrating an X-autosome translocation in cattle. To date this has only been found in females.

Photograph by Eldridge.

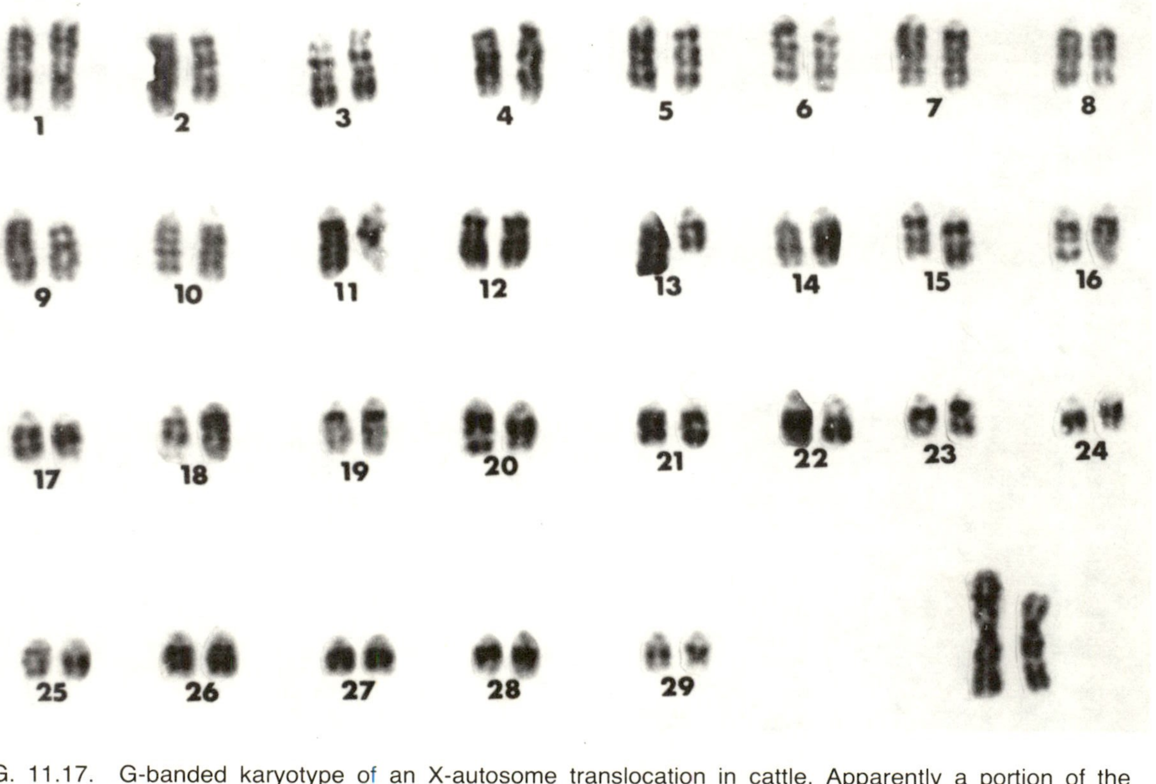

FIG. 11.17. G-banded karyotype of an X-autosome translocation in cattle. Apparently a portion of the telomeric end of chromosome 13 became translocated to the p (short) arm of the X chromosome.

Photograph by Eldridge.

cluding two G-bands, has become attached to the short arm of one X chromosome. The X-autosome translocated chromosome was found consistently in all cells from lymphocyte cultures. Perhaps it was coincidence, but as in Gustavsson's (1971B) case the dam carried a 1/29 translocation and the sire had a normal X as determined from chromosome studies of his other female offspring. Differing from the Gustavsson case, this female did not receive the 1/29 translocation. One apparently normal male calf has been born which, without G-banding, was normal in its chromosome complement. At 1 year of age the calf had normal libido and was normal in all other respects. The second calf from this cow was a female, which also resulted from conception at the first service. It has one normal X chromosome and one translocation chromosome. A third calf, female, had apparently normal chromosomes. The original X-autosome translocation female and her X-autosome daughter have each produced one more calf, neither of which has the X-autosome translocation. Further analysis is being made. A connective tissue fibroblast culture from the original cow showed the same chromosome pattern. This culture is currently being used in somatic cell hybridization with a hamster culture by G. Echard in Toulouse, France, for chromosome mapping studies.

Another set of observations on an X-autosome translocation was made by Basrur *et al.* (1982). They found the translocation in a Limousin-Jersey crossbred. In the herd, several cattle carried the aberration, including one cow, her paternal half-sister, and her daughter. The specific autosome involved was not identified, but it appeared to be different from the one reported by Gustavsson (1971B). The level of fertility was lowered by this translocation.

DOUBLE TRANSLOCATIONS IN CATTLE

Since chromosomal aberrations do not occur at a high frequency in cattle generally, the discovery of two translocations in one animal, if translocations occur at random, would be expected to happen very infrequently. It is possible that the tendency for an animal to have one translocation may increase the probability of a second translocation. In other words, the inherent tendency for an animal to have chromosome breakages which result in translocations may increase the probability of a second breakage–translocation event. However, it is also possible that the discovery of one translocation in a breed or herd may have resulted in more extensive studies of that group of animals than in cattle generally, and the probability of the discovery of a second translocation has been increased. Several cases have been reported in which an animal had two translocations.

The first double translocation found in an animal was reported by Gustavsson *et al.* (1968; Gustavsson 1971B). While studying animals with a 1/29 translocation in SRB cattle in Sweden one female was discovered to have an X-autosome translocation. The animal had a low level of fertility and did not transmit the translocation chromosome to the one offspring produced. It is interesting to note that the only other X-autosome translocation reported to date came from a dam with a 1/29 translocation, but the 1/29 translocation was not transmitted.

The second double translocation reported (Queinnec *et al.* 1974) was only identified in a subsequent paper (Darre *et al.* 1975) as one between the chromosomes 7–11 and 20–25. It also occurred in conjunction with a 1/29 Robertsonian translocation. The male carrying these two translocations transmitted the 1/29 to a son and the other to one daughter.

The third report (DiBerardino *et al.* 1979) identified, by R-banding, the 1/29 Robertsonian translocation and a new Robertsonian translocation, the 14/24 (1976 Reading Karyotype Conference) in a Podolian-type heifer near Salerno, Italy.

A fourth report on double translocations in cattle came from Japan in the Black Japanese (Wagyu) cattle (Masuda *et al.* 1980). Two heifers from mating a 1/29 heterozygous sire to 5/21 heterozygous dams produced heifers with both translocations (see Fig. 11.9).

It is interesting to speculate how far the reduction in chromosome numbers in cattle could go if all Robertsonian translocations could be accumulated in the homozygous condition. The following nine translocations would not duplicate any chromosomes: 1/29 + 3/4 + 5/21 + 6/28 + 2/8 + 11/16 + 14/24 + 25/27 + 10/15 (Table 11.2). This could reduce the number to 42, instead of 60, and would leave chromosomes 7, 9, 12, 13, 17, 18, 19, 20, 22, 23, and 26 as the 11 acrocentric pairs. Each translocation individually has caused no measurable effect upon phenotype, except on fertility in the heterozygous condition. Presumably full fertility would be restored in the homozygous condition since only bivalents occur at synapsis in meiosis. Such an accumulation, however, might make crossing with cattle with the normal 60 chromosomes very difficult, especially reducing the fertility of the hybrid offspring.

OTHER TRANSLOCATIONS

Prior to the development of banding techniques, a translocation resulting in one very long acrocentric chromosome was reported in the Red Danish Milk Breed by Hansen (1969). He called this a tandem translocation. Apparently a break occurred very near the centromere in one chromosome and at the telomere in another, with subsequent fusion and a loss of the one centromere, since the animals had only 59 chromosomes. The translocation was transmitted to somewhat less than half of the progeny and decreased fertility by about 10% in the males. Both males and females were clinically normal.

Further study by Hansen (1979) has indicated that the tandem translocation was between autosomes 1 and 18. The centromeric region of chromosome 18 had become fused to the telomeric region of chromosome 1 without losing all of its centromere. C-banding revealed two heterochromatin regions in this translocation chromosome. The translocation is described as 59,XY, −1, −18, + tandic (1 : 18). Although heterochromatin areas have been found distributed throughout the karyotype of some species, the heterochromatin is normally localized in the centromeric region of cattle chromosomes. There is some similarity between this translocation and the 27/29 translocation found by Bongso and Basrur in a Guernsey, described next.

Bongso and Basrur (1976) reported a translocation in a Guernsey bull used in artificial insemination, which had a record of fertility slightly higher than the breed average. The translocation appeared to involve the fusion of chromosomes 27 and 29 and produced an apparently dicentric chromosome. This dicentric condition, which has now been reported in five cases, did not seem to affect meiosis. They did not report on the frequency of this chromosome in the offspring of the bull.

TRISOMY

Within a period of 4 years, starting in 1968, three groups of researchers in three different countries found a trisomic condition associated with extreme brachygnathia (very short lower jaw) in cattle: Herzog and Höhn (1968, 1970) in Germany, Mori *et al.* (1969) in a Holstein male in Japan, and Dunn and Johnson (1972) in a Brown Swiss male in the United States. Herzog *et al.* (1977) found that the chromosome involved in the trisomic condition was number 18, although this chromosome has been identified by the Standardization Conference of Reading as number 17 (Figs. 11.18 and 11.19). Mori *et al.* (1969) made no attempt to identify the chromosome pair which was trisomic. Dunn and Johnson counted only 30 cells, 17 of which had 61 chromosomes. They thought the extra chromosome was one of the largest, designating it as 61,XY,?A, following the human system of grouping chromosomes. The condition may not have been caused by trisomy of the same chromosome in each case. As new cases were found by Herzog *et al.* (1977) the chromosomes were banded and positively identified. They have called this characteristic, phenotypically, the "lethal brachygnathia trisomy syndrome," or LBTS. Figures 11.18 and 11.19 illustrate this anomaly. Not all cases had 100% of the cells trisomic.

Two other cases of congenital abnormalities associated with a trisomic condition were reported by Tschudi *et al.* (1975, 1977). One of these had an umbilical hernia and heart defects; the other was an intersex with arthrogryposis and microophthalmia. Both were Simmental calves.

In 1967 Dunn *et al.* cultured cells from a markedly deformed mass that was born twin to a male. It had no heart, head, limbs, etc., but did have distinguishable parts of the digestive tract. Blood was not obtainable, but tissues from parts of the mass were cultured. Two cell lines were found, one 60,XX? and the other 61,XX?. Apparently only one member of the number 1 pair of chromosomes was in each line and there were one or two extra, small, acrocentric chromosomes. Most cases of anidian monsters (shapeless masses of tissue, usually without a heart, and dependent upon the circulation of the normal twin) have been thought to be developed from a piece of the early embryo which had broken off and developed as a parasite of the normal twin. This case, because it was a female born twin to a male, was obviously a case of dizygotic twins.

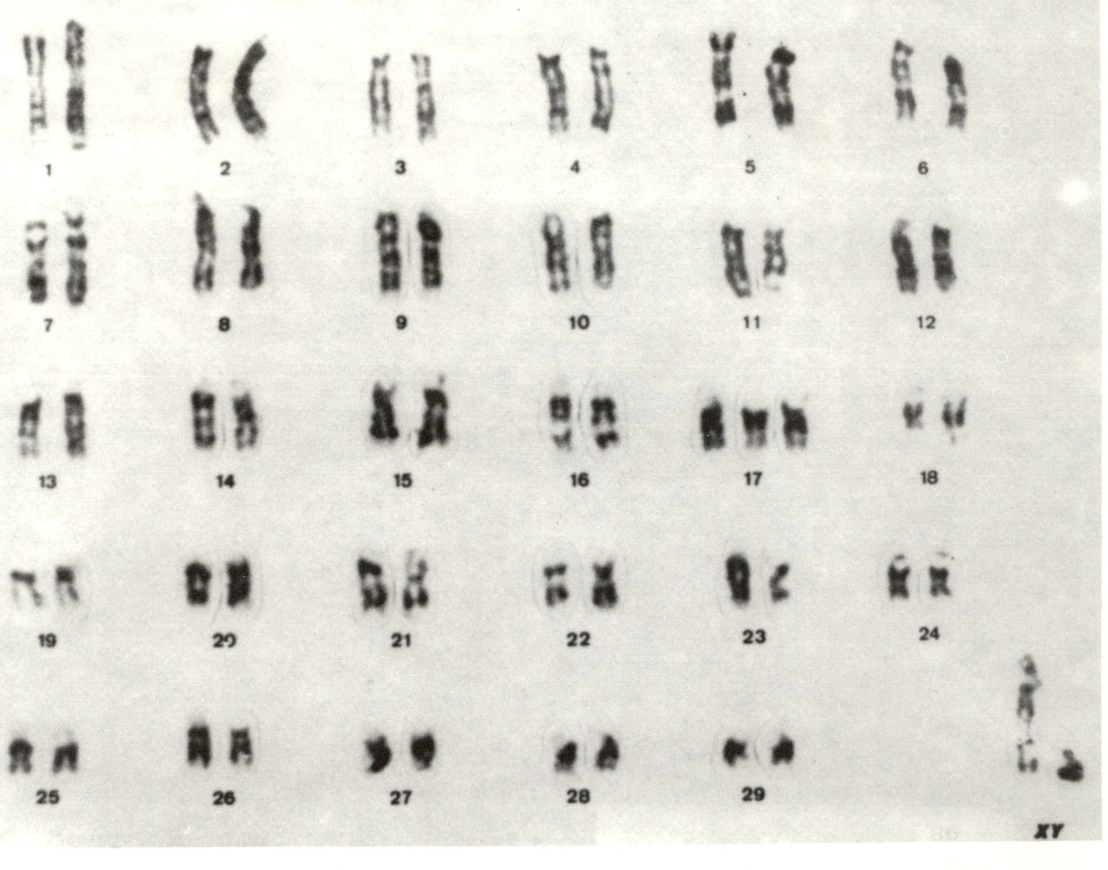

FIG. 11.18. G-banded karyotype of trisomy-17. In early reports this was designated trisomy-18. It has been reported in Germany, Japan, and the United States. Phenotypically, the animals have severe brachygnathia and some other associated defects.

From Herzog et al. (1977).

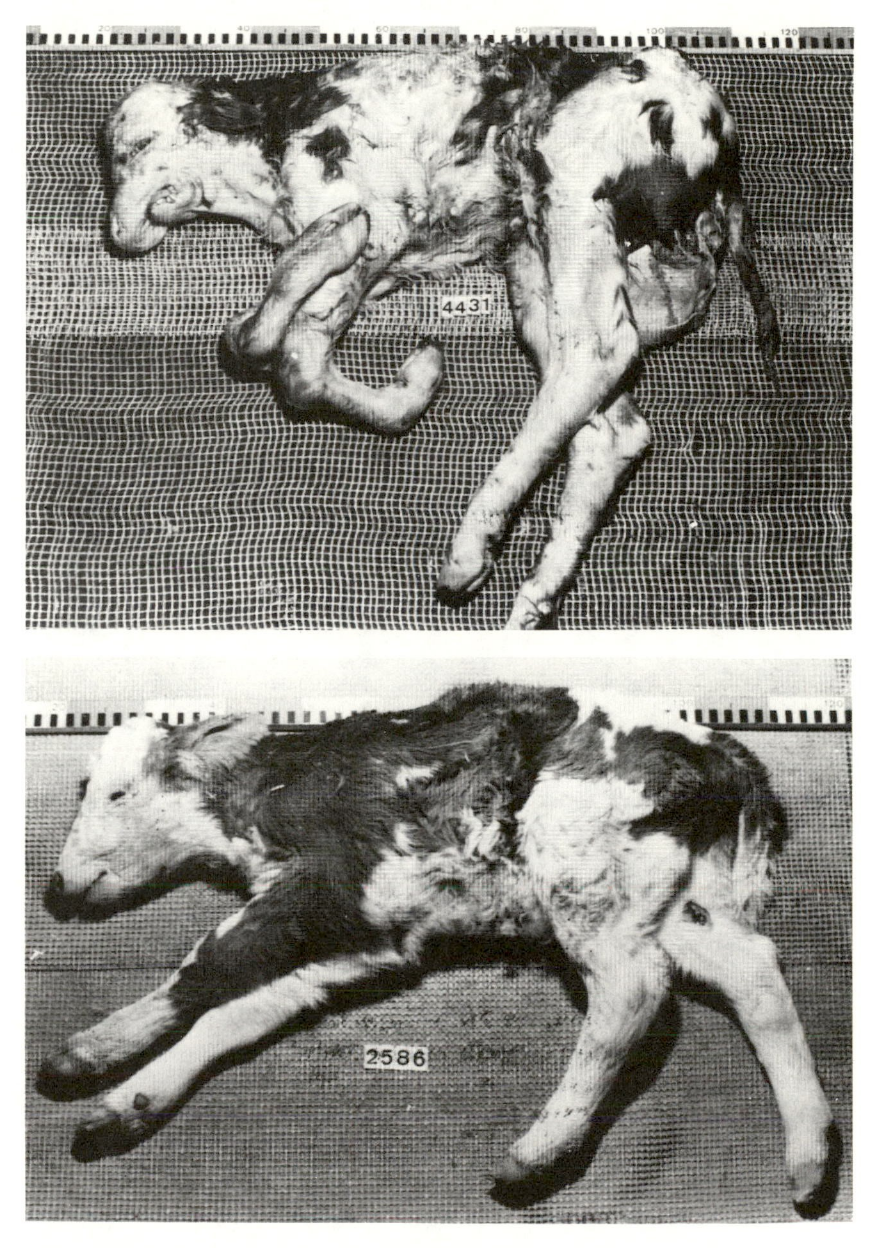

FIG. 11.19. Two animals with trisomy-17. Phenotypic variation is found among these affected animals, similar to the variation that occurs among animals with defects resulting from single-gene mutations, probably reflecting variation in the basic genotypes.

From Herzog et al. (1977).

At least three cases of XXY males have been found in cattle (Logue *et al.* 1979; Finger *et al.* 1969; Scott and Gregory 1965) (Fig. 11.20). This is comparable to the Klinefelter syndrome in humans. The testicles are usually underdeveloped and the animal is sterile (Fig. 11.21). No estimate has been made of the frequency of occurrence of such animals. Rieck *et al.* (1970) indicated that the sire of a

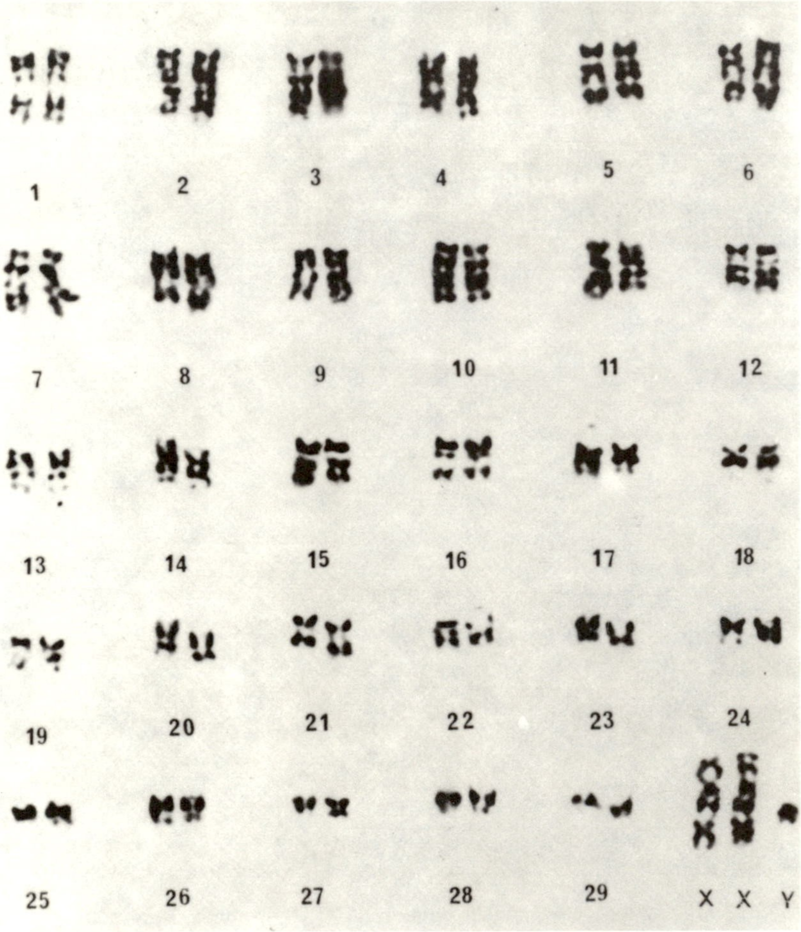

FIG. 11.20. G-banded karyotype of XXY British Friesian bull.
From Logue et al. (1977).

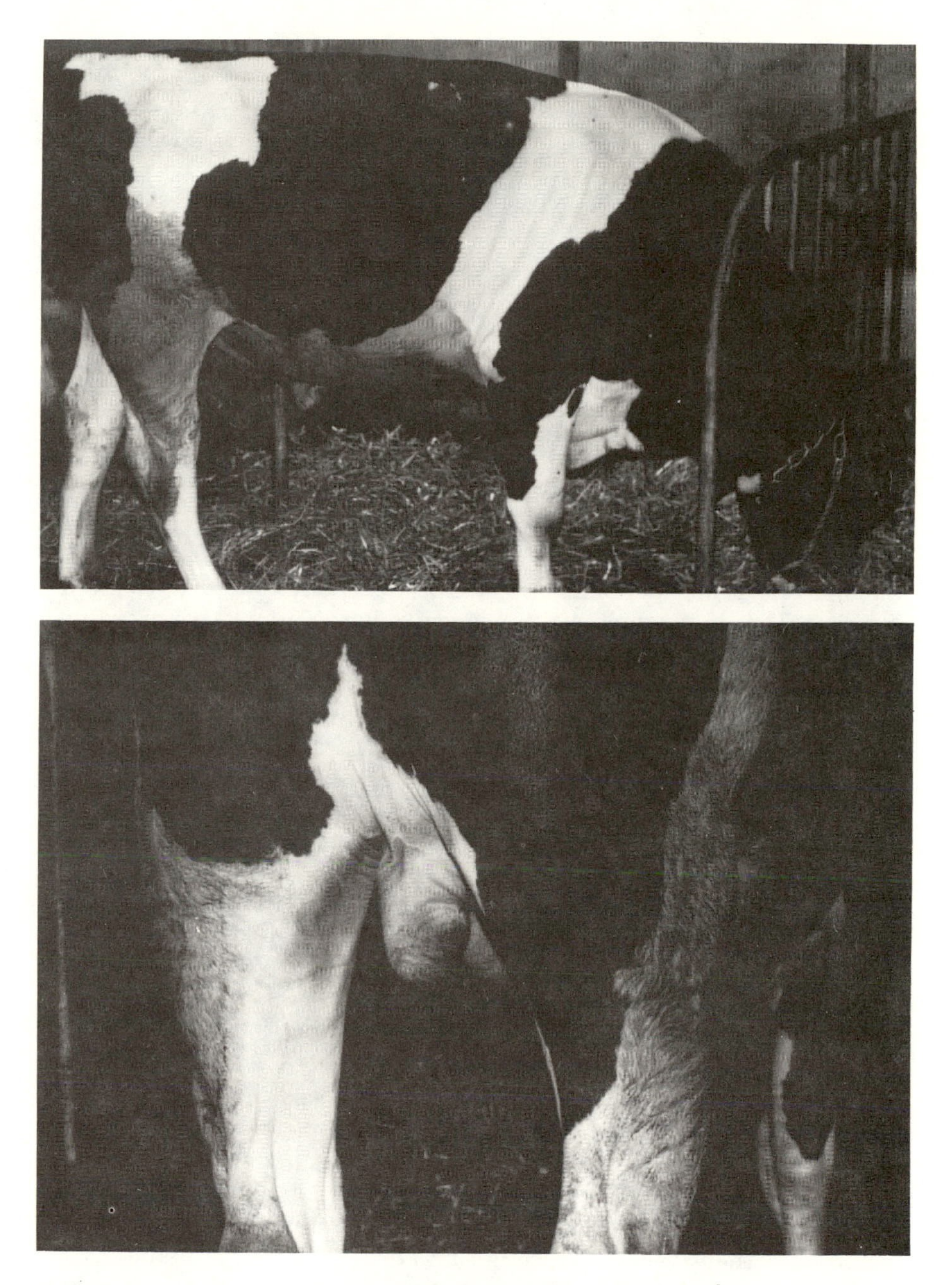

FIG. 11.21. XXY British Friesian bull at Glasgow Veterinary College. This bull was selected as a calf for artificial insemination, but when testis growth was observed to be below normal, the lymphocytes were cultured for chromosome studies.

Photograph by Eldridge.

61,XXX female was also the grandsire of a 61,XXY male, so there could be a familial tendency in certain strains of cattle toward sex-chromosome trisomy.

Rieck *et al.* (1970) and Norberg *et al.* (1976) have each reported one case of a 61,XXX female in cattle, and Swartz and Vogt (1983) and Linares *et al.* (1981) have each reported two, for a total of six. In one case reported by Rieck *et al.* (1970) the sex organs and sexual function in a Fleckvieh heifer showed nothing peculiar. In another case (Norberg *et al.* 1976) in a Norwegian Red heifer, heat was observed at 13 months, but no further heat periods were observed, up to 18 months of age. The animal had a small uterus and underdeveloped ovaries with one corpus luteum. The dam and maternal half-sibs were normal 60,XX. Linares *et al.* (1981) found one of two Swedish Red and White heifers to be anestrous with small ovaries, and the other cycled but did not conceive. Swartz and Vogt (1983) found one Pinzgauer and one Red Polled heifer to be trisomic-X among 71 heifers which did not conceive in two successive breeding seasons. The reproductive tracts were not examined.

Chromosomes from tumors may vary frequently from the normal number and morphology of the species. Basrur *et al.* (1964) found a lymphosarcoma in a bovine to have 61,XXX as the dominant type.

It is not uncommon to find trisomic cells existing in a mosaic, or chimeric, condition in animals. Probably the normal cells provide for the viability of the animal and protect it from the deleterious effects which might occur if the animal were 100% trisomic. Some of these cases are reported in a later section of this chapter on chimerism and mosaicism.

PERICENTRIC INVERSION

Two cases of a pericentric inversion have been reported. Short *et al.* (1969) found a small metacentric chromosome in the XY cell line of a Charolais × Guernsey freemartin and its male co-twin, which they assumed to be a pericentric inversion. Popescu (1976) first reported the occurrence of a pericentric inversion in a Normandy bull, and then was able through breeding to study 27 female offspring. The inversion was transmitted to 16 of the 27, which indicated that it was transmitted in the same way as a simple Mendelian dominant. The daughters of the bull had low fertility. The

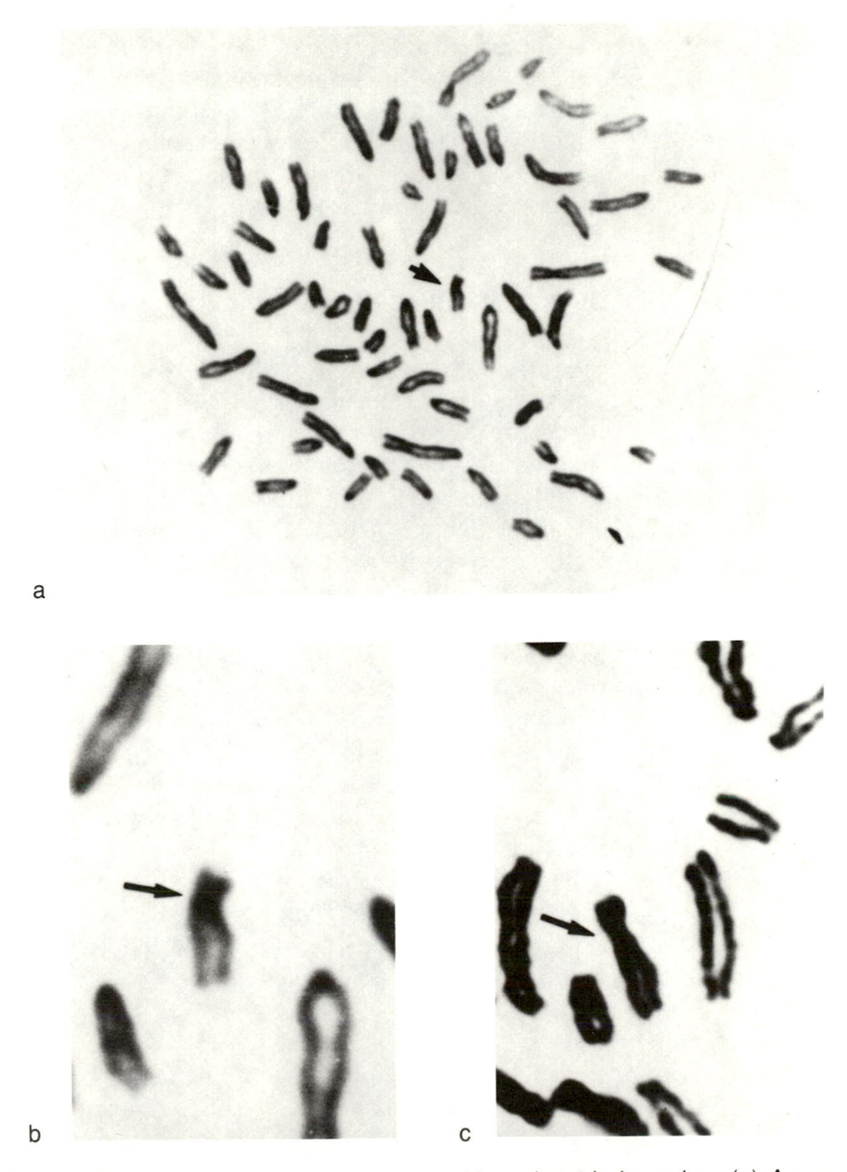

FIG. 11.22. Bovine female metaphase with pericentric inversion. (a) Arrow identifies the aberrant chromosome. (b,c) Enlarged pericentric inversion in two different cells.

From Popescu (1976).

low fertility may or may not have been due to the inversion. By banding it was identified as chromosome 14. No other phenotypic characteristics were associated with the inversion (see Fig. 11.22).

Pericentric inversions may occur occasionally in somatic cells (Eldridge and Blazak 1977). The aberration may be found as a mosaic condition, occurring at a low frequency within an animal, or as a chromosomal modification appearing rarely during somatic mitosis.

Y-CHROMOSOME POLYMORPHISMS

The one chromosomal characteristic that distinguishes *Bos indicus* from *Bos taurus* is the variation in morphology of the Y chromosome. In *Bos taurus* the Y chromosome is submetacentric, the centromere being located at a point about 42–25% of the length from one end (Cribiu and Popescu 1974; Gustavsson 1969). Kieffer and Cartwright (1968) first reported that *Bos indicus* males did not have a submetacentric Y chromosome, but the Y chromosome is an acrocentric similar to the smallest autosomes. This variation has been confirmed by numerous persons, and has been used (Eldridge and Blazak 1977) as a method for indicating possible evolutionary relationships within breeds. It is interesting that Africander cattle which are frequently referred to as *Bos indicus* have the submetacentric Y chromosome, indicating the use of *Bos taurus* bulls in establishing the breed (Märki *et al.* 1984; Megev, 1984). Gupta *et al.* (1974) confirmed the acrocentric nature of *Bos indicus* in bulls of two Indian breeds, Sahiwal and Red Sindhi.

Pinheiro *et al.* (1980) found both the *Bos taurus* type and *Bos indicus* type of Y chromosomes in the Ibage breed in Brazil. This breed was developed from Angus and Nellore cattle, and still reflects the use of both breeds of bulls in the development of this new breed.

Since the Y chromosome is short, although not the shortest chromosome, and since many published photographs of cattle chromosomes have been rather condensed, it has been difficult for many to distinguish whether the Y chromosome was submetacentric or metacentric. Therefore, the term metacentric has frequently been used in its description.

In Ayrshire (Fechheimer 1973) and Charolais bulls (Cribiu and Popescu 1974) one case each has been reported to have an unusually long Y chromosome. In each case no phenotypic variation from normal was observed in the animals with the long Y chromosome. Charolais and Montbeliard cattle were found to have significantly longer Y chromosomes (Cribiu 1975) than Maine–Anjou, Normandy, and FFPN cattle. Hansen and Elleby (1975) also found some Charolais bulls with longer Y chromosomes.

According to Jorge (1974) the Y chromosome of Jersey cattle is metacentric and smaller than in other breeds. This has also been observed by others (James 1980; Gupta 1976). All 23 Jersey bulls studied by Eldridge *et al.* (1983) were observed to have metacentric Y chromosomes. Selected for measurement were six bulls which had different sires in the sire line back four generations. Their centromeric index and total length relative to the X chromosome were significantly different from Holstein, Brown Swiss, and Simmental. The evidence is, therefore, quite conclusive that the Jersey has a Y chromosome that differs from many other breeds.

Variation in the length of the Y chromosomes has been found in humans and other mammals. No apparent phenotypic variation seems to be associated with these chromosomal variations. Certain morphological characteristics of the Y chromosome may be typical of specific breeds, but variation in length has been observed among individuals within several breeds.

CHIMERISM AND MOSAICISM IN CATTLE

Mosaic is a rather broad term which describes an individual having cells of more than one genotype. The cells may have become different through mutational changes, such as a Robertsonian translocation occurring in some somatic cells, or through a transfer of cells such as occurs in freemartinism. Chimera is a more precise term for an individual composed of cells from two or more zygotes, and therefore is a more specific term for the cell mixture typical of freemartins. The relationship between freemartinism and chimerism is described in detail in Chapter 9.

In the Bulgarian Brown breed, Dobryanov and Konstantinov (1970) found a male calf with 12.35% of its cells 61,XYY and 1.68% 59,XO. They indicated that this was the first report of such a mosaic in cattle. There were no phenotypic characteristics indicated to be in association with this chromosomal aberration.

Eldridge (1974) reported on a Dexter cow in which approximately 10% of its cells did not carry the Robertsonian translocation found in the remainder of its cells.

Dunn *et al.* (1970) found a bovine true hermaphrodite in which both diploid XX and triploid XXY cells were found. It is interesting to note in this case that observation of the first 300 metaphases from cultured lymphocytes yielded no triploid cells. Other cultured tissues yielded triploid cells up to a 50 : 50 proportion, which led to further observation of the lymphocytes. Only two triploid cells were found in 700 lymphocytes. There were also statistically significant differences between cultures made from mesodermal tissues of the right and left sides of the animal, indicating asymmetry.

Lojda *et al.* (1976) found a mosaic 60,XY/61,XXY/59,XO bull which produced 23% gonadal hypoplasia in his male calves. One of these calves was a 60,XY/60,XX/59,XO mosaic.

Dain and Bridge (1977) found a Friesian heifer born twin to a dead bull calf. The heifer had abnormal genitalia and was an XX/XY/XXY chimera. The skin cells cultured were XX/XXY, and a small number of cells had a translocation chromosome also.

Among 71 beef heifers which did not conceive in two successive breeding seasons, Swartz and Vogt (1983) found two mixoploid mosaic animals: 59,XO/60,XX/61,XXX and 59,XO/60,XX/61,XO). The first was a Pinzgauer × Angus heifer and the second was a Charolais.

Halnan (1976) reported briefly on three apparent cases of mosaic trisomy in Hereford cattle, two bulls with arthropathy and a related female. Only about 12% of the cells were found with 61,XX or 61,XY. Since the chromosomes were not banded, the chromosomes involved were not specifically identified.

Miyake *et al.* (1984) studied the semen and overall fertility characteristics of a mosaic XY/XYY Holstein bull. It was unilaterally cryptorchid, produced less than the normal volume of semen, but had a higher conception rate than normal. Nine paternal half-sibs and 11 daughters all had normal karyotypes. The ratio of XYY cells to XY cells in lymphocytes decreased with age from 48% at 1 year to 11% at 5 years of age. No abnormal characteristics were reported in any of his calves. The cell cultures from skin and kidney gradually lost the XYY cells, but the culture from the spleen retained its original ratio at least to the 29th passage, at 171 days.

Both bulls and heifers from mixed sex twin births have been

found to be sex chromosome chimeras, 60,XX/60,XY, from lymphocyte cultures. This chimerism has also been found in cells cultured from other tissues; kidney, lung, bone marrow, connective tissue, and gonads, but with the frequency of the donor cells usually lower than in the lymphocytes. The suggestion has been made that if a chimeric bull had XX cells among his spermatogonia, then there would be an increased number of X-bearing spermatozoa as compared with a normal 60,XY bull, and consequently an increased frequency of female calves among his progeny. Three chimeric bulls have been reported to have a significant deviation in sex ratio favoring the female. Dunn *et al.* (1979) reported one with 71% females in 48 calves, Lojda (1972) one with 60% females in 168 progeny, and DeGiovanni *et al.* in 1975 with 78% females in 168 calves. Many other chimeric bulls have had no significant deviation in the sex ratio of their progeny.

Ohno *et al.* (1962) found 60,XX cells in the testes of two newborn males, twin to females. One had twice as many 60,XX cells as 60,XY cells. They were not certain that these were precursors of spermatogonia. No 60,XX cells were found in testes of 11 chimeric bulls, ranging in age from 1 to 9 years, by Dunn *et al.* (1979). Since 58.3% of the 12 chimeric AI bulls were culled for reproductive problems, as compared with 5.4% of the 128 control bulls, they concluded that problems which could occur in 60,XX spermatogonial cells might contribute to a reduction in reproductive fitness later in life.

If spermatozoa could be produced from 60,XX spermatogonia in chimeric bulls, then these progeny would possess the hereditary characteristics, including blood type, of the female twin instead of the bull. In all cases where blood types have been determined, no such evidence has been found (Dunn *et al.* 1979).

HOMOLOGY WITH SHEEP AND GOATS AND OTHER MEMBERS OF THE SUPERFAMILY BOVIDEA

Wurster and Benirschke (1968) pointed out the remarkable similarity in the number of chromosomes in the superfamily Bovidea as expressed by the NF (*nombre fondamentale*). NF is the number of chromosome arms in a karyotype, in which acrocentric chromosomes are counted as one and metacentric or submetacentric are counted as two. In the 50 species reviewed, the NF varied only from 58 to 62, with only 3 exceptions, even though the diploid number varied from 30 to 60. From this study, they state: "This indicates an almost exclusive use of the Robertsonian fusion mechanism of karyotype evolution in this group of species which represent 30 different genera." Nothing has been published since this report that would contradict this statement. The majority of the cases with 62 NF had metacentric or submetacentric sex chromosomes. The NF for *Bos taurus* is 62 and that for *Bos indicus* is 61.

The G-banded homology (Evans *et al.* 1973) apparent among cattle, goats, and sheep is further evidence of the similarity of karyotypes in these species. The goat and sheep differ only by the three pairs of submetacentric chromosomes in sheep compared to all acrocentric chromosomes in goats. The metacentric chromosomes of the sheep were made up of chromosomes 1 and 3, 2 and 8, and 4 and 9 of the goat as determined by G-banding. However, cattle have two pairs of chromosomes, which Evans *et al.* identified as 11 and 12, which had no counterparts in the other two species.

CATTLE HYBRIDS

Gray (1971) published a very complete list of reports on hybrids of livestock. Crosses of *Bos taurus* with *Bos indicus* are highly viable and are completely fertile in both sexes. The development of the breeds of Santa Gertrudis, Brangus, and Braford in the United States is evidence of the absence of any serious fertility problems from the crossing of these two species. There is extensive crossbreeding of Zebu cattle in India with European dairy breeds, in order to increase milk production, and the relatively low frequency of fertility problems within these crosses is further evidence of compatability. The fact that the Y chromosome of *Bos taurus* is submetacentric whereas in *Bos indicus* it is acrocentric apparently has no major, deleterious effects among hybrids of these species.

Both *Bos taurus* and *Bos indicus* reciprocal crosses with *Bos* (Poephagus) *grunniens* (Yak) have been made. The females are usually fertile and the F_1 males usually sterile, although the external genitalia appear normal and their libido is not less than the parent males.

Crosses of the American buffalo, *Bison bison,* with domestic cattle, *Bos taurus,* have been made in considerable numbers according to Basrur (1969). Backcrosses have also been made to each parent. Intercrosses have been made using F_1 females and backcross females to males after two or more backcrosses have reestablished their fertility. The F_1 males are sterile. The chromosomes of *Bison bison* are nearly identical with *Bos taurus,* all autosomes being acrocentric and the X chromosome being a large submetacentric.

Lenoir and Lichtenberger (1980) compared the X chromosomes of *Bos taurus, Bison bison,* and the Basolo hybrid of the two species and demonstrated that the centromere is located at a different point in the hybid than in the *B. taurus* and *B. indicus.* The centromeric index, which is either the short (p) or long (q) arm divided by the total length, was used to determine the difference. They appear generally similar, but not exactly the same. However, the Y chromosome is a small acrocentric, as in *Bos indicus.* The reason for the male sterility in the hybrids is not established, but the high frequency of sterility is similar to the general situation in species hybrids, that the heterogametic sex is sterile more frequently than the homogametic sex. The sex ratio of F_1 hybrids also favors the females in the cattle bison cross. The chromosomes and karyotype of a female F_1 hybrid from mating a bison male to a Hereford female are shown in Figure 11.23.

Shumov (1980) found that the only obvious chromosomal difference between *Bison bison* and *Bos taurus* parents and their hybrids was the difference in Y chromosomes. Abnormal meiotic figures occurred in the F_1 hybrids.

The European bison, *Bison bonasus,* has also been crossed with *Bos taurus* and *Bos indicus* (Betancourt *et al.* 1974), with results somewhat similar to those obtained with the American bison. *Bison bonasus* has chromosomes similar in number and morphology to *Bos taurus,* with the Y chromosome submetacentric.

Natural matings have been observed between cattle and many other different species of Bovidae, but living offspring have seldom been found. Many reports have indicated that spermatozoa from widely differing species can penetrate the ovum, but development does not proceed beyond the early stages.

Problems with viability and fertility of species crosses have been recognized for many years. The cause of these problems is still obscure. Certainly the difficulty inherent in synapsis of chromosomes during meiosis could be one of the reasons for lack of viability of the

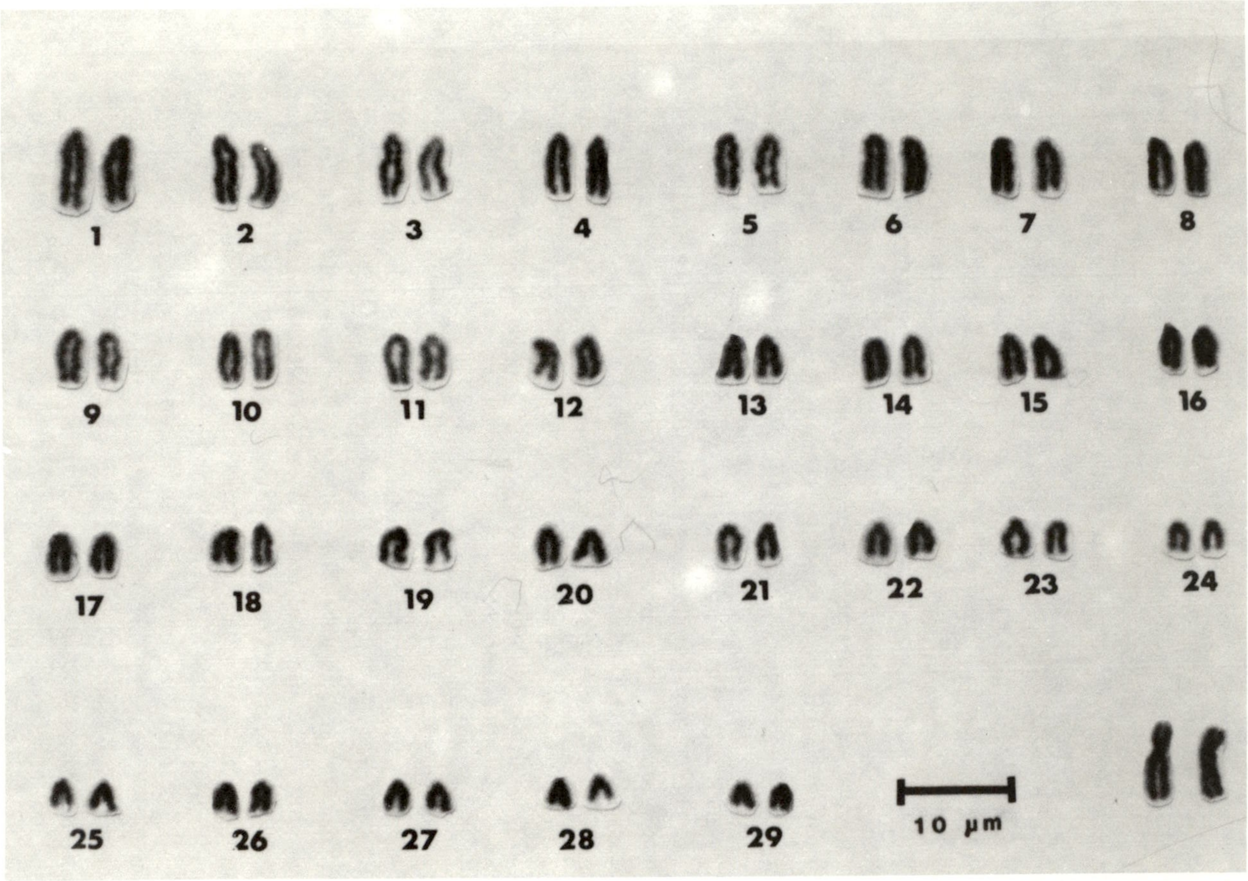

FIG. 11.23. Female F_1 hybrid from *Bison bison* male and a grade Hereford cow, Giemsa stained.
Photograph by Eldridge.

germ cells, but also many physiological barriers undoubtedly exist and affect both germ cell development and embryo viability. The general tendency for the heterogametic sex in hybrids to be less fertile than the homogametic sex is also true in cattle.

POPULATION STUDIES

Studies to determine the frequency of specific chromosomal aberrations within a breed of cattle, or the frequency of general chromosomal aberrations, are not numerous. This probably reflects the relatively recent development of cytogenetic studies of livestock, but also the tendency to pursue more detailed studies on the specific effects of each aberration rather than on population studies. In a few cases, large enough numbers of animals within a breed or a geographical area have been investigated so that estimates of the frequency of aberrant chromosomes could be made.

Gustavsson (1971A) observed the chromosomes from 2989 SRB cattle in Sweden and found 13% to be heterozygous for the 1/29 translocation and .5% homozygous. If one uses the Hardy-Weinberg equation the frequency of the 1/29 translocation in the breed would have been about 7%. Within nearly 3000 animals in this study only one other chromosomal abnormality was discovered, an X-autosome translocation. In 1979, Gustavsson reported that more than 2600 bulls and a total of over 6000 SRB cattle had been investigated chromosomally.

Fechheimer (1973) observed the chromosomes from 743 bulls from three artificial insemination (AI) organizations. He found 1 elongated Y chromosome in an Ayrshire bull, and 13 chimeric animals with both XX and XY types of lymphocytes. No other chromosomal aberrations were found.

Pollock (1974) studied chromosomes from 421 bulls used in AI in Britain and found one Robertsonian translocation and eight XX–XY chimeric animals among 330 British Friesians. The Robertsonain translocation was 2/8 as identified from later banding, but was tentatively identified as 2/4 in his 1974 publication. Although the population frequency of a chromosomal aberration cannot be firmly established by one case, the frequency in this sample would be $\frac{1}{2} \div$ 330, or .15%. When one considers the fact that Fechheimer (1973) found no chromosomal aberrations in 537 Holstein bulls, the .15%

might be considered the top limit of frequency for this breed, based on data to date.

Blazak and Eldridge (1977) calculated a frequency of 2.4% for the 1/29 translocation in the American Brown Swiss breed based upon 299 animals and the number of animals in the breed known to be sired by a popular AI bull, Hycrest Royal Jester, which had a 50% frequency of heterozygotes among his progeny.

The Channel Island breeds, Jersey and Guernsey, were studied by James and Eldridge (1978) and no aberrations were found in 202 Jerseys and 300 Guernseys. Statistical analyses of these data show that, if the frequency of chromosomal abnormalities is as high as 1% in the Guernsey breed, the sample of 302 animals was large enough that the probability of missing such an abnormality was only 5%. In other words, the probability was 95% that such an abnormality would have been found if the random frequency in the population were 1%. For Jerseys, the smaller sample would have had a probability of 95% that an abnormality would have been found if the frequency of abnormal chromosomes were at least 1.43%.

The British White breed has the highest frequency of a chromosomal aberration among those studied to date. The aberration was the 1/29 Robertsonian translocation, and the frequency in the first herd studied was 60% (Eldridge 1975). A study of animals in additional herds by Peter Wilkes (1979) indicated a frequency of 40% in the entire breed.

Amrud (1969) observed 430 head of Norwegian Red cattle after discovering the 1/29 Robertsonian translocation in a sterile heifer. He found 18, or 4.2%, of the animals with only 59 chromosomes, including the translocation. This calculated to a chromosomal aberration frequency of $18 \div (2 \times 430)$, or 2.1%.

Sixty-five Danish AI bulls were found by Hansen and Elleby (1975) to have no chromosomal aberrations.

From the Ivory Coast of Africa 136 animals from 4 breeds and some crossbreeds were studied by Popescu *et al.* (1979), and 5 were heterozygous for the 1/29 Robertsonian translocation. No other deviations from normal were found. Two breeds had submetacentric Y chromosomes as in *Bos taurus,* and two breeds had acrocentric Y chromosomes as in *Bos indicus.*

Several other studies have been made of smaller numbers of animals, mostly reflecting the 1/29 Robertsonian translocation. The relatively low frequency of aberrations when compared with humans, where 1 out of 160 newborns has chromosomal aberrations (Thompson and Thompson 1980), is probably due to the intensity of selection practiced in livestock husbandry, but it could be a species difference.

Population studies of chromosomes from breeds of livestock would be useful to establish the frequencies of chromosomal aberrations. Results could lead to hypotheses about evolutionary development of livestock. Since most livestock species have relatively large numbers of chromosomes, and in many species only very small differences exist in length and morphology, detailed banding studies would also be useful in discovering inversions and deletions. Although neonatal studies of human chromosomes have revealed a number of chromosomal aberrations, few neonatal studies have yet been published on livestock other than poultry.

Herzog and co-workers (1977) did a karyotype analysis on 847 newborn calves with congenital abnormalities, the most extensive study to date of the relationship between phenotypic abnormalities and chromosomal aberrations. Among these, 141 were found to have some type of chromosomal aberration. They did not feel that their studies were sufficient to determine the frequency of chromosomal aberrations in the population, particularly since their studies did not include a sizable sample of normal newborn calves. The most frequent aberration was gonosomal (sex chromosomal) chimerism, a total of 92 animals, 2 of which were single-born or autonomous chimeras and 3 of which were associated with the 1/29 Robertsonian translocation. The second most frequent aberration was trisomy-17, a total of 19 animals, but 9 of them were autosomal chimeras in which the trisomic condition was found in one of two cell lines, both cell lines being of the same sex, either XX or XY. The third most frequent aberration was the 1/29 Robertsonian translocation, found in 17 animals including the 3 gonosomal chimeras. Twelve animals had chromosome breaks, some with abnormal chromosome associations at a frequency significantly higher than the usual amount. The other four animals were gonosomal trisomics, one XXX, two XXY, and one a diploid–triploid chimera XX/XXY.

Another case of chromosome breaks in a bull associated with a phenotypic abnormality, foreleg paralysis, was reported by Sitko and Dianovsky (1981). The progeny of this bull had 31% showing the defect.

Moustafa *et al.* (1983) studied 1331 bulls of 9 breeds and 3 crossbred types, using G-banding for individuals in which chromosomal aberrations were first found by Giemsa staining. No chromosomal aberrations were found among 45 Hereford, 37 Limousin, 14 Charolais, 14 German Red Pied, 10 "Hungarofriz," 6 Jersey, 16 Holstein–Friesian × Jersey, or 6 Blonde Aquitaine × Simmental. Among 63 Podolian bulls, 6 had the 1/29 Robertsonian translocation. Out of 398 Simmental bulls, 13 had the 1/29, 1 had a 5/18, and 1 had the 14/21 Robertsonian translocations. Fourteen of the 661 Holstein bulls and 5 of the 398 Simmental bulls were sex chromosome chimeras, reflecting the higher frequency of twinning and larger numbers studied in these breeds. The freedom from Robertsonian translocations of animals in the Holstein–Friesian breed was typical of other studies on this breed of dairy cattle.

BLASTOCYST AND EMBRYO CHROMOSOMES

Embryonic mortality is one of the causes of lowered fertility in livestock. Problems related to embryo implantation, crowding in the uterus in litter-bearing species, nutritional status of pregnant females, hormone imbalance, and genetic causes such as recessive lethal genes affecting early embryonic survival are all causes of embryonic mortality. The effects of chromosomal aberrations on early embryonic survival have also been postulated, and have been studied in a small number of cases.

In the first study of early cattle embryos (McFeely and Rajakoski 1968) 1 out of 12 blastocysts had a diploid–tetraploid mosaicism in a ratio of 11 diploid to 7 tetraploid cells. Although the numbers are quite small, if this study were found to be typical of cattle, then 1 out of 12, or 8% of all embryos, would probably be lost and could be attributed to chromosomal anomalies, since such an embryo would not have been expected to survive.

Among 13 early blastocysts of cattle Eldridge *et al.* (1978) found one blastocyst with only triploid cells. Figure 11.24 shows the chromosomes of this blastocyst, which was from a 1/29 Robertsonian translocation female. The three X chromosomes and the translocation chromosome are identified among the 89 chromosomes in the cell. The triploid condition probably resulted from fertilization of the ovum by two X-bearing sperm. Incorporation of the second polar body into the nucleus would have been expected to result in *two*

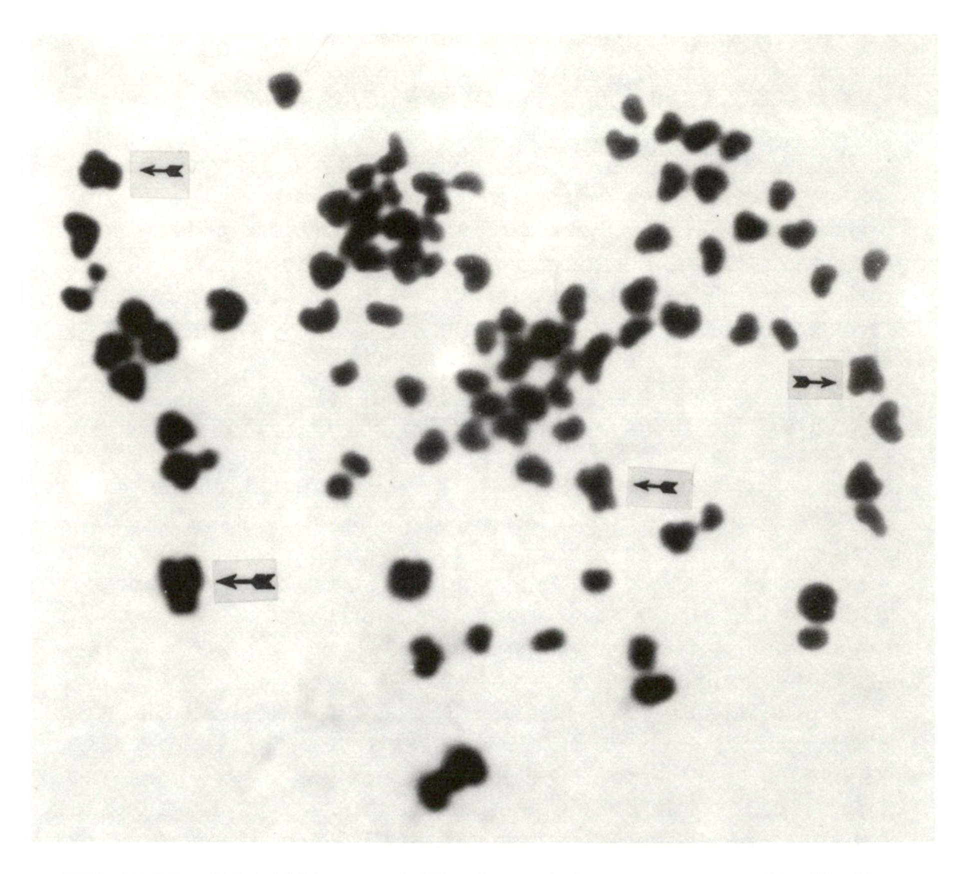

FIG. 11.24. Triploid blastocyst. The three X chromosomes are identified by the small arrows, the 1/29 translocation chromosome by the large arrow. Since the cow was a 1/29 heterozygote it was assumed that triploidy was the result of fertilization by two sperm.

Photograph by Eldridge.

translocation chromosomes and only 88 chromosomes in total, so this explanation was rejected. Prior to the harvesting procedures, the blastocyst was noted to appear abnormal, being somewhat cylindrical in shape rather than spherical. In another blastocyst one tetraploid cell was the only countable set of chromosomes. Since tetraploid cells at a low frequency are rather common in cattle (Hare and Singh 1980), the finding of one such cell as the only one in a blastocyst cannot be accepted as definitive of that blastocyst, so it probably should be ignored. Combining the frequency of 1 out of 13 with McFeely and Rajakoski's (1968) data, 2 out of 25 blastocysts were found abnormal, an estimate of about 8%.

Generally only 70–80% of the potential blastocysts in cattle are recovered about 12 ± 2 days after breeding in the most successful, nonsurgical procedures used for embryo transplanting. This recovery rate is the ratio of the number of blastocysts to the number of corpora lutea. This can be compared with the nearly 100% surgical recovery of 6-day blastocysts from rabbits, for example. Chromosomal studies of cattle blastocysts are hampered by this less than perfect recovery rate. Even if acceptable metaphase spreads of chromosomes are obtained from all recovered blastocysts, the important question about the fate of the unrecovered blastocysts still remains unanswered. The ova may not have been fertilized, chromosomal imbalances may have resulted in early disintegration, the blastocysts may have not been recovered by flushing, or some other factor may have reduced the potential number of blastocysts. More definitive information might be obtained on embryo mortality associated with chromosomal aberrations if the blastocysts were recovered at about the eight-cell stage, or earlier.

In an extensive study of chromosomes from 224 cattle fetuses after implantation, and therefore probably over 20–24 days of age, Fechheimer and Harper (1980) found no aneuploid or polyploid individuals, and no unusual chromosomal configurations. They did find several chimeric fetuses, which are discussed in Chapter 9. The absence of chromosomal aberrations at these later stages of gestation is further evidence that embryonic loss associated with chromosomal aberrations occurs at very early stages, prior to implantation.

MEIOTIC STUDIES

Chiasma Frequency

Early work on cattle chromosomes was done with testicular tissue. With the newer techniques of tissue culture and harvesting of cells, the more recent studies have concentrated upon mitotic cell division of lymphocytes in blood cultures and in cultures of other tissues.

Certain types of studies, however, can only be done with cells undergoing meiosis. These studies include estimates of chiasma frequencies, estimates of rates of nondisjunction in animals with Robertsonian translocations, and identification of deviations from normal chromosomal divisions for other types of chromosomal aberrations (see Fig. 11.3). Only a limited number of meiotic studies of cattle have been published.

Logue and Harvey (1978) and Popescu (1971), using testicular tissue, reported approximately 50 chiasma per cell. Logue found 49.5 (SD = 4.1) per cell, from 325 cells, and Popescu found 53.73 per cell, from 38 cells, a reasonably close agreement.

On the other hand, Jagiello *et al.* (1974, 1976) reported a mean frequency of chiasma per bivalent in studies of oocytes at approximately 1.2, which, converted to a per-cell basis, would be 36, smaller than the spermatocyte counts. This discrepancy may reflect a difference in the criteria used for identifying chiasmata, or perhaps a sex difference. It is interesting to note that the mean number of chiasma counted per bivalent, from oocytes studied in Jagiello's laboratory, was significantly lower than that found in three primate species, including humans.

Since Hulten (1974) found 50.61 (SD = 3.87) chiasmata in a human male and Jagiello *et al.* (1974) found 1.89 chiasma per bivalent in human oocytes it appears quite possible that there is a sex difference. Sex differences in number and location of chiasma have been reported in mice (Polani 1972).

Oocyte Studies

Another approach to determining the effect of chromosomal aberrations on early embryonic loss would be a study of the chromosomes of oocytes. It is quite well established that nearly 100% of all ova which reach the oviduct are fertilized (Bearden *et al.* 1956; Kidder *et al.* 1954) when highly fertile bulls are used. Chromosomal abnormalities which are present in the ova that could cause early embryonic mortality include duplications, deficiencies, a diploid pronucleus, or perhaps some major structural change in a chromosome, although if the gamete were balanced the embryo would presumably be viable. Ooctye studies would not detect aberrations of a similar nature in the spermatozoa, fertilization by two spermatozoa, or early mitotic errors in the blastocysts. Studies of ooctye chromo-

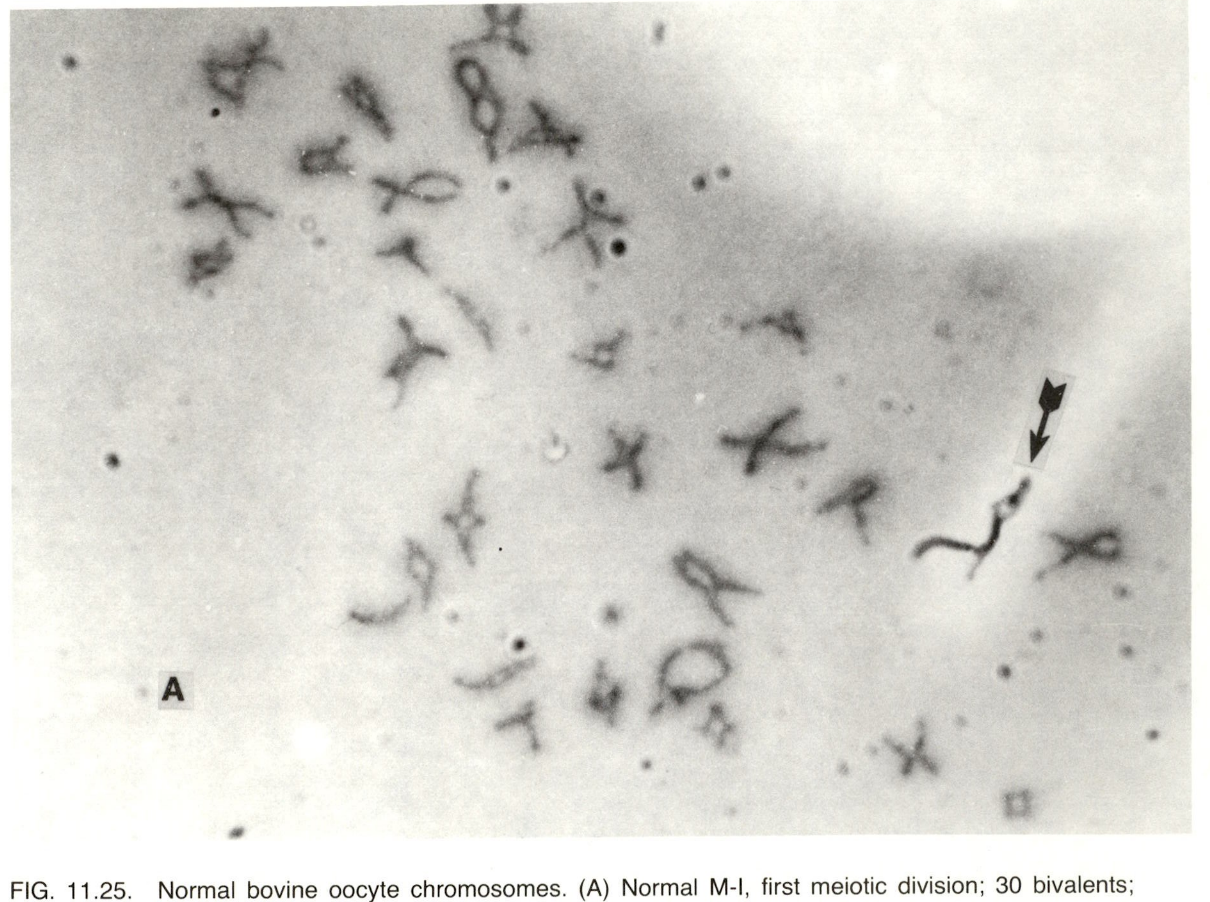

FIG. 11.25. Normal bovine oocyte chromosomes. (A) Normal M-I, first meiotic division; 30 bivalents; arrow identifies artifact.

From Koenig (1982).

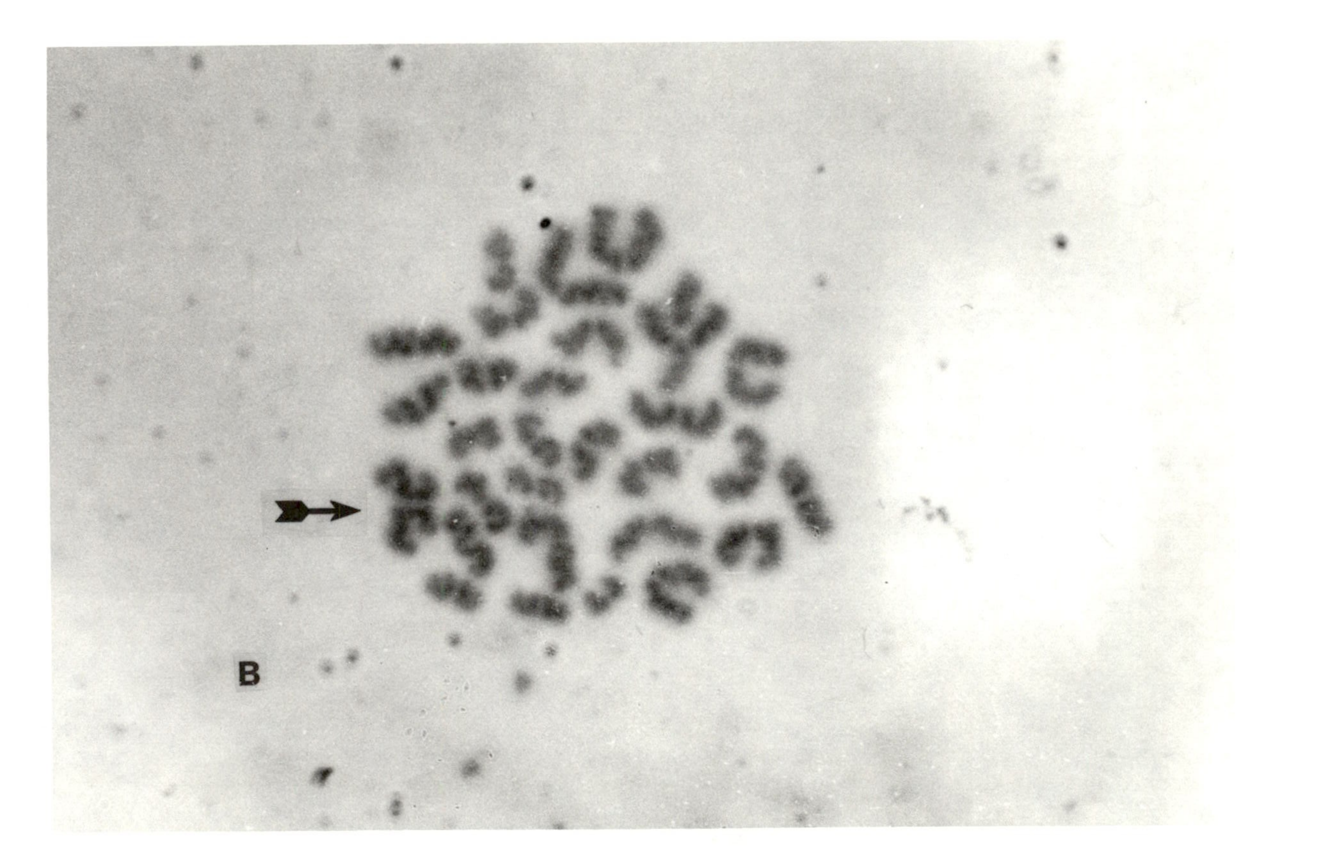

FIG. 11.25. (*Continued*). (B) Normal M-II, second meiotic division; 30 chromosomes; arrow indicates the X chromosome.
From Koenig (1982).

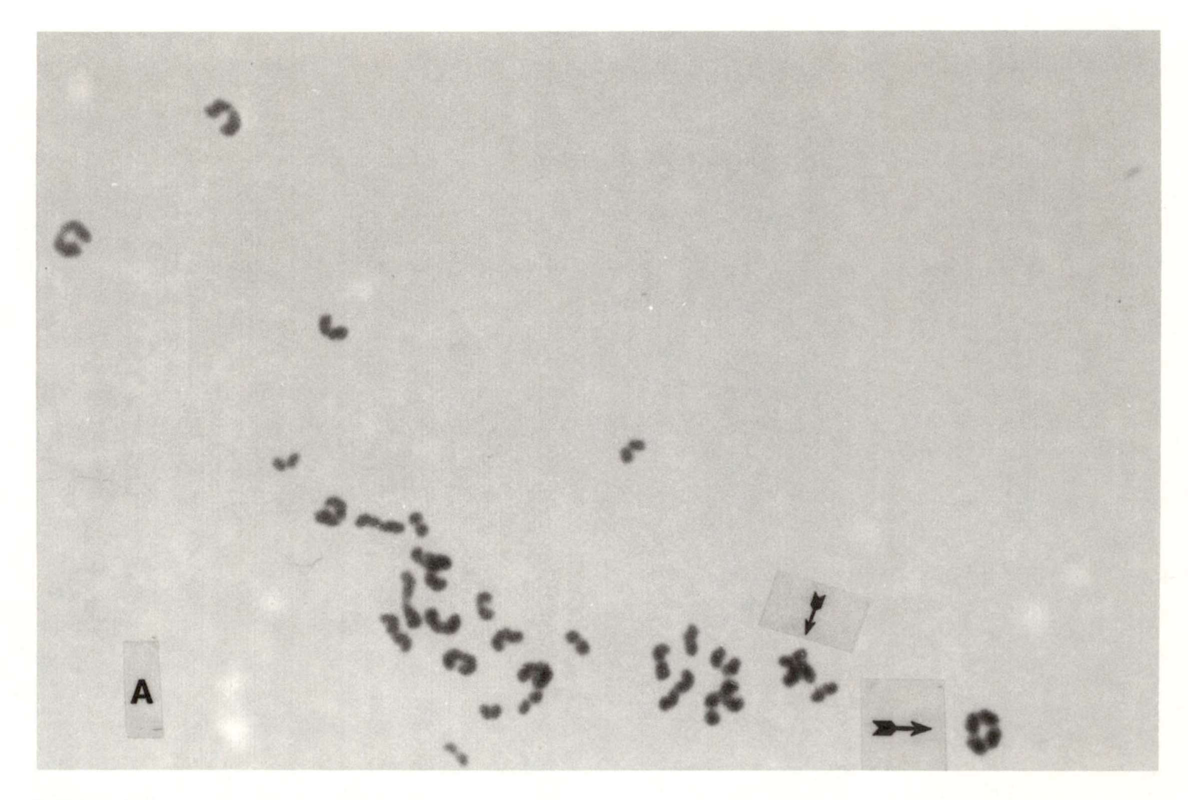

FIG. 11.26. Abnormal bovine oocyte chromosomes. (A) Aneuploid M-II; note the one bivalent (marked with the large arrow) which had not separated in metaphase I; The X chromosome is indicated by the small arrow.

From Koenig (1982).

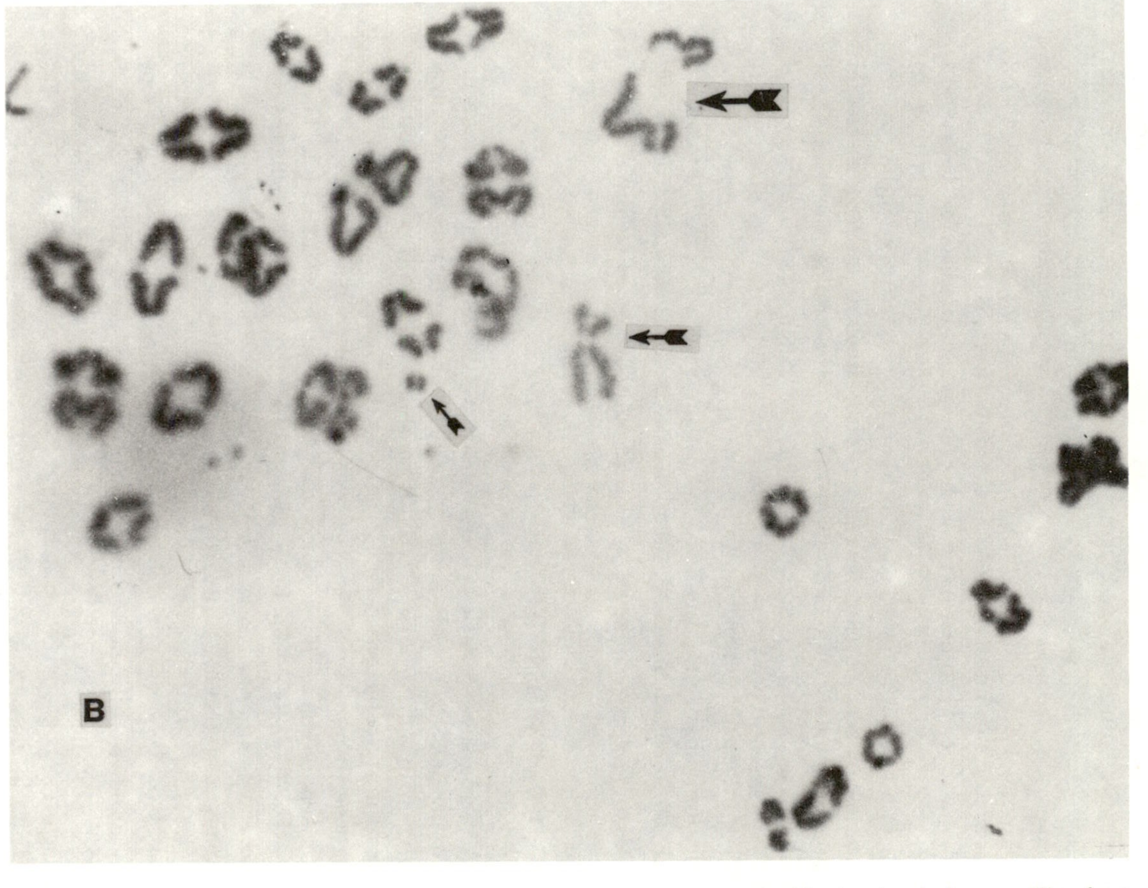

FIG. 11.26. (*Continued*). (B) Composite photograph of chromosomes with structural abnormalities from one oocyte in M-I; the large arrow indicates a break, the small arrow indicates a pair of fragments, and the intermediate-sized arrow shows an unexplained, but aberrant, chromosome.

From Koenig (1982).

somes also would have to be based upon the assumption that any ovum recovered from an ovarian follicle would have an equal opportunity for being ovulated. Some follicles become atretic after each ovulation—do not rupture and then regress—and it is possible that follicles with ova that are chromosomally aberrant may have a higher possibility of becoming atretic than do follicles containing ova with a normal chromosome constitution.

A number of studies have been done on maturation of oocytes in vitro, on possible fertilization of oocytes matured in vitro, on chiasma frequency in oocytes, on the effects of certain chemicals on chromosomes of cattle oocytes, on the effect of follicular size on maturation of oocytes, and on the effect of the presence or absence of cumulus cells on chromosomes of oocytes. Koenig *et al.* (1983) found 23% of the oocytes to be abnormal, from heifer ovaries obtained at slaughter (Figs. 11.25 and 11.26). This estimate is conservative since only hyperploid cells were counted as abnormal, hypoploid cells were counted as indeterminate. Also, some structural abnormalities resulted in oocytes being designated as abnormal.

SEX IDENTIFICATION

Altering the sex ratio of calves born would be especially valuable to dairymen since so few bulls are needed in relationship to cows. Bulls for breeding purposes could be obtained in sufficient numbers from highly selected cows if 1 calf in 100 births were a male. The increase in number of females would also permit greater selection intensity in the females. Numerous techniques for treatment of semen to alter sex ratios have been tried with almost no success. In those experiments when some success was indicated there was a greatly decreased fertility rate.

Two successful methods for determining the sex of an embryo have been developed. One is by amniocentesis at 70–100 days of pregnancy and the other by culturing a small piece of the blastocyst before the blastocyst is transferred to a recipient cow. Bongso and Basrur (1977) reported a method of amniocentesis where fluid from the amniotic sac was obtained by puncture through the wall of the rectum. Singh *et al.* (1977) reported on a similar method with variable degrees of success. After determing the sex, if the fetus is a male, abortion can be induced and the cow bred again. The disadvantage of this method lies in the increased calving interval for the cows in which an abortion is induced. The gestation must be at 70–100 days of pregnancy and at least 21 days must elapse before the next estrus. This extends the calving interval by at least 3 months, which decreases the lifetime milk production of the cow. In the subsequent breeding the sex ratio is still 1 : 1, so with this method the sex ratio of females to males is moved only to 3 : 1. Routine use therefore would probably be uneconomical. Sex identification by amniocentesis might be useful in certain specific situations.

The other method was developed by Hare *et al.* (1976). They obtained a fragment of the trophoblast from 14- and 15-day-old embryos as far from the inner cell mass as possible and, after treatment lasting about 3.5 hours, observed the chromosomes. The embryos were then transferred to recipient females, where 37.5% of them developed. The success rate of nonbiopsied embryos transferred at 14 to 15 days was 45.8%, so they concluded that the biopsy did not markedly affect the success of transfer considering the small numbers involved and other variables. Correct diagnoses of sex was 87.5%, the only error being in an embryo where determination of sex had been questionable. This method in conjunction with embryo transplants appears to have promise in the selection of sex of calves. See Chapter 6 for information on varying the sex ratio by breeding.

SISTER-CHROMATID EXCHANGE

The separation of each chromosome during mitosis results in two sister chromatids, each made up of two strands of DNA. It has been generally thought that under normal conditions there is little exchange of chromatin between the pair of sister chromatids. However, some relatively recent work has shown that some exchanges do occur either between the two strands making up the chromatid or between the two chromatids, the "daughter" chromosomes not being completely the original strand or the newly formed one. By use of tritium-labeled DNA or by exposure of cell or tissue cultures to 5-bromodeoxyuridine (BUdR) and subsequent staining with Giemsa, Hoechst 33258, acridine orange, or some other techniques, an exchange of chromatin between sister chromatids has been demonstrated. The treatment with BUdR or other material has been recognized as a cause for some of the exchanges, but within similar treatments significant variation has been found in the response of individuals. In addition, certain chemicals, biological infections, and physiological conditions have been found to be associated with higher frequencies of sister-chromatid exchanges (SCE) (Yaeger *et al.* 1983).

Most studies of SCE in animals have been done on human or laboratory animal chromosomes. However, DiBerardino and Shoffner (1979) found that exchanges do occur in cattle (*Bos taurus*). They cultured lymphocytes from 8 males and 23 females and found 5.4 ± 2.1 SCE per cell. There were no differences between the sexes, but a highly significant difference in frequency was found between individuals. Most SCE were terminal, with many "fewer interstitial (double) and triple exchanges." The late-replicating X chromosome had 2.5 times as many SCE as the early-replicating homologue. They postulated that high SCE frequency might be symptomatic of physiological or environmental disturbances, but that much more study is needed to establish norms before the frequency could be used diagnostically.

Popescu (1978) found statistically significant differences between animals in the number of SCE per cell. He also found in 17 animals a mean of 12.2 SCE per cell, higher than found by DiBerardino and Shoffner (1979).

MITOCHONDRIAL VARIATION IN BOVINE DNA

Not all hereditary variation is transmitted through the chromosomes. Laipis *et al.* (1982) reported differences among maternal lineages of Holstein cattle in the loss of the HAE III restriction enzyme site. This type of research has opened up the possibility of a better understanding of maternal inheritance in cattle.

REFERENCES

AMRUD, J. 1969. Centric fusion of chromosomes in Norwegian Red cattle (NRF). Hereditas *62,* 293–302.

BASRUR, P. K. 1969. Hybrid sterility. *In* Comparative Mammalian Genetics, K. Benirschke, Editor, pp. 107–131. Springer-Verlag, New York.

BASRUR, P. K., GILMAN, J. P. W., and McSHERRY, B. J. 1964. Cytological observations on a bovine lymphosarcoma. Nature (London) *201,* 368–371.

BASRUR, P. K., REYES, E. R., and BAIRD, J. 1982. Chromosome anomaly in a subfertile cow. Am. Soc. Anim. Sci., Abstr. *47.*

BEARDEN, H. J., HANSEL, W., and BRATTON, R. W. 1956. Fertilization and embryonic mortality rates of bulls with histories of either low or high fertility in artificial breeding. J. Dairy Sci. *39,* 312–318.

BEGIMKULOV, B. K., BAKAI, A. V., and KRASOTA, V. F. 1980. A Robertsonian type translocation in a Zebu bull and two of his sons. Dokl. Vses. Akad. Sh. Nauk *11,* 31–33.

BETANCOURT, A., GUTIERREZ, C., and SANCHEZ, A. 1974. The chromosomes of "*Bos taurus,*" "*Bos indicus,*" "*Bison bonasus*" and their hybrids. 1st World Congr. Genet. Appl. Livestock Prod. *3,* 173–176.

BIGGERS, J. D., and McFEELY, R. A. 1963. A simple method for the display of chromosomes from cultures of white blood cells with special reference to the ox. Nature (London) *199,* 718–719

BLAZAK, W. F., and ELDRIDGE, F. E. 1977. A Robertsonian translocation and its effect upon fertility in Brown Swiss cattle. J. Dairy Sci. *60,* 1133–1142.

BONGSO, A., and BASRUR, P. K. 1976. Chromosome anomalies in Canadian Guernsey bulls. Cornell Vet. *66,* 476–488.

BONGSO, A., and BASRUR, P. K. 1977. Bovine fetal cells in vitro: Fate and fetal sex prediction accuracy. In Vitro *13,* 769–776.

BRUERE, A. N., and CHAPMAN, H. M. 1973. Autosomal translocations in two exotic breeds of cattle in New Zealand. Vet. Rec. *92,* 615–618.

CHIARELLI, B., DECARLI, L., and NUZZO, F. 1960. Morphometric analysis of chromosomes of *Bos taurus* L. Caryologia *13,* 766–772.

CRIBIU, E. P. 1975. Inter-racial variation in the length of the Y chromosome of *Bos taurus* L. Ann. Genet. Sel. Anim. *7,* 139–144.

CRIBIU, E. P., and POPESCU, C. P. 1974. One case of an abnormally long Y chromosome in *Bos taurus* L. Ann. Genet. Sel. Anim. *6,* 387–390.

CRIBIU, E. P., and POPESCU, C. P. 1980. Distribution of the 1/29 Robertsonian translocation in France. Proc. 4th Eur. Colloq. Cytogenet. Domest. Anim. 130–135.

CROSSLEY, R., and CLARK, E. G. 1962. The application of tissue culture techniques to the chromosomal analysis of *Bos taurus.* Genet. Res. *3,* 167–168.

DAIN, A. R., and BRIDGE, P. S. 1977. A freemartin calf with XX/XXY mosaicism. Ann. Genet. Sel. Anim. *9,* 533.

DARRE, R., QUEINNEC, G., and BERLAND, H. M. 1972. The 1/29 translocation in cattle. General study on its importance in southeast France. Rev. Med. Vet. *123* (New Ser. 35), 477–494.

DARRE, R., BERLAND, H. M., and QUEINNEC, G. 1975. A new Robertsonian translocation in cattle. Ann. Genet. Sel. Anim. *6,* 297–303.

DEGIOVANNI, A., POPESCU, C. P., and SUCCI, G. 1975. First cytogenetic study from an Italian AI center. Ann. Genet. Sel. Anim. *7,* 311–315.

DEGIOVANNI, A., SUCCI, G., MOLTENI, L., and CASTIGLIONI, J. 1979. A new autosomal translocation in "Alpine Grey Cattle." Ann. Genet. Sel. Anim. *11,* 115–120.

DEGIOVANNI, A., POPESCU, C. P., SUCCI, G., and MOLTENI, L. 1980. Meiotic study of a new autosomal translocation in "Alpine Grey Cattle." Proc. 4th Eur. Colloq. Cytogenet. Domest. Anim. 158–163.

DiBERARDINO, D., and SHOFFNER, R. N. 1979. Sister chromatid exchange in chromosomes of cattle (*Bos taurus*). J. Dairy Sci. *62,* 627–632.

DiBERARDINO, D., IANNUZZI, L., FERRARA, L., and MATASSINO, D. 1979. A new case of Robertsonian translocation in cattle. J. Hered. *70,* 436–438.

DOBRYANOV, D., and KONSTANTINOV, G. 1970. A case of mosaicism of the 58AXY, 58AXYY type in a male calf of the Bulgarian Brown Breed. C. R. Acad. Sci. Agricoles Bulg. *3,* 271–276.

DUNN, H. O., and JOHNSON, R. H., JR. 1972. A 61,XY cell line in a calf with extreme brachygnathia. J. Dairy Sci. *55,* 524–526.

DUNN, H. O., LEIN, D. H., and KENNY, R. M. 1967. The cytological sex of a bovine anidian (amorphous) twin monster. Cytogenetics *6,* 412.

DUNN, H. O., McENTEE, K., and HANSEL, W. 1970. Diploid–triploid chimerism in a bovine true hermaphrodite. Cytogenetics *9,* 245–249.

DUNN, H. O., McENTEE, K., HALL, C. E., JOHNSON, R. H., JR., and STONE, W. H. 1979. Cytogenetic and reproductive studies of bulls born co-twin with freemartins J. Reprod. Fertil. *57,* 21–30.

DYRENDAHL, I., and GUSTAVSSON, I. 1979. Sexual functions, semen characteristics and fertility of bulls carrying the 1/29 chromosome translocation. Hereditas *90,* 281–289.

ELDRIDGE, F. E. 1974. A dicentric Robertsonian translocation in a Dexter cow. J. Hered. *65,* 353–355.

ELDRIDGE, F. E. 1975. High frequency of a Robertsonian translocation in a herd of British White cattle. Vet. Rec. *96,* 71–73.

ELDRIDGE, F. 1980. X-autosome translocation in cattle. Proc. 4th Eur. Colloq. Cytogenet. Domest. Anim. 23–30.

ELDRIDGE, F. E., and BALAKRISHNAN, C. R. 1977. C-band variations in Robertsonian translocations in cattle. Nucleus *20,* 28–30.

ELDRIDGE, F. E., and BLAZAK, W. F. 1977. Comparison between the Y chromosomes of Chianina and Brahma crossbred steers. Cytogenet. Cell Genet. *18,* 57–60.

ELDRIDGE, F. E., LARSON, L. L., JAMES, R. C., and COWAN, C. W. 1978. Chromosome studies of bovine blastocysts. J. Dairy Sci. *61* (Suppl. 1), 87.

ELDRIDGE, F. E., KOENIG, J. L. F., and HARRIS, N. 1983. Y chromosome variation in the bovine. J. Dairy Sci. *66* (Suppl. 1), 245.

ELLSWORTH, S. M., PAUL, S. R., and BUNCH, T. D. 1979. A 14/28 dicentric Robertsonian translocation in a Holstein cow. Theriogenology *11,* 165–171.

EVANS, H. J., BUCKLAND, R. A., and SUMNER, A. T. 1973. Chromosome homology and heterochromatin in goat, sheep and ox studied by banding techniques. Chromosoma *42,* 383–402.

FECHHEIMER, N. S. 1973. A cytogenetic survey of young bulls in the USA. Vet. Rec. *93,* 535–536.

FECHHEIMER, N. S., and HARPER, R. L. 1980. Karyological examination of bovine fetuses collected at an abattoir. Proc. 4th Eur. Colloq. Cytogenet. Domest. Anim. 194–199.

FINGER, K. A., HERZOG, A., HOHN, H., and RIECK, G. W. 1969. Progress in skeleton cytogenetics in veterinary medicine. Giessener Beitr. Erbpathol. Zuchthyg. *2/3,* 13–30.

FISCHER, H. 1971. Chromosome analysis of native Thailand Cattle. Z. Tierz. Zuchtungsbiol. *88,* 215–221.

FORD, C. E., POLLOCK, D. L., and GUSTAVSSON, I. 1980. Proceedings of the First International Conference for the Standardisation of Banded Karyotypes of Domestic Animals. Hereditas *92,* 145–162.

FROGET, J., COULTON, J., NAIN, M. C., and DUABIEZ, J. M. 1972. Chromosomal anomaly of the centric fusion type in a Charolais calf. Bull. Soc. Sci. Vet. Med. Comp. Lyon *74,* 131–135.

GRAY, A. P. 1971. Mammalian Hybrids, 262 pp. Commonwealth Agricultural Bureaux of Farnham Royal, England.

GUPTA, P. 1976. Personal communication.

GUPTA, P., SINGH, L., and RAY-CHAUDHURI, S. P. 1974. Chromosomes of Indian breeds of cattle. Nucleus *17,* 129–132.

GUSTAVSSON, I. 1969. Cytogenetics, distribution and phenotypic effects of a translocation in Swedish cattle. Hereditas *63,* 67–169.

GUSTAVSSON, I. 1971A. Distribution of the 1/29 translocation in the A.I. bull population of Swedish Red and White cattle. Hereditas *69,* 101–106.

GUSTAVSSON, I. 1971B. Early DNA replication patterns of the normal sex chromosomes and a presumptive X-autosome translocation in cattle (*Bos taurus* L.). Nature (London) *229,* 339–341.

GUSTAVSSON, I. 1971C. Chromosomes of repeat breeder heifers. Hereditas *68,* 331–332.

GUSTAVSSON, I. 1975. Some comments on the eradication of the 1/29 translocation in Sweden. Eur. Kolloq. Zytogenet. Vet., Tierzucht. Saugetierkunde 263–268.

GUSTAVSSON, I. 1979. Distribution and effects of the 1/29 Robertsonian translocation in cattle. J. Dairy Sci. *62,* 825–835.

GUSTAVSSON, I., and HAGELTORN, M. 1976. Staining technique for definite identification of individual cattle chromosomes in routine analysis. J. Hered. *67,* 175–178.

GUSTAVSSON, I., and RENDEL, J. 1971. A translocation in cattle and its association to polymorphisms in red cell antigens, transferrins and carbonic anhydrases. Hereditas *67,* 35–38.

GUSTAVSSON, I., and ROCKBORN, G. 1964. Chromosome abnormality in three cases of lymphatic leukaemia in cattle. Nature (London) *203,* 990.

GUSTAVSSON, I., FRACCARO, M., TIEPOLO, L., and LINDSTEN, J. 1968. Presumptive X-autosome translocation in a cow: Preferential inactivation of the normal X chromosome. Nature (London) *218,* 183–184.

HALNAN, C. R. E. 1976. A cytogenetic survey of 1,101 Australian cattle of 25 different breeds. Ann. Genet. Sel. Anim. *8,* 131–139.

HANADA, H., MURAMATSU, S., ABE, T., and FUKUSHIMA, T. 1981. Robertsonian chromosome polymorphism found in a local herd of the Japanese Black cattle. Ann. Genet. Sel. Anim. *13,* 205–211.

HANSEN, K. M. 1969. Bovine tandem fusion and infertility. Hereditas *63,* 453–454.

HANSEN, K. M. 1972. Bovine chromosomes identified by quinacrine mustard and fluorescent microscopy. Hereditas *70,* 225–234.

HANSEN, K. M. 1973. Heterochromatin (C bands) in bovine chromosomes. Hereditas *73,* 65–70.
HANSEN, K. M. 1979. Bovine autosomal tandem fusion translocation—A unique translocation in the animal kingdom. 1st Int. Symp. Res. Cytogenet. Resistance Anim. and Man, Vienna.
HANSEN, K. M., and ELLEBY, F. 1975. Chromosome investigation of Danish A. I. beef bulls. Nord. Vet.-Med. *27,* 102–106.
HARE, W. C. D., and SINGH, E. 1980. Chromosomal analysis of early bovine embryos. Proc. 4th Eur. Colloq. Cytogenet. Domest. Anim. 172–180.
HARE, W. C. D., MITCHELL, D., BETTERIDGE, K. J., EAGLESOME, M. D., and RANDALL, G. C. B. 1976. Sexing two-week-old bovine embryos by chromosomal analysis prior to surgical transfer: Preliminary methods and results. Theriogenology *5,* 243–253.
HARVEY, M. J. A., and LOGUE, D. N. 1975. Studies on the 13/21 Robertsonian translocation in Swiss Simmental cattle. Proc. 2nd Eur. Colloq. Cytogenet. Domest. Anim. 155–161.
HERSCHLER, M.S., and FECHHEIMER, N. S. 1966. Centric fusion of chromosomes in a set of bovine triplets. Cytogenetics *5,* 307–312.
HERSCHLER, M. S., FECHHEIMER, N. S., and GILMORE, L. O. 1962. Somatic chromosomes of cattle. J. Anim. Sci. *21,* 972–973.
HERZOG, A., and HÖHN, H. 1968. Autosomal trisomy in a calf with brachygnathia inferior and congenital edema. Dtsch. Tieraerztl. Wochenschr. *75,* 604–606.
HERZOG, A., HÖHN, H., and RIECK, G. W. 1977. Survey of recent situation of chromosome pathology in different breeds of German cattle. Ann. Genet. Sel. Anim. *9,* 471–491.
HÖHN, H., and HERZOG, A. 1970. Two additional cases of autosomal trisomy in calves with brachygnathia inferior and other aberrations. Giessener Beitr. Erbpathol. Zuchthyg. *3,* 1.
HSU, T. C., and POMERAT, C. M. 1953. Mammalian chromosomes in vitro. II. A method for spreading the chromosomes of cells in tissue culture. J. Hered. *44,* 23–29.
HULTEN, M. 1974. Chiasma distribution at diakinesis in the normal human male. Hereditas *76,* 55–78.
JAGIELLO, G. M., MILLER, W. A., DUCAYEN, M. B., and LIN, J. S. 1974. Chiasma frequency and disjunctional behavior of ewe and cow oocytes matured in vitro. Biol. Reprod. *10,* 354–363.
JAGIELLO, G., DUCAYEN, M., FANG, J.-S., and GRAFFEO, J. 1976. Cytogenetic observations in mammalian oocytes. Chromosomes Today *5,* 43–63.
JAMES, R. C. 1980. A study of U.S. registered Jersey and Guernsey cattle chromosomes M.S. Thesis, Univ. of Nebraska.
JAMES, R. C., and ELDRIDGE, F. E. 1978. Incidence of chromosomal aberrations in Guernsey and Jersey cattle. J. Dairy Sci. *61*(Suppl. 1), 86.
JORGE, W. 1974. Chromosome study of some breeds of cattle. Caryologia *27,* 325–329.
KIDDER, H. E., BLACK, W. G., WILTBANK, J. N., ULBERG, L. C., and CASIDA, L. E. 1954. Fertilization rates and embryonic death rates in cows bred to bulls of different levels of fertility. J. Dairy Sci. *37,* 691–697.

KIEFFER, N. M., and CARTWRIGHT, T. C. 1968. Sex chromosome polymorphism in domestic cattle. J. Hered. *59,* 34–36.

KING, W. A., LINARES, T., GUSTAVSSON, I., and BANE, A. 1980. Presumptive translocation type trisomy in embryos sired by bulls heterozygous for the 1/29 translocation. Hereditas *92,* 167–169.

KNUDSEN, O. 1956. Chromosomal investigations in bulls. Fortpflanz. Zuchthyg. Haustierbesamung. (Suppl. to Dtsch. Tieraerztl. Wochenschr.) *6,* 5–8.

KOENIG, J. L. F., ELDRIDGE, F. E., and HARRIS, N. 1983. A cytogenetic analysis of bovine oocytes cultured in vitro. J. Dairy Sci. *66* (Suppl 1), 253.

KOENIG, J. L. F. 1982. A cytogenetic analysis of bovine oocytes cultured in vitro. M.S. Thesis, Univ. of Nebraska.

KOVACS, A. 1976. A new autosomal translocation in Hungarian Simmental. Veterinaermed. Gesellsch. 162–167.

KOVACS, A., and PAPP, M. 1977. Report on chromosome examination of A.I. bulls in Hungary. Ann. Genet. Sel. Anim. *9,* 528.

KOVACS, A., MESZAROS, I., SELLYEI, M., and VASS, L. 1973. Mosaic centromeric fusion in a Holstein-Friesian bull. Acta Biol. Acad. Sci. Hung. *24,* 215–220.

KRALLINGER, H. F. 1931. Cytological studies on some domestic animals. Arch. Tierernaehr. Tierzucht. Abt. B *5,* 127–187.

LAIPIS, P. J., WILCOX, C. J., and HAUSWIRTH, W. W. 1982. Nucleotide sequence variation in mitochondrial deoxyribonucleic acid from bovine liver. J. Dairy Sci. *65,* 1655–1662.

LENOIR, F., and LICHTENBERGER, J. M. 1980. Comparison of sex chromosomes of Bosolo hybrid Beefalo, American buffalo and domestic cattle. Proc. 4th Eur. Colloq. Cytogenet. Domest. Anim. 260–271.

LIN, C. C., NEWTON, D. R., and CHURCH, R. B. 1977. Identification and nomenclature for G-banded bovine chromosomes. Can. J. Genet. Cytol. *19,* 271–282.

LINARES, T., KING, W. A., and GUSTAVSSON, I. 1980. Morphological appearances of bovine embryos sired by bulls heterozygous for the 1/29 translocation. Proc. 4th Eur. Colloq. Cytogenet. Domest. Anim. 188–192.

LINARES, T., KING, W. A., GUSTAVSSON, I. and LARSSON, K. 1981. Trisomy-X in Swedish Red and White breed heifers. 23rd Ann. Mtg. ASAS Abstr., 162.

LOGUE, D. N., and HARVEY, M. J. A. 1978. Meiosis and spermatogenesis in bulls heterozygous for a presumptive 1/29 Robertsonian translocation. J. Reprod. Fertil. *54,* 159–165.

LOGUE, D. N., HARVEY, M. J. A., ELDRIDGE, F. E., and POLLOCK, D. 1977. Identification of some chromosomal anomalies in cattle. Personal Communication.

LOGUE, D. N., HARVEY, M. J. A., MUNRO, C. D., and LENNOX, B. 1979. Hormonal and histological studies in a 61,XXY bull. Vet. Rec. *104,* 500–503.

LOJDA, L. 1972. Chromosomal chimerism of the sire as a cause of the shift in sex ratio in his offspring. Proc. 7th Int. Congr. Anim. Reprod. Artif. Insem., Munich *2,* 1110–1113.

LOJDA, L., RUBES, J., STAIKSOVA, M., and HAVRANDSOVA, J. 1976. Chromosomal findings in some reproductive disorders in bulls. Proc. 8th Int. Congr. Anim. Reprod. Artif. Insem., Krakow, July 12–16 *1,* 158.

LYON, M. F. 1961. Gene action in the X-chromosome of the mouse. Nature (London) *190,* 372–373.

MAKINO, S. 1944. Karyotypes of domestic cattle, zebu and domestic water buffalo (Chromosome studies in domestic mammals, IV). Cytologia *13,* 247–264.

MAKINO, S. 1956. A Review of the Chromosome Numbers in Animals, 300 pp. Hokuryukan, Tokyo.

MAKINO, S., and NISHIMURA, I. 1952. Water pretreatment squash technique. Stain Technol. *27,* 1.

MÄRKI, U., ROBINSON, T., and OSTERHOFF, D. R. 1984. Y chromosome dimorphism in African cattle. Proc. 6th Eur. Colloq. Cytogenet. Domest. Anim. (To be published.)

MASUDA, H., and WAIDE, Y. 1980. Chromosome aberrations in a low fertility bull. Jpn. J. Anim. A.I. Res. *2,* 1–4.

MASUDA, H., TAKAHASHI, T., SOEJIMA, A., and WAIDE, Y. 1978. Centric fusion of chromosomes in a Japanese Black bull and its offspring. Jpn. J. Zootech. Sci. *49,* 853–858.

MASUDA, H., SHIOYA, Y., and FUKUHARA, R. 1980. Robertsonian translocation in Japanese Black cattle. Jpn. J. Zootech. Sci. *51,* 26–32.

MAYR, B., KOPP, E., and SCHLEGER, W. 1980. Identification of a 1/29 Robertsonian translocation in Austrian Grey cattle. Wien. Tieraerztl. Wochenschr. *67,* 292–294.

McFEELY, R. A., and RAJAKOSKI, E. 1968. Chromosome studies on early embryos of the cow. Proc. 6th Int. Congr. Reprod. Anim. Insem. Artif., Paris.

MELANDER, Y. 1959. The mitotic chromosomes of some cavicorn mammals (*Bos taurus* L., *Bison bonasus* L., and *Ovis aries* L.). Hereditas *45,* 649–664.

MELANDER, Y., and KNUDSEN, O. 1953. The spermiogenesis of the bull from a karyological point of view. Hereditas *39,* 505–517.

MEYER, E. H. H. 1984. Classification of *Bos indicus* cattle breeds in Southern and Central Africa as Sanga or Zebu type by means of Y chromosome morphology Proc. 6th Eur. Colloq. Cytogenet. Domest. Anim. (To be published.)

MIYAKE, Y. I., KANAGAWA, H., and ISHIKAWA, T. 1984. Further chromosomal and clinical studies on the XY/XYY mosaic bull. Jpn. J. Vet. Res. *39,* 9–21.

MOORHEAD, P. S., NOWELL, P. C., MELLMAN, W. J., BATTIPS, D. M., and HUNGERFORD, D. S. 1960. Chromosome preparations of leukocytes cultured from human peripheral blood. Exp. Cell Res. *20,* 613–616.

MORI, M., SASAKI, M., MAKINO, S., ISHIKAWA, T., and KAWATA, K. 1969. Autosomal trisomy in a malformed newborn calf. Proc. Jpn. Acad. Sci. *45,* 955–959.

MOUSTAFA, A. R., IBRAHIM, H., RAHMAN, H. A., and KOVACS, A. 1983. Quality, freezability and fertility of the semen of pre-selected A.I. bulls carrying various chromosome aberrations. Anim. Reprod. Sci. *6,* 167–175.

NICHOLS, W. W., LEVAN, A., and LAWRENCE, W. C. 1962. Bovine chromosomes by the peripheral blood method. Hereditas *48,* 536–538.

NIEHBUHR, E. 1972. Dicentric and monocentric Robertsonian translocation in man. Humangenetik *16,* 217–226.

NORBERG, H. S., REFSDAL, A. O., GARM, O. N., and NES, N. 1976. A case report on X-trisomy in cattle. Hereditas *82,* 69–72.

NOWELL, P. C. 1960. Phytohemagglutinin: An initiator of mitosis in cultures of normal human leukocytes. Cancer Res. *20,* 462–466.

OHNO, S., TRUJILLO, J. M., STENIUS, C., CHRISTIAN, L. C. and TEPLITZ, R. L. 1962. Possible germ cell chimeras among newborn dizygotic twin calves (*Bos taurus*). Cytogenetics *1,* 258–265.

OKAMOTO, A., YOSHIZAWA, M., and MURAMATSU, T. 1981. Studies of the Robertsonian translocation in Japanese Black cattle. Bull. Coll. Agric. Utsunomiya Univ. *11,* 1–8.

PAPP, M., and KOVACS, A. 1980. 5/18 dicentric Robertsonian translocation in a Simmental bull. Proc. 4th Eur. Colloq. Cytogenet. Domest. Anim. 51–54.

PATHAK, S., and WURSTER-HILL, D. H. 1977. Distribution of constitutive heterochromatin in carnivores. Cytogenet. Cell Genet. *18,* 245–254.

PINHEIRO, L. E. L., and FERRARI, I. 1980. A new type of Robertsonian translocation in cattle. Resumos 5th Encontro de Pesquisas Veterinarias Nov. 6–7.

PINHEIRO, L. E. L., MORAES, J. C. F., MATTEVI, M. S., ERDTMANN, B., SALZANO., F. M., and FILHO, A. M. 1980. Two types of Y chromosome in a Brazilian cattle breed. Caryologia *33,* 25–32.

POLANI, P. E. 1972. Centromere localization at meiosis and the position of chiasmata in the male and female mouse. Chromosoma *36.* 343–374.

POLLOCK, D. 1972. A chromosome abnormality in Friesian cattle in great Britain. Vet. Rec. *90,* 309–310.

POLLOCK, D. L. 1974. Chromosome studies in artificial insemination sires in Great Britain. Vet. Rec. *95,* 266–277.

POLLOCK, D. L., and BOWMAN, J. C. 1974. A Robertsonian translocation in British Friesian cattle. J. Reprod. Fertil. *40,* 423–432.

POPESCU, C. P. 1971. The meiotic chromosomes of cattle (*Bos taurus* L.). Ann. Genet. Sel. Anim. *3,* 125–143.

POPESCU, C. P. 1976. New data on pericentric inversion in cattle (*Bos taurus* L.). Ann. Genet. Sel. Anim. *8,* 443–448.

POPESCU, C. P. 1977. A new type of Robertsonian translocation in cattle. J. Hered. *68,* 139–142.

POPESCU, C. P. 1978. Sister chromatid exchanges (S.C.E.) in normal and abnormal bovine (*Bos taurus* L.) karyotypes. Cytologia *43,* 533–540.

POPESCU, C. P. 1980. Cytogenetics study on embryos sired by a bull carrier of 1/29 translocation. Proc. 4th Eur. Colloq. Cytogenet. Domest. Anim. 180–186.

POPESCU, C. P., CRIBIU, E. P., POIVEY, J. P., and SEITZ, J. L. 1979. Cytogenetic study of cattle from the Ivory Coast. Rev. Elev. Med. Vet. Pays Trop. *32,* 81–84.

QUEINNEC, G., DARRE, R., BERLAND, H. M., and RAYNAUD, J. C. 1974. Study of the 1/29 translocation in cattle of southeast France: Zootechnical results. 1st World Cong. Genet. Appl. Livestock Prod., Madrid. *3,* 131–151.

REFSDAL, A. O. 1976. Low fertility in daughters of bulls with 1/29 translocation. Acta Vet. Scand. *17,* 190–195.

RIECK, G. W., HOHN, H., and HERZOG, A. 1970. X-trisomy in cattle with familial tendency toward meiotic non-disjunction. Cytogenetics *9,* 401–409.

SAMARINEANU, M., LIVESCU, B.-E., and GRANCIU, I. 1976. Identification of a Robertsonian translocation in one breed of Romanian cattle. Lucr. Stiint. Taurine *3,* 53–60.

SASAKI, M. S., and MAKINO, S. 1962. Revised study of the chromosomes of domestic cattle and the horse. J. Hered. *53,* 157–162.

SCOTT, C. D., and GREGORY, P. W. 1965. An XXY trisomic in an intersex of *Bos taurus.* Genetics *52,* 473–474.

SHORT, R. V., SMITH, J., MANN, T., EVANS, E. R., HALLETT, J., FRYER, A., and HAMERTON, J. L. 1969. Cytogenetic and endocrine studies of a freemartin heifer and its bull co-twin. Cytogenetics *8,* 369–388.

SHUMOV, A. V. 1980. Chromosomes of the sub-species *Bos* (bison) and hybrids with cattle. Vopr. Gibridizatsii Kopytnykh. 91–94.

SINGH, E. L., EAGLESOME, M. D., HARE, W. C. D., and MITCHELL, D. 1977. Chromosomal studies on cultured fetal fluids collected by transacrostatic amniocentesis from heifers during the third month of pregnancy. Ann. Genet. Sel. Anim. *9,* 536.

SITKO, M., and DIANOVSKY, J. 1981. Genetic analysis in commercial herds of cattle in Slovakia. Veterinarstvi *31,* 155–157. (Anim. Breed Abstr. *49,* 6298.)

SOLDATOVIC, B., PROLIC, Z., and CRETKOVIC, M. 1977. The importance of cytogenetics for the diagnosis of congenital abnormalities in cattle. Vet. Glas. *3,* 33–40.

STRANZINGER, G. F., and FORSTER, M. 1976. Autosomal chromosome translocation of Piebald cattle and Brown cattle. Experientia *32,* 24–26.

SUCCI, G., MOLTENI, L., and DEGIOVANNI, A. 1980. Cytogenetical study of some Italian cattle breeds in decreasing or in way of extinction. Proc. 4th Eur. Colloq. Domest. Anim. 136–140.

SWARTZ, H. A., and VOGT, D. W. 1983. Chromosome abnormalities as a cause of reproductive inefficiency in heifers. J. Hered. *74,* 320–324.

THOMPSON, J. S., and THOMPSON, M. W. 1980. Genetics in Medicine, 3rd Edition. Saunders, Philadelphia, PA.

TSCHUDI, P., UELTSCHI, G., MARTIG, J., and KUPFER, U. 1975. Autosomal trisomy as the cause for defects of the interventricular septum in a Simmental calf. Schweiz. Arch. Tierheilkd. *117,* 335–340.

TSCHUDI, P., ZAHNER, B., KUPFER, U., and STAMPFLI, G. 1977. Chromosome investigation of Brown Swiss cattle. Schweiz. Arch. Tierheilkd. *119,* 329–336.

ULBRICH, F., WEINHOLD, E., and PFEIFFER, R. A. 1963. Preparation of bovine chromosomes. Nature (London) *199,* 719.

WILKES, P. R. 1979. Personal communication.

WODSEDALEK, J. E. 1920. Studies on the cells of cattle with special reference to spermatogenesis, oogonia and sex determination. Biol. Bull. *38,* 290–316.

WURSTER, D. H., and BENIRSCHKE, K. 1968. Chromosome studies in the superfamily Bovoidea. Chromosoma *25,* 152–171.

YAEGER, J. W., HINES, C. J., and SPEAR, R. C. 1983. Exposure to ethylene oxide at work increases sister chromatid exchanges in human peripheral lymphocytes. Science *219,* 1221–1223.

YOSIDA, T. H., and LAMONTAIN, E. J. 1964. Chromosomes of normal and dwarf cattle. Jpn. J. Genet. *38,* 351–355.

12

Sheep and Goat Chromosomes

INTRODUCTION AND EARLY STUDIES

R. O. Berry in 1941 and S. Makino in 1943 presented good review papers on the early work done with sheep and goat chromosomes. Sheep chromosomes will be considered first.

Apparently Wodsedalek in 1922 reported the first counts in sheep. He reported 33 as the $2n$ number for the male and 34 for the female, assuming an XO type of sex determination.

Krallinger in 1931 found 50–60 chromosomes in sheep, with 30 in the primary spermatocytes. Novikov (1935), Butarin (1934), Bruce (1935), and Pchakadze (1936) all reported the $2n$ number in sheep as 60, and reported no V-shaped chromosomes in anaphase. Somehow they all overlooked the submetacentric chromosomes and decided that the chromosomes were all rod-shaped. The centromeric regions apparently were so indistinct in their preparations that they interpreted each of the three pairs as two chromosomes.

Shiwago (1931), Berry (1938, 1941), and Ahmed (1940) all found 54 chromosomes. They recognized the 6 submetacentric chromosomes as single chromosomes. Thus, it appears that Shiwago was the first to establish the true diploid number of 54 in sheep, the number which has now been corroborated by many authors. In the earlier paper by Berry (1938), he observed only four U-shaped, or J-shaped, large chromosomes. In his later paper he identified six U-shaped chromosomes in the sheep, which correspond to what are now known as six large submetacentric chromosomes. Makino (1943) compared the chromosomes of the sheep with the goat and noted the marked similarities. He postulated that the two species may have descended from a common ancestor and that the six chromosomes of the sheep that were V-shaped at anaphase could have arisen by a fusion of goat-type, rod-shaped chromosomes, at their ends. This is the same conclusion that Evans *et al.* (1973) came to in their study of banded chromosomes. The banding of the two species is similar enough that each of the two arms of the submetacentric sheep chromosomes was identified as specific goat chromosomes. The longest sheep metacentric was made up by centric fusion of goat chromosomes 1 and 3. The second-longest sheep metacentric was made up from centric fusion of goat chromosomes 2 and 8. The third sheep metacentric was the result of fusion of goat chromosomes 4 and 9. These combinations were confirmed by Bruere *et al.* (1974).

Zartman and Bruere (1974) published a more detailed idiogram of the banding patterns of sheep, from karyotypes that show greater detail than those of Evans *et al.* (1973). The arrangement of chromosomes by length, however, differed from Evans *et al.* (1973), and the karyotypes are therefore not strictly comparable.

Figures 12.1 and 12.2 illustrate the karyotypes of male and female sheep by Giemsa staining. The X chromosome cannot be identified with certainty in Giemsa-stained preparations as it can be with G-banded chromosomes, but it is the longest acrocentric, so it can be identified with a reasonable degree of confidence. The Y chromosome is very definitely the smallest chromosome, but, as with other species, the length may be found to be variable if larger numbers of males are observed (Fig. 12.3). The Y chromosome, in excellent preparations, also is seen to be metacentric. C-banding (Fig. 12.4) shows an absence of bands in the centromeric region of the sex chromosomes as in cattle, but a weak C-band appears in the long arm of the X chromosome. Weak C-bands are found in the 1 and 3 metacentrics with a stronger C-band in the second metacentric.

Bunch and Foote (1977) concluded that the mouflon (*Ovis musimon* or *Ovis orientalis*) or a mouflon-like species ($2n = 54$) was probably the ancestor of domesticated sheep. They do not exclude the contribution of some other species, even those with $2n = 52$ or 56, to domestic sheep, however. They base their conclusions partially on crosses of an argali (*O. ammon nigrimontana*) ram with mouflon ewes and subsequent backcrosses and inter-se crosses.

Studies of Iranian wild sheep by Valdez *et al.* (1978) showed variation of diploid chromosome numbers from 54 to 58. Animals hybridized freely without regard to chromosome numbers. The freedom

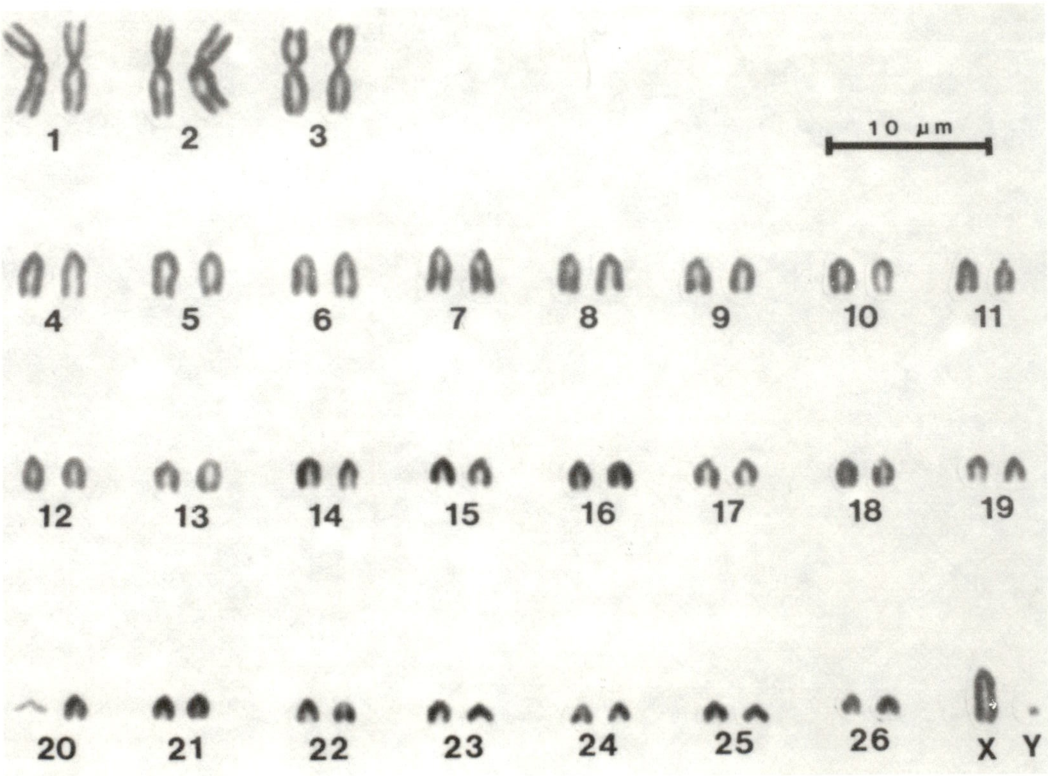

1 2 3

10 μm

4 5 6 7 8 9 10 11

12 13 14 15 16 17 18 19

20 21 22 23 24 25 26 X Y

FIG. 12.1. Giemsa-stained karyotype of a male sheep. The X chromosome is the longest acrocentric and the Y chromosome is metacentric and is considerably smaller than any other chromosome.

Photograph by Eldridge.

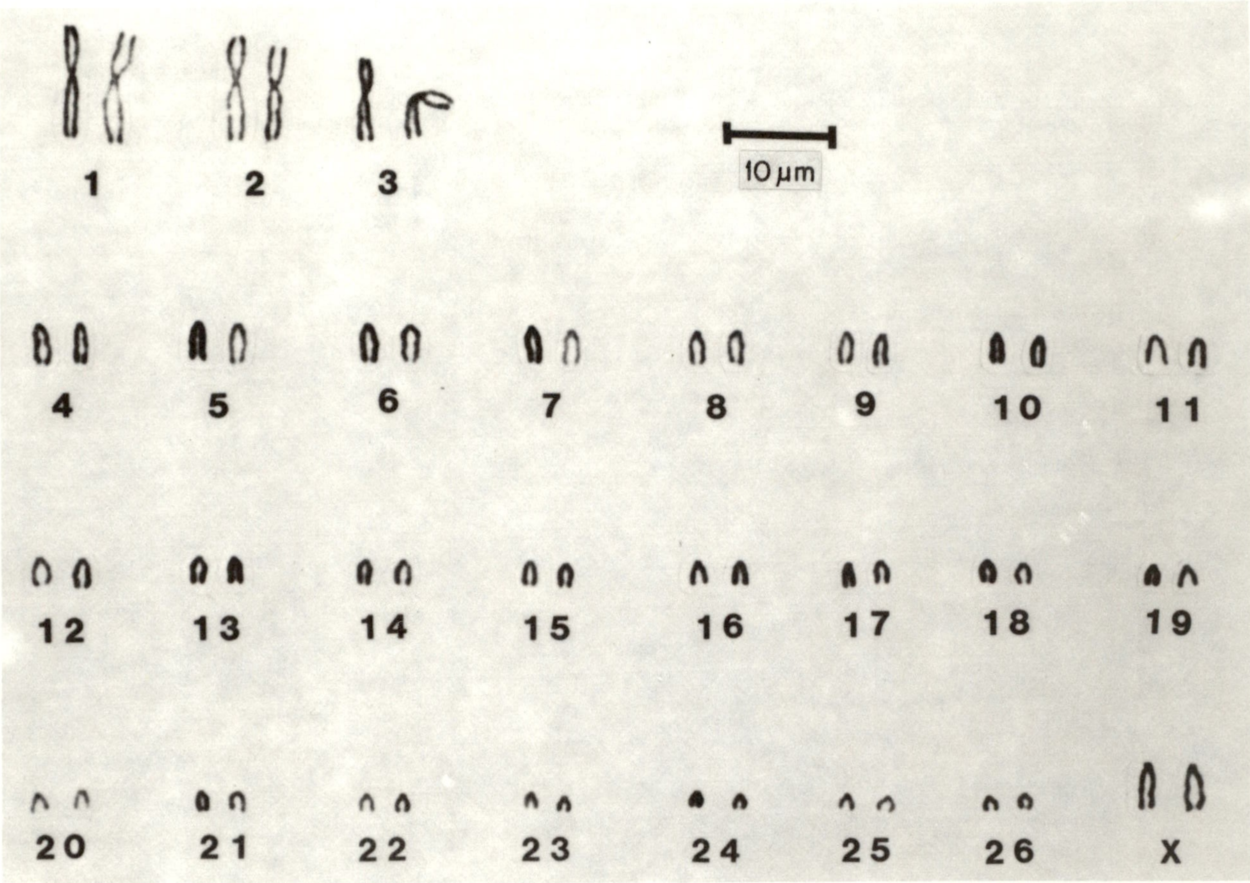

FIG. 12.2. Giemsa-stained karyotype of a female sheep. The X chromosome is the longest acrocentric.
Photograph by Eldridge.

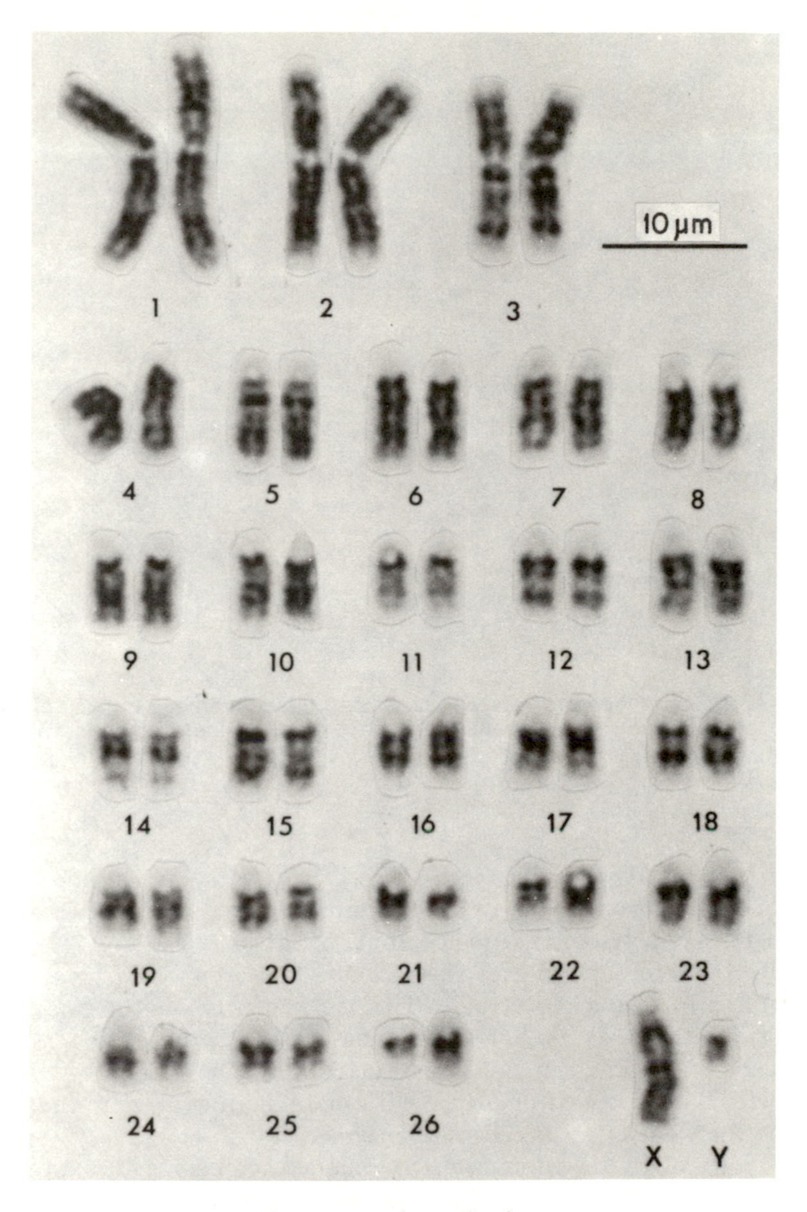

FIG. 12.3. G-banded karyotype of a male sheep.
Photograph courtesy of R. A. Buckland; Ford et al.

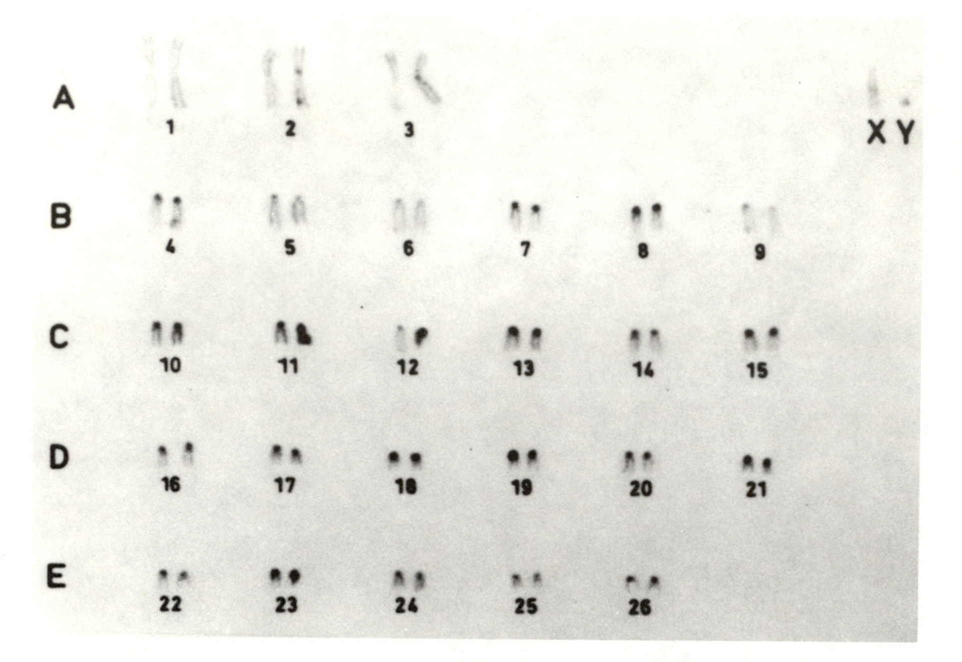

FIG. 12.4. C-banded male sheep karyotype. The centromeric regions of the gonosomes do not show C-bands, and metacentric autosomes 1 and 5 are weak.

Photograph by B. Glahn-Luft.

from problems with infertility in crosses with varying chromosome numbers in sheep seems to differ from the problems encountered with 1/29 Robertsonian translocations in cattle.

The X chromosome of sheep is the longest acrocentric chromosome according to Evans *et al.* (1973). The Y chromosome is the smallest chromosome, so small that little structure can be observed, but it appears to be metacentric (Figs. 12.1 and 12.3). Khavary (1974) reported a longer Y chromosome with more distinct arms in the Zel breed than has been found in other sheep.

No differences in chromosome numbers have been found among breeds of domestic sheep, although some variations have been found within breeds. Studies have been reported on Rambouillet, Suffolk, Merino, Karakul, Corriedale, Merino, Southdown, Ryeland, Dorset

Horn, Border Leicester, Romney Marsh, Cheviot, Zel, and numerous others.

ROBERTSONIAN TRANSLOCATIONS IN SHEEP

In the superfamily Bovoidea, centric fusion, or Robertsonian translocation, has been widely recognized to be the major chromosomal mechanism involved in the evolutionary development of different species. Domestic sheep (*Ovis aries*) have a diploid number of 54, including three pairs of submetacentric chromosomes, which by G-banding have been determined to be centric fusions of six pairs of acrocentric chromosomes found in the goat, the species considered most representative of the primitive karyotype (Evans *et al.* 1973; Bunch *et al.* 1976B). These three pairs of submetacentric chromosomes are sometimes referred to as M1, M2, and M3.

In addition to the three pairs of metacentrics typical of the sheep, four more Robertsonian translocation chromosomes have been studied. Three of these, referred to as Massey I or t_1, Massey II or t_2, and Massey III or t_3, have been extensively studied in New Zealand by Bruere and co-workers. The fourth was described first by Korobitsyna *et al.* (1974) in the Siberian snow sheep, *Ovis nivicola,* and was studied in more detail by Nadler and Bunch (1977).

In addition to the three pairs of submetacentrics normally found in sheep, Bruere *et al.* (1976) described three Robertsonian translocations. They are identified as t_1 which is 5/26, t_2 which is 8/11, and t_3 which is 7/25, the chromosome numbers corresponding to the sheep karyotype. The G-banded karyotype published by Evans *et al.* (1973) and the analysis by Bunch (1978) do not correspond directly to the G-banded karyotypes of Bruere and co-workers, so care must be used in identifying the chromosomes which make up the Robertsonian translocations. These have been more completely described in earlier papers, and have been designated by different terminologies as the research has progressed. In the 1976 paper Bruere *et al.* described the origin of the Perendale breed, which had 26% of the animals studied carrying the t_3 translocation. The Drysdale breed also had a high frequency of the t_3 translocation. The New Zealand Romney had both the t_1 and t_2 translocations. All appeared to have obtained the translocations from the Romney Marsh breed imported from Kent in the south of England. The New Zealand Romney is a dominant breed in New Zealand based upon importations of the Romney Marsh breed in 1853 and additional importations over the next 60 years. Subsequent to the discovery of these translocations in New Zealand, the t_1 translocation was found at the frequency of 7.6% in 158 Romney Marsh sheep in England (Bruere *et al.* 1978).

Bruere and Ellis (1979) analyzed chromosomally 741 lambs from matings of rams with several different combinations of translocation chromosomes to 856 ewes with various combinations. Most of the rams had at least two translocation chromosomes. Chi-square analysis of various matings consistently showed no significant deviation from expected numbers. Combining this new information with the results of many earlier matings of different combinations (Bruere and Chapman 1974) led to their conclusion that in sheep these three Robertsonian translocations had no deleterious effect upon fertility.

Further matings of triple-heterozygous rams to ewes with normal karyotypes resulted in significant deficiencies of $52,t_1t_3$ and $53,t_2$ lambs among 83 born and analyzed (Bruere *et al.* 1981). Also, more lambs were born with no translocation chromosomes than was expected. Meiotic studies showed more metaphase II figures with 27 chromosomes and no translocations and fewer $25,t_1t_3$ figures than expected in a total of 287. They suggest that the frequency of nondisjunction may be determined genetically and not by the translocation per se. Overall fertility still may not be affected, since the abnormal germ cells may be selected against. The general level of fertility in sheep with various combinations of t_1, t_2, and t_3 does not seem to be decreased. The similarity in genetic base for the animals studied may permit this maintenance of fertility, as contrasted to the crosses of the tobacco mouse, which has many Robertsonian translocation pairs of chromosomes, with the house mouse, which has none (Cattanach and Moseley 1973). Geographical isolation of these mouse species may have resulted in many other genetic differences which affected fertility when the two species were crossed.

The fourth naturally occurring Robertsonian translocation in a wild sheep species has been referred to as M4 by Bunch (1978). By G-banding he identified this submetacentric chromosome as 11/17 based upon his idiogram of goat chromosomes. He also identified t_3 based upon this same idiogram as 11/29, so M4 and t_3 would appear to have one arm in common, the number 11 chromosome. This might

TABLE 12.1. Known Chromosome Involvement in Centric Fusions of Domestic and Wild Sheep (*Ovis*) with the Numerical Identity of Acrocentrics Based on the Idiogram of Goats (*Capra*, $2n = 60$)

			Chromosome identity		
Species examined	Diploid number	Translocations	Present study	Evans *et al.* (1973)	Bruere *et al.* (1974)
Ovis vignei	58	M1	1/5	1/5[a] (1/4)	
Ovis ammon	56	M2	3/10	3/9 (2/8)	
Ovis aries	54	M3	4/9	5/10 (5/10)	
	54	T1	8/30		7/30[a] (5/26)
	54	T2	12/15		12/15 (8/11)
	54	T3	11/29		11/29 (7/25)
Ovis nivicola	52	M4	11/17		

[a] Adjusted chromosomes positioning to follow the genus *Capra* idiogram $2n = 60$; differences in ranking are due to original chromosome positioning: members within the parentheses are the author's original description. [From Bunch (1978).]

cause problems in fertility if animals with t_3 were mated with the Siberian snow sheep, which possesses M4. Table 12.1 compares the numbering systems of Bunch (1978) with Evans *et al.* (1973) and Bruere *et al.* (1974) (Figs. 12.5 and 12.6, pp. 198 and 199).

CHIMERISM IN SHEEP

Freemartinism in sheep has been recognized at least since 1928 when Roberts and Greenwood described a Southdown × Welsh Mountain crossbred, which was remarkably similar to freemartin heifers. Studies of heterosexual twin pairs and higher multiple mixed littermates with respect to skin graft tolerance led to some further discoveries of freemartin-type females. Stormont *et al.* (1953) identified erythrocyte mosaicism in a pair of mixed twins, and by combining some previous reports with his data concluded that .8% of ewe lambs may be freemartins. This conclusion was based on the assumption that one third of sheep births are twins. They thought this estimate might be too high, but subsequent work has not reduced it. Freemartin ewes that overtly display enlarged clitorises and male-type behavior are not numerous, but some more covert types of infertile females may be freemartins. However, Slee

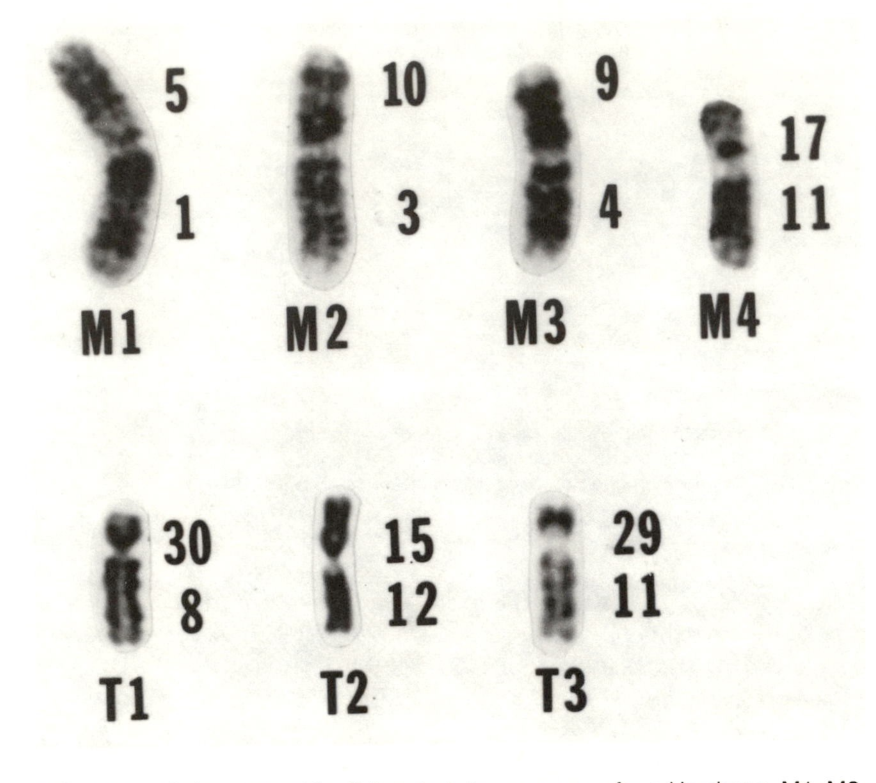

FIG. 12.5. Submetacentric, G-banded chromosomes found in sheep. M1, M2, and M3 are the three submetacentric chromosomes common to domestic sheep. M4 is found in the Siberian snow sheep. T1, T2, and T3 are the Robertsonian translocations discovered by Bruere in domestic sheep. The numbering of the arms is that of Bunch (1978)

Photograph courtesy of Bunch and with permission of the Journal of Heredity 69, 77 (1978), Fig. 2.

(1963) compared 187 ewes born twin to females with 167 born twin to males and found no significant difference in fertility.

Dain (1971) found 2 females in 161 sets of unlike sexed multiple births, or a 1.2% frequency of accurately identified sex chromosome chimerism. Wilkes *et al.* (1978) found a 54,XX/XY female chimera which shared the right horn of the uterus with a dead, semi-mummified male. The chimeric ewe had external female genitalia, but internally had male reproductive organs similar to other reported sheep freemartins. The male born from the left horn was a normal 54,XY animal.

Numerous other reports have been made over the years of individual or small numbers of freemartin sheep. Since the mid-'60s these reports have usually included chromosomal analyses. In all of these cases studied cytologically XX/XY cell chimerism has been found in the blood lymphocytes. No cases have been reported of XY cells in tissue cultures from skin or other organs.

Since sheep have a multiple-birth frequency much higher than cattle, it would appear that in their evolutionary development some

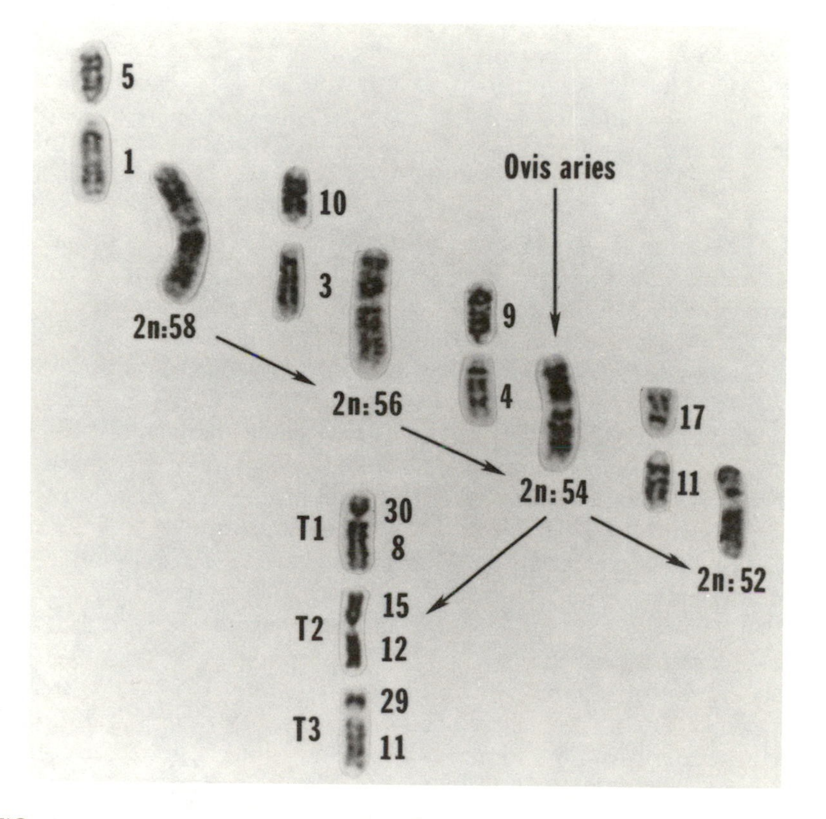

FIG. 12.6. Proposed evolutionary development of *Ovis aries*, related to the Siberian snow sheep, and Robertsonian translocations reported by Bunch. *Photograph by T. D. Bunch.*

protective mechanism against anastomosis of the placental vascular system has developed.

OTHER CHROMOSOMAL ABERRATIONS IN SHEEP

Other than Robertsonian translocations and sex chromosome chimerism, a very limited number of chromosomal aberrations have been reported in sheep.

Glahn-Luft and Wassmuth (1977, 1980) mentioned animals with a karyotype 54,XYt($1p^-$;$24q^+$) in which the aberration had been transmitted as expected. In the second report the second chromosome involved was identified as chromosome 20 rather than 24. When heterozygous rams were mated to heterozygous ewes a decrease in multiple births was found, compared to mating with normal ewes. They ascribed this decrease to death of fertilized ova or embryos (Figs. 12.7, 12.8, and 12.9).

In 121 rams of four breeds, Moraes *et al.* (1980) found mosaicism in 3 animals. Two of these had a small extra fragment or marker chromosome in approximately 25% of the cells observed. In the other case 15% of the cells were 55,XYY. No results on fertility were reported on the latter case, but the other two rams appeared nearly normal in fertility.

A number of sheep with different congenital abnormalities have been studied chromosomally; no relationship has been found between the abnormalities and chromosomal aberrations (Zhapbasov and Baikenova 1978). Two different cases of anidian monsters in sheep had different analyses. Dunn and Roberts (1972) found a fourfold increase in the number of cells with fewer than 53 chromosomes. Eldridge (1980), however, found a normal pattern of 54,XX in the case he studied.

Six rams were found by Bruere and Kilgour (1974) to have karyotypes of 55,XXY, similar to the Klinefelter syndrome in humans. The animals were not chimeric; apparently had no mental retardation, which is typical of humans with this syndrome; and showed no reduction in libido when compared to normal rams. Testicular and scrotal size were markedly smaller than in normal rams. All were from the Romney breed in New Zealand, but in a previous report one case from the Cheviot breed was found (Bruere *et al.* 1969). About

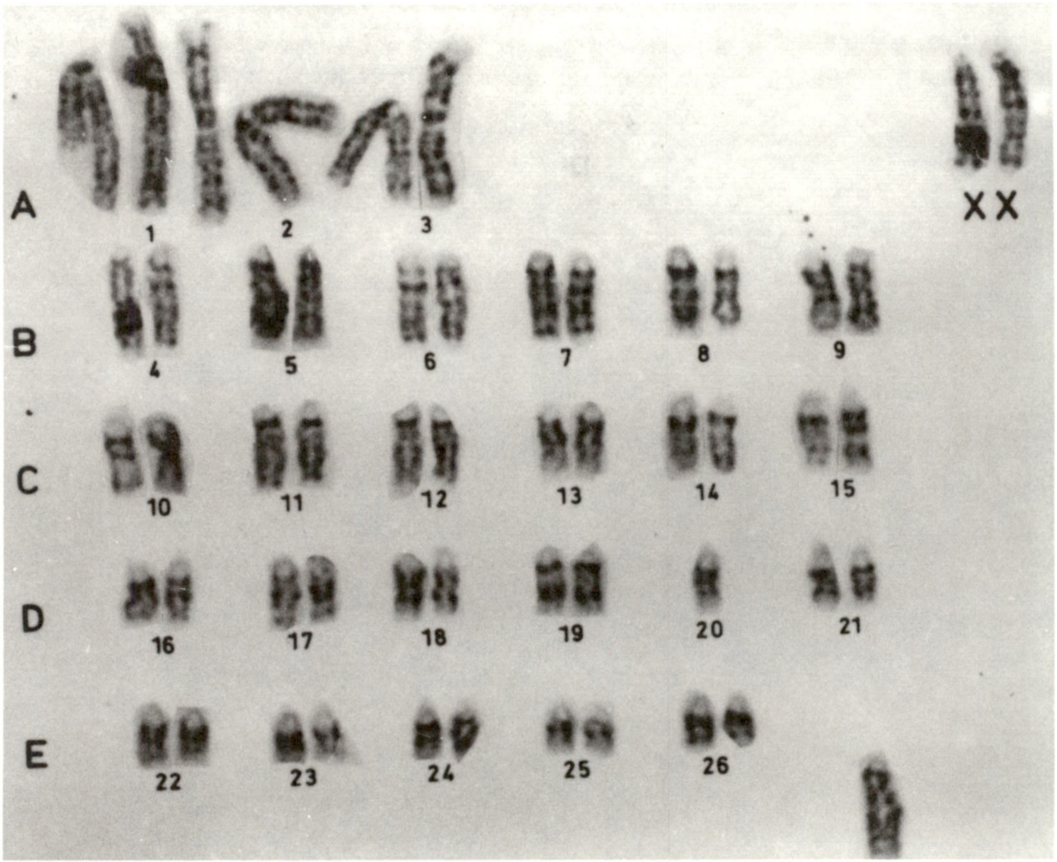

FIG. 12.7. G-banded karyotype of a heterozygous female sheep with the balanced translocation 54,XXt($1p^-$;$20q^+$).

Photograph by Glahn-Luft and Wassmuth (1980).

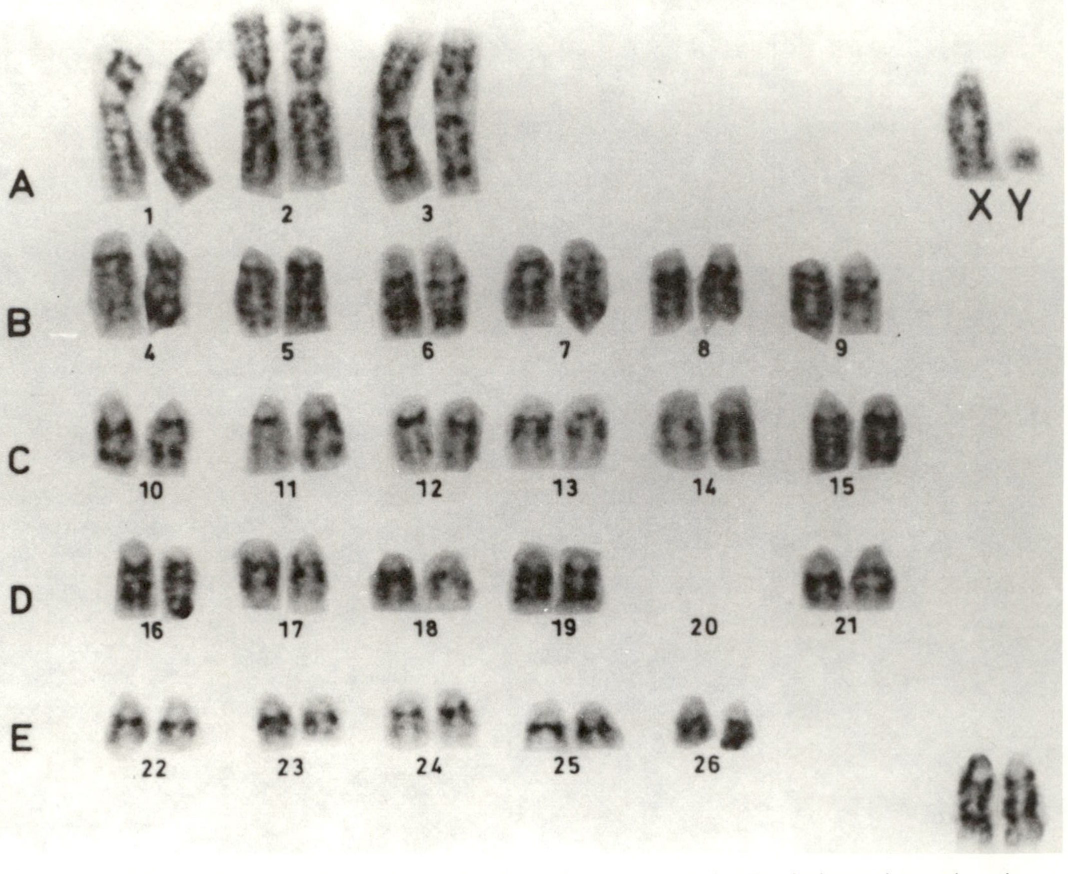

FIG. 12.8. G-banded karyotype of a male sheep homozygous for the balanced translocation 54,XYt($1p^{-}$;$20q^{+}$).

Photograph by Glahn-Luft and Wassmuth (1980).

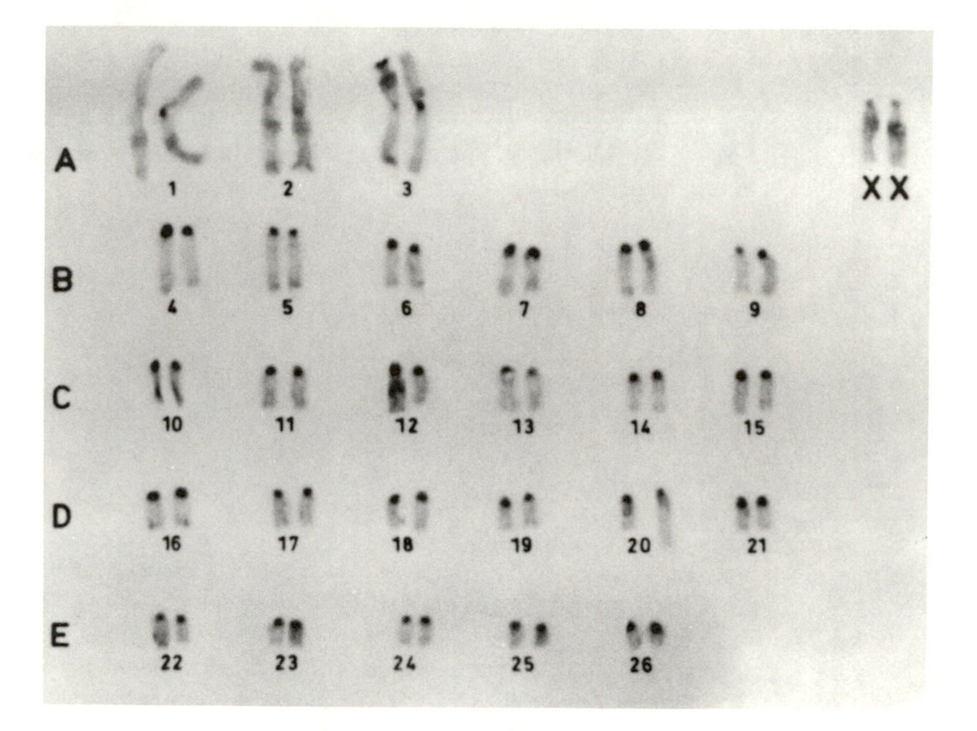

FIG. 12.9. C-banded karyotype of a female sheep heterozygous for the balanced translocation 54,XXt(1p$^-$;20q$^+$).
Photograph by Glahn-Luft and Wassmuth (1980).

3.5% of rams have testicular hypoplasia, but no population study has been made to determine whether or not the major cause of this condition is this chromosomal aberration.

OTHER CAUSES OF CHROMOSOMAL ABERRATIONS

Where studies are being planned either to determine the normal chromosome complement of animals or to establish a relationship between some characterstic and the karyotype, it is highly desirable to control the environmental conditions as competely as possible. Age of animals can affect their chromosome constitution, as has been shown by Bruere (1967): he found that there were more hypermodal and hypomodal cells in sheep 4–7+ years old than in 2-year olds. Crenshaw *et al.* (1974) found that cyclophosphamide (CPA) fed to sheep at oral doses of 25 or 75 mg/kg decreased the mitotic index (the number of cells in metaphase related to the number of cells in interphase) and increased aneuploidy. Other drugs may have specific effects upon chromosomes in lymphocytes.

EARLY STUDIES ON GOAT CHROMOSOMES

Goats have 60 chromosomes as the diploid number, all acrocentric except the Y, which is the smallest element and is metacentric or submetacentric. Makino (1943), in a complete historical review of early chromosome studies on goats, as well as sheep, indicated that Sokolov (1930) was the earliest author to report on goat chromosomes, and he reported a $2n$ number of 60. All subsequent reports have agreed on the number, in contrast to the variable reports and disagreements among early workers on most other species.

The X chromosome is one of the longest acrocentric chromosomes. Depending upon its location in the metaphase spread, whether it is straight, bent, or has its chromatids crossed, and the same factors in the longest autosomes, it may be longest or second or third in length. Without G-banding it cannot be identified with certainty. The Y chromosome of the goat is the smallest and the only metacentric (or possibly submetacentric) chromosome (Basrur and Stoltz 1967; Makino *et al.* 1967) (Figs. 12.10 and 12.11).

The goat karyotype has been accepted by numerous authors as the "primitive" karyotype, since all closely related species studied appear to be derived from it through the occurrence of Robertsonian translocations. Banding patterns, either G, Q, or R, have demonstrated remarkable homology between the chromosomes of goats and related species. Since it is not possible to actually reconstruct the evolution of these animals, it is not claimed that other species descended from goats, but descent of goats and the other species from a common ancestor with a goatlike karyotype is generally agreed upon. Evans *et al.* (1973) used the G-banded goat chromosomes as the basic type for describing homology with sheep and cattle.

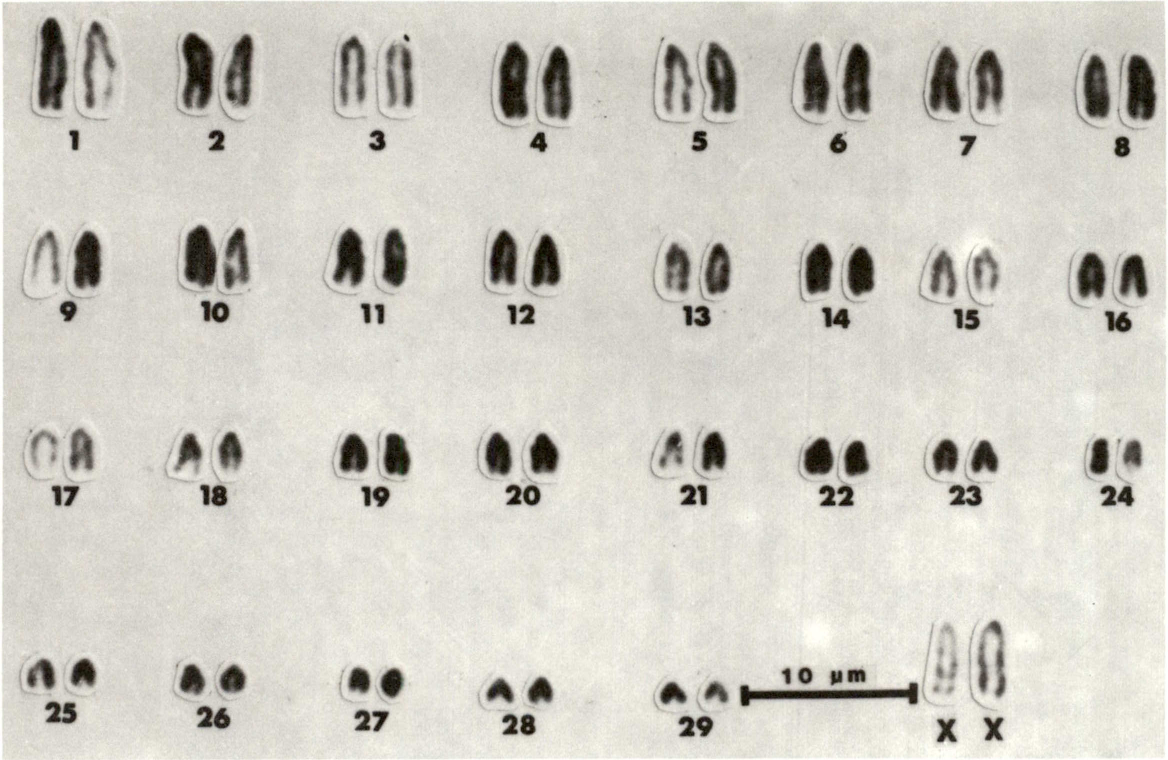

FIG. 12.10. Giemsa-stained karyotype of a female goat.
Photograph by Eldridge.

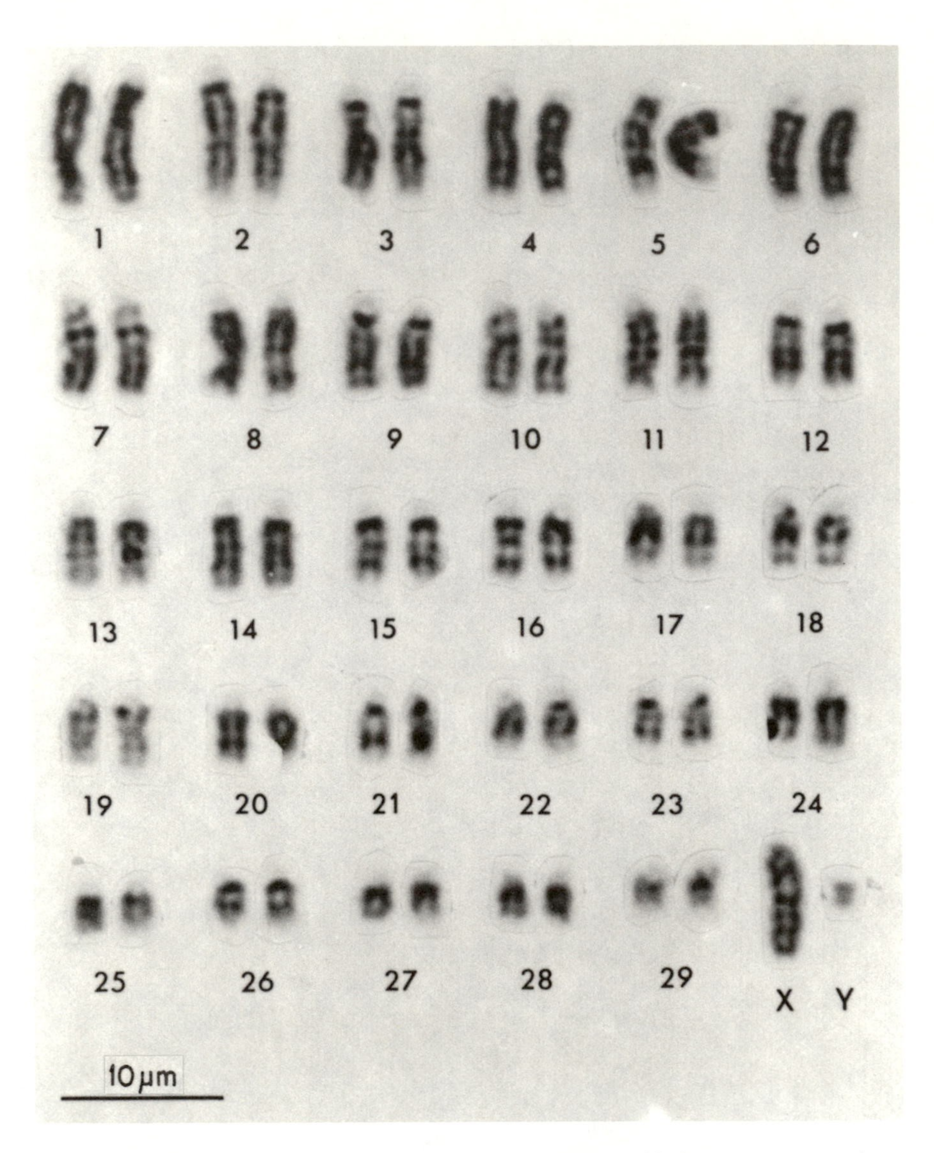

FIG. 12.11. G-banded karyotype of a male goat. The Y chromosome can be seen to be metacentric.

Photograph courtesy of R. A. Buckland; from Ford et al. (1980).

ROBERTSONIAN TRANSLOCATIONS IN GOATS

Apparently the first report of a Robertsonian translocation in goats was by Padeh *et al.* (1965), discovered while studying an XX/XY intersex animal. Padeh *et al.* (1971) reported the results of mating animals carrying a Robertsonian translocation that they had first reported in 1966, which was unrelated to the first chimeric case. No phenotypical variations from normal were found in heterozygous or six homozygous animals. All were found in the Saanen breed. The translocation chromosome was studied over four generations. The specific chromosomes involved in the translocation were not identified.

Hulot (1969) found a Robertsonian translocation in another Saanen goat which Popescu (1972) bred to determine transmission. The animals did not deviate from normal phenotypically, and the chromosome was transmitted as expected. No G-banding was reported, but by length the chromosomes were suggested to be numbers 2 and 13. The karyotype appears similar, if not identical, to the photograph of the karyotype reported by Padeh *et al.* (1971).

Sohrab *et al.* (1973), while studying goat × sheep hybrids, found one goat with 531 out of 545 metaphases which contained a Robertsonian translocation chromosome, with a $2n$ number of 59 in those cells. The animal was a crossbred Toggenburg-type female. The photograph showed the translocation chromosome to be very similar to the ones reported by Padeh *et al.* (1971) and Popescu (1972). No breeding results were given.

GOAT INTERSEXES

For several decades the relationship between the polled condition and intersexuality has been known in the Saanen and some other goat breeds. The dominant gene which causes the polled condition acts as a recessive gene causing intersexuality when homozygous. This is a good example of the pleiotropic effect of a gene. It could also be the result of two very closely linked genes.

Development of techniques for accurate observation of chromosomes has permitted identification of the genetic sex of intersex individuals. The Y chromosome in goats is rather easily identified since it is only half as long as the shortest autosomes. Even without identifying the X chromosome, the presence or absence of the Y can allow accurate identification of the sex of each cell. The polled condition also was found to be associated with a modification of the sex ratio, fewer females than expected being born in matings of polled × polled animals. Classification of the genetic (chromosomal) sex of the intersexes as either males or females allowed a more accurate determination of the sex ratio.

Soller *et al.* (1969) cytologically studied 30 goats, 17 of which were classified as pseudohermaphrodites; 13 were infertile males, either with testicular hypoplasia (3 animals) or with epididymal sperm granuloma (10 animals). They found that 16 of the 17 pseudohermaphrodites were genetically XX, and one was a sex chromosome chimera, with 30 cells XX and 47 cells XY. All of the cells from tissue cultures of muscle, kidney, lung, liver, and testis were XX, similar to the situation that is usually found in freemartin cattle. The first report of a sex chromosome chimera in goats according to Popescu (1972) was by Padeh *et al.* (1965) in an intersex Saanen goat which also had a Robertsonian translocation.

The 10 infertile males with epididymal sperm granuloma were all found to be XY, genetically males. However, the males with testicular hypoplasia were XX, genetically females. The expression of pseudohermaphroditism is highly variable, so the animals with testicular hypoplasia represent the extreme effect of the polled gene on genetic females. The epididymal sperm granuloma represents the effect of the polled gene on genetic males. If all the pseudohermaphrodites and the infertile "males" showing testicular hypoplasia are classified as genetic females the apparently abnormal sex ratio can be accounted for.

Ilberry *et al.* (1967) reported a similar set of results, finding two out of six intersex goats to be sex chimeras, XX/XY, and the other six to be genetically female. One of the chimeras had a ratio of XX cells to XY of 34 : 69 and the other 98 : 1. These were Saanen and Toggenburg goats.

Hamerton *et al.* also in 1969 investigated 35 polled intersex goats anatomically, histologically, and with several biochemical analyses, as well as chromosomally. One out of the 32 whose chromosomes were studied was 60,XX/60,XY and the other 31 were 60,XX. They indicated that their findings were in complete agreement with Soller *et al.* (1969). These three reports were based upon 70 animals which had been selected because of an intersex condition. Four of these were sex chromosome chimeras and 66 were genetically female. Therefore the evidence is quite clear that within the polled breeds intersexuality is caused predominantly by the recessive action of the gene for polledness, but that about 1 case in 17 may also involve sex chromosome chimerism.

Several other sex chromosome chimeras have been reported in goats, including one examined in our laboratory but not reported in the literature. Three reports of special interest were reported by Padeh *et al.* (1965), Smith and Dunn (1981), and Bon Durant *et al.* (1980), since they were all horned. In these cases the intersexuality was probably the result of chimerism and not the effect of the polled gene. Therefore, intersexuality in the goat is not the result of one factor only, but may be caused by chimerism, or by the gene for the polled condition acting upon both genetic males and genetic females.

SHEEP × GOAT HYBRIDS

Goats have 58 acrocentric autosomes, an acrocentric X and a small metacentric Y chromosome. Sheep have 52 autosomes, of which 6 (3 pair) are submetacentric, 46 are acrocentric, the X is the longest acrocentric chromosome, and the Y is a small submetacentric chromosome. Hybrids should therefore have 57 chromosomes, of which 3 autosomes should be submetacentric. The X and Y chromosomes may not be distinguishable between the species.

For numerous reasons the production of sheep × goat hybrids has been of interest. As early as 1932 researchers attempted to produce such hybrids (Berry 1938). Fertilization of ewes with goat semen seldom results in conception, but the reverse—breeding does with ram semen—is much more frequently successful. The resulting embryos usually live until about 30–36 days after insemination, although occasionally one may survive into the 6th week, and the others may die by about 20 days. The hybrid embryos die at an earlier stage in goat females which have had a hybrid fetus previously. Imperfect development of the fetal–maternal placental union appears to be the final cause of embryonic death.

Examination of chromosomes from amnion cells of 30-day embryos by Berry (1938) verified the expected number of 57 in the hybrids. These results have been corroborated by Hancock and Jacobs (1966), by Buttle and Hancock (1966), and by Ilberry *et al.*

(1967). Through examination of cultured cells and of liver cells exposed immediately to colchicine without culture they found the three submetacentrics, 57 chromosomes, and both male and female hybrid embryos. The embryos were 30 to 33 days into gestation.

Transfer of sheep embryos into goats, goat embryos into sheep, and hybrid embryos into both goats and sheep as twins with embryos of maternal origin failed (Hancock and McGovern 1968). Also, goats treated parenterally with sheep semen failed to keep the hybrid embryos alive beyond the 6th week (McGovern 1973).

Successful production of living sheep × goat hybrids has been accomplished by Bratanov *et al.* (1980), another case was reported by Bunch *et al.* (1976A), and one more case was reported by Eldridge *et al.* (1983). In all three cases positive identification of the hybrid was made through chromosome studies. Bratanov *et al.* (1980) found the male hybrids to be unable to fertilize females, even though sperm production was relatively normal. Females with 57 chromosomes were fertile and in backcrosses with sheep produced offspring with 56 chromosomes. The second backcross generation had 54 chromosomes.

Bunch *et al.* (1976A) examined the chromosomes of a hybrid from natural mating of a Barbados ram, considered to be one of the most primitive types of sheep in the United States, with a "Spanish" goat female. The female hybrid (Fig. 12.12) had 57 chromosomes, as expected. The male twin hybrid to the female was lost before chromosomal examination could be made. The crosses were made by Mr. William P. Moore of Hunt, Texas. The hybrid female was then mated to a Barbados ram and produced twins, a male and female, and both had 55 chromosomes, including 5 metacentric and 48 acrocentric autosomes (Figs. 12.13 and 12.14).

An apparent sheep × goat hybrid was born twin to a female goat and both twins were examined cytologically by Eldridge *et al.* (1983). The dam of the twins was a goat owned and bred by Dr. Sarah Pratt, Sedgwick, Kansas. The apparent hybrid (Fig. 12.15)

FIG. 12.12. (Top) Female sheep × goat hybrid.

Photograph by Bunch (1976A), courtesy of Theriogenology.

FIG. 12.13. (Bottom) Female backcross of a sheep × goat hybrid female mated to a Barbados ram.

Photograph by Bunch (1976A), courtesy of Theriogenology.

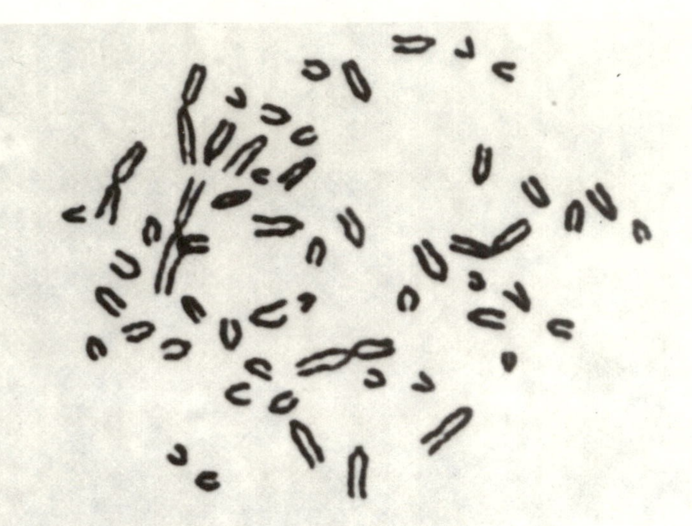

FIG. 12.14. Metaphase and karyotype of the chromosomes from a female backcross of a sheep × goat hybrid female mated to a Barbados ram.
Photograph by Bunch (1976A), courtesy of Theriogenology.

FIG. 12.15. Female sheep × goat hybrid.
Photograph by Leipold.

was found to have 57 chromosomes, including 3 submetacentrics (Fig. 12.16). The goat twin had 60 chromosomes and a normal goat karyotype (Fig. 12.10). The hybrid conceived once from mating to a ram and aborted near term a stillborn fetus with a heart defect. She was bred again, unsuccessfully, but during a pregnancy examination the rectum was perforated and she died from the subsequent infection.

More closely related species have been successfully hybridized. Bunch *et al.* (1977) mated a domestic goat doe to an Auodad (*Ammotragus lervia*) ram and produced an F_1 female hybrid. The hybrid was then mated to a domestic goat buck and produced a single backcross female. The Aoudad has 58 chromosomes, including 1 pair of metacentric autosomes. Both the hybrid and backcross females had 59 chromosomes with 1 metacentric autosome. Dain (1980)

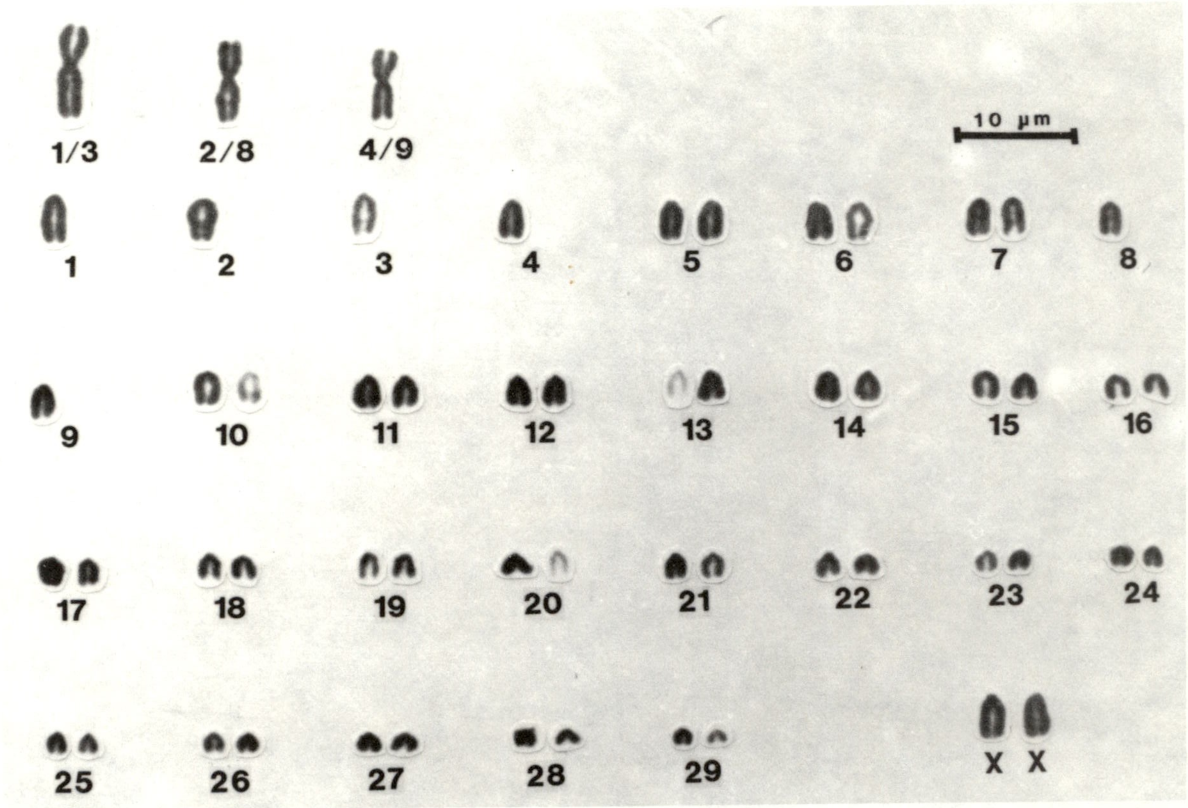

FIG. 12.16. Karyotype of a sheep × goat hybrid female.
Photograph by Eldridge.

made the same cross and produced a pair of stillborn twins. Chromosomes from tissue cultures were somewhat variable, with only a quarter of the cells having 59 chromosomes.

Hauschteck-Jungen and Meili (1967) crossed domestic goats with the ibex (*Capra ibex*) and were successful. The two species appear to be closely related.

Chromosome studies are invaluable in determining the possible production of sheep × goat hybrids. Eldridge and Doane (1979) examined the chromosomes of 7 lambs which appeared at birth to be quite different from usual lambs of Suffolk-type ewes. All 20 animals had 54 chromosomes including 3 pairs of metacentric and submetacentric autosomes. It was clearly established that these were not sheep × goat hybrids. Possibly the sire which had been mated to these ewes, purchased as bred ewes, was a Barbados-type ram.

REFERENCES

AHMED, I. A. 1940. The structure and behaviour of the chromosomes of the sheep during mitosis and meiosis. Proc. Roy. Soc. Edinburgh *60B*, 260–270.

BASRUR, P. K., and STOLTZ, D. R. 1967. The Y chromosome of the goat. J. Hered. *58*, 261–262.

BERRY, R. O. 1938. Comparative studies on the chromosome numbers in sheep, goat, and sheep × goat hybrids. J. Hered. *29*, 343–350.

BERRY, R. O. 1941. The chromosome complex of domestic sheep (*Ovis aries*). J. Hered. *32*, 261–267.

BON DURANT, R. H., McDONALD, M. C., and TROMMERSHAUSEN-BOWLING, A. 1980. Probable freemartinism in a goat. J. Am. Vet. Med. Assoc. *177*, 1024–1025.

BORLAND, R. 1964. The chromosomes of domestic sheep. J. Hered. *55*, 61–64.

BRATANOV, K., DIKOV, V., SOMLEV, B., and EFREMOVA, V. 1980. Chromosome complement and fertility of sheep × goat hybrids. Proc. 4th Eur. Colloq. Cytogenet. Domest. Anim. 262–266.

BRUCE, H. A. 1935. The spermatogenetic history in sheep. Ph.D. Thesis, Univ. of Pittsburgh Library.

BRUERE, A. N. 1967. Evidence of age aneuploidy in the chromosomes of the sheep. Nature (London) *215*, 658–659.

BRUERE, A. N., and CHAPMAN, H. M. 1974. Double translocation heterozygosity and normal fertility in domestic sheep. Cytogenet. Cell Genet. *13*, 342–351.

BRUERE, A. N., and ELLIS, P. M. 1979. Cytogenetics and reproduction of sheep with multiple centric fusions (Robertsonian translocations). J. Reprod. Fertil. *57*, 363–375.

BRUERE, A. N., and KILGOUR, R. 1974. Normal behaviour patterns and libido in chromatin-positive Klinefelter sheep. Vet. Rec. *95*, 437–440.

BRUERE, A. N., MARSHALL, R. B., and WARD, D. P. J. 1969. Testicular hypoplasia and XXY sex chromosome complement in two rams: The ovine counterpart of Klinefelter's syndrome in man. J. Reprod. Fertil. *19,* 103–108.

BRUERE, A. N., ZARTMAN, D. L., and CHAPMAN, H. M. 1974. The significance of the G-bands and C-bands of three different Robertsonian translocations of domestic sheep (*Ovis aries*). Cytogenet. Cell Genet. *13,* 479–488.

BRUERE, A. N., CHAPMAN, H. M., JAINE, P. M., and MORRIS, R. M. 1976. Origin and significance of centric fusions in domestic sheep. J. Hered. *67,* 149–154.

BRUERE, A. N., EVANS, E. P., BURTENSHAW, M. D., and BROWN, B. B. 1978. Centric fusion polymorphism in Romney Marsh sheep of England. J. Hered. *69,* 8–10.

BRUERE, A. N., SCOTT, I. S., and HENDERSON, L. M. 1981. Aneuploid spermatocyte frequency in domestic sheep heterozygous for three Robertsonian translocations. J. Reprod. Fertil. *63,* 61–66.

BUNCH, T. D. 1978. Fundamental karyotype in domestic and wild species of sheep. J. Hered. *69,* 77–80.

BUNCH, T. D., and FOOTE, W. C. 1977. Evolution of the $2n$ = 54 karyotype of domestic sheep (*Ovis aries*). Ann. Genet. Sel. Anim. *9,* 509–515.

BUNCH, T. D., FOOTE, W. C., and SPILLETT, J. J. 1976A. Sheep–goat hybrid karyotypes. Theriogenology *6,* 379–385.

BUNCH, T. D., FOOTE, W. C., and SPILLETT, J. J. 1976B. Translocations of acrocentric chromosomes and their implications in the evolution of sheep. Cytogenet. Cell Genet. *17,* 122–136.

BUNCH, T. D., ROGERS, A., and FOOTE, W. C. 1977. G-band and transferrin analysis of aoudad–goat hybrids. J. Hered. *68,* 210–212.

BUTARIN, N. S. 1935. The chromosome complex of Arkar (*Ovis ploii karelini*), kurdiuchny ram (*Ovis steato pyga*) and their F_1 hybrid. C. R. Acad. Sci. U.R.S.S. N.S. *4,* 287–290.

BUTTLE, H. L., and HANCOCK, J. L. 1966. The chromosomes of the goats, sheep and their hybrids. Res. Vet. Sci. *7,* 230–231.

CATTANACH, B. M., and MOSELEY, H. 1973. Non-disjunction and reduced fertility caused by the tobacco mouse metacentric chromosomes. Cytogenet. Cell Genet. *12,* 264–287.

CRENSHAW, D. B., LASLEY, J. F., and KINTNER, L. D. 1974. In vivo cytogenetic effects of CPA in the ovine. J. Anim. Sci. *39,* 142.

DAIN, A. R. 1971. The incidence of freemartinism in sheep. J. Reprod. Fertil. *24,* 91–97.

DAIN, A. R. 1980. A cytogenetic study of a Barbary sheep (*Ammotragus lervia*) × domestic goat (*Capra hircus*) hybrid. Experientia *36,* 1358–1360.

DUNN, H. O., and ROBERTS, S. J. 1972. Chromosome studies of an ovine acephalic–acardiac monster. Cornell Vet. *62,* 425–431.

ELDRIDGE, F. E. 1980. Chromosomes of *Acardius amorphus* lamb. Proc. 4th Eur. Colloq. Cytogenet. Domest. Anim. 17–22.

ELDRIDGE, F., and DOANE, T. D. 1979. Genetic study shows the basic difference. Farm, Ranch Home Quart., Fall, Univ. Nebr. *26,* 26–28.

ELDRIDGE, F., LEIPOLD, W. H., and HARRIS N. 1983. Sheep × goat hybrid: An additional case. J. Dairy Sci. *66* (Suppl. 1), 253.

EVANS, H. J., BUCKLAND, R. A., and SUMNER, A. T. 1973. Chromosome homology and heterochromatin in goat, sheep and ox studied by banding techniques. Chromosoma *42,* 383–402.

FORD, C. E., POLLOCK, D. L., and GUSTAVSSON, I. 1980. Proceedings of the first International Conference for the Standardisation of Banded Karyotypes of Domestic Animals. Univ. of Reading. Hereditas *92,* 145–162.

GLAHN-LUFT, B., and WASSMUTH, R. 1977. C-banding in sheep with balanced homolog interchromosomal translocation. Ann. Genet. Sel. Anim. *9,* 540.

GLAHN-LUFT, B., and WASSMUTH, R. 1980. Effect of a 1/20 translocation on the reproductive performance in sheep. Proc. 31st Ann. Mtg. Eur. Assoc. Anim. Prod. 1–2.

HAMERTON, J. L., DICKSON, J. M., POLLARD, C. E., GRIEVES, S. A., and SHORT, R. V. 1969. Genetic intersexuality in goats. J. Reprod. Fertil. Suppl. *7,* 25–51.

HANCOCK, J. L., and JACOBS, P. A. 1966. The chromosomes of goat × sheep hybrids. J. Reprod. Fertil. *12,* 591–592.

HANCOCK, J. L., and McGOVERN, P. T. 1968. Transfer of goat × sheep hybrid eggs to sheep and reciprocal transfer of eggs between sheep and goats. Res. Vet. Sci. *9,* 411–415.

HAUSCHTECK-JUNGEN, E., and MEILI, R. 1967. Chromosome study of the wild (*Capra ibex*) and domestic (*Capra hircus*) goat. Chromosoma *21,* 198–210.

HULOT, F. 1969. A new case of centric fusion in the domestic goat (*Capra hircus* L.). Ann. Genet. Sel. Anim. *1,* 175–176.

ILBERRY, P. L. T., ALEXANDER, G., and WILLIAMS, D. 1967. The chromosomes of sheep × goat hybrids. Aust. J. Biol. Sci. *20,* 1245–1247.

KHAVARY, H. 1974. The normal karyotype of the Zel breed of sheep. 1st World Cong. Genet. Appl. Livestock Prod., Madrid *3,* 219–222.

KOROBITSYNA, K. V., NADLER, C. F., VORONTSOV, N. N., and HOFFMAN, R. S. 1974. Chromosomes of the Siberian snow sheep, *Ovis nivicola,* and implications concerning the origin of amphiberingian wild sheep (subgenus *Pachyceros*). Quart. Res. (NY) *4,* 235–245.

KRALLINGER, H. F. 1931. Cytological studies in domestic animals. Wiss. Arch. Landwirtsch. Abt. B *5,* 127–187.

MAKINO, S. 1943. The chromosome complexes in goat (*Capra hircus*) and sheep (*Ovis aries*) and their relationship: Chromosome studies in domestic mammals, II. Cytologia *13,* 39–54.

MAKINO, S., SHIMBA, H., SOFUNI T., and IKEUCHI, T. 1967. A revised study of the chromosomes in the goat and the sheep. Proc. Jpn. Acad. Sci. *43,* 913–917.

McGOVERN, P. T. 1973. The effect of maternal immunity on the survival of goat × sheep hybrid embryos. J. Reprod. Fertil. *34,* 215–220.

MORAES, J. C. F., MATTEVI, M. S., and FERREIRA, J. M. M. 1980. Chromosome studies in Brazilian rams. Vet Rec. *107,* 489–490.

NADLER, C. F., and BUNCH, T. D. 1977. G-band patterns of the Siberian snow sheep (*Ovis nivicola*) and their relationship to chromosomal evolution in sheep. Cytogenet. Cell Genet. *19,* 108–117.

NOVIKOV, J. I. 1935. Chromosomes in the spermatogenesis of interspecific hybrids of the European Mouflon and the domestic sheep (Merino). C. R. Acad. Sci. U.R.S.S. N.S. *4,* 93–94.

PADEH, B., WYSOKI, M., AYALON, N., and SOLLER, B. 1965. An XX/XY hermaphrodite in the goat. Isr. J. Med. Sci. *1,* 1008–1012.

PADEH, B., WYSOKI, M., and SOLLER, B. 1971. Further studies on a Robertsonian translocation in the Saanen goat. Cytogenetics *10,* 61–69.

PCHAKADZE, G. M. 1936. A new data about the chromosome number in domestic sheep. C. R. Acad. Sci. U.R.S.S. N.S. *3,* 333–334.

POPESCU, C. P. 1972. The mode of transmission of a centric fusion to the offspring of a buck (*Capra hircus* L.). Ann. Genet. Sel. Anim. *4,* 355–361.

ROBERTS, J. A. F., and GREENWOOD, A. W. 1928. An extreme freemartin and freemartin-like condition in the sheep. J. Anat. *63,* 87–94.

SHIWAGO, P. I. 1931. Karyotype studies on ungulates. 1. The chromosome complex of sheep and goats. Z. Zellforsch. Mikroskop. Anat. *13,* 511–522.

SLEE, J. 1963. Immunological tolerance between litter-mates in sheep. Nature (London) *200,* 654–656.

SMITH M. C. and DUNN, H. O. 1981. Freemartin condition in a goat (clinical report). J. Am. Vet. Med. Assoc. *178,* 735–737.

SOHRAB, M., McGOVERN, P. T., and HANCOCK, J. L. 1973. Two anomalies of the goat karyotype. Res. Vet. Sci. *15,* 77–81.

SOKOLOV, I. 1930. The chromosomes in spermatogenesis of the goat (*Capra hircus*). Bull. Bur. Genet. Leningrad Akad. Nauk SSSR *8,* 63–76.

SOLLER, M., PADEH, B., WYSOKI, M., and AYALON, N. 1969. Cytogenetics of Saanen goats showing abnormal development of the reproductive tract associated with the dominant gene for polledness. Cytogenetics *8,* 51–67.

STORMONT, C., WEIR, W. C., and LANE, I. L. 1953. Erythrocyte mosaicism in a pair of sheep twins. Science *118,* 695–696.

VALDEZ, R., NADLER, C. F., and BUNCH, T. D. 1978. Evolution of wild sheep in Iran. Evolution *32,* 56–72.

WILKES, P. R., MUNRO, I. B., and WIJERATNE, W. V. S. 1978. Studies on a sheep freemartin. Vet. Rec. *102,* 140–142.

WODSEDALEK, J. R. 1922. Studies on the cells of sheep with reference to spermatogenesis, oogenesis, and sex determination. Anat. Rec. *23,* 103.

ZARTMAN, D. L., and BRUERE, A. N. 1974. Giemsa banding of the chromosomes of the domestic sheep (*Ovis aries*). Can. J. Genet. Cytol. *16,* 555–564.

ZHAPBASOV, R., and BAIKENOVA, S. D. 1978. Some results of cytogenetic investigation of lambs with inborn anomalies of development and study of chromosome replication in sheep. 14th Int. Congr. Genet. Moscow, Pt. 1, Sect. 1–12, 284.

13

Swine Chromosomes

ESTABLISHMENT OF CHROMOSOME NUMBER

In 1913 Wodsedalek made the first recorded investigation of pig chromosomes, reporting a diploid number of 18 in males and 20 in females. The next report on chromosomes of the pig was in 1917 when Hance, studying several tissues, arrived at 40 as the diploid number. He was quite critical of the quality of preparations used by Wodsedalek. The first report giving the diploid number as 38, the number known now to be correct, was by Krallinger in 1931. He was a careful observer who first established the correct number as 60 in cattle also. In spite of this report, and agreement by Bryden (1933), Hillebrand (1936), and Crew and Koller (1939), Makino (1944) published a report in which he stated that the diploid number was 40 "without any exception and with no obscurity." Subsequently Sachs (1954), Spalding and Berry (1956), and Aparicio (1960) reported 40 chromosomes. The number 10 autosome (1976 Reading Conference) in pigs is quite narrow at the centromeric region and in many metaphases almost appears to be two acrocentric chromosomes closely associated at the centromeres. This chromosome probably was the one that led to the confusion. In 1962 Makino *et al.*, in a footnote, acknowledged that the correct number was 38. With the use of bone marrow cells for studying chromosomes, Gimenez-Martin *et al.* (1962) clearly showed the diploid number to be 38 in domestic swine. Subsequent work with cultured blood lymphocytes and other cultured tissues has confirmed the diploid number of 38 for domestic swine. Stone (1963) and McConnell *et al.* (1963) apparently were the first to report chromosomes of the pig from lymphocyte cultures, but

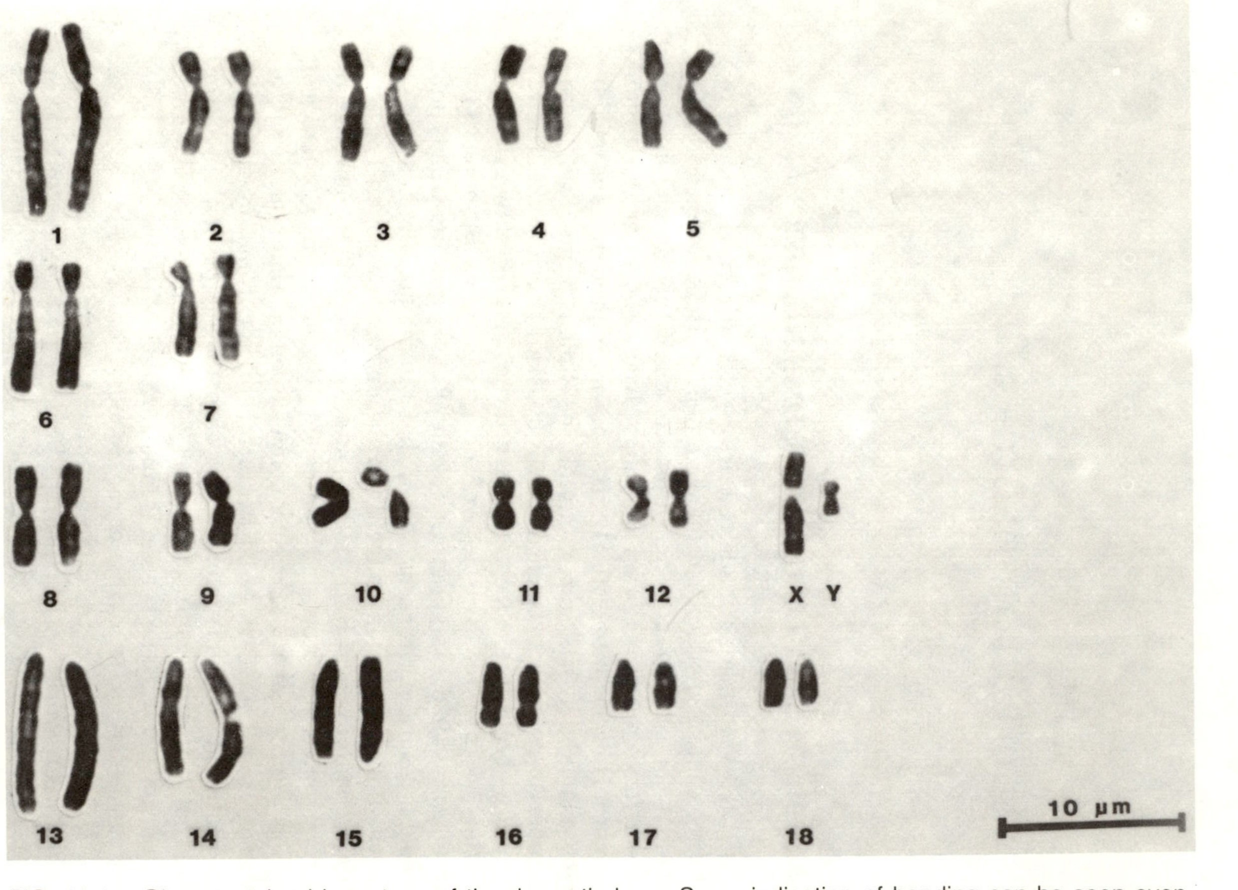

FIG. 13.1. Giemsa-stained karyotype of the domestic boar. Some indication of banding can be seen even though the slides were not treated to produce bands.

Photograph by Eldridge.

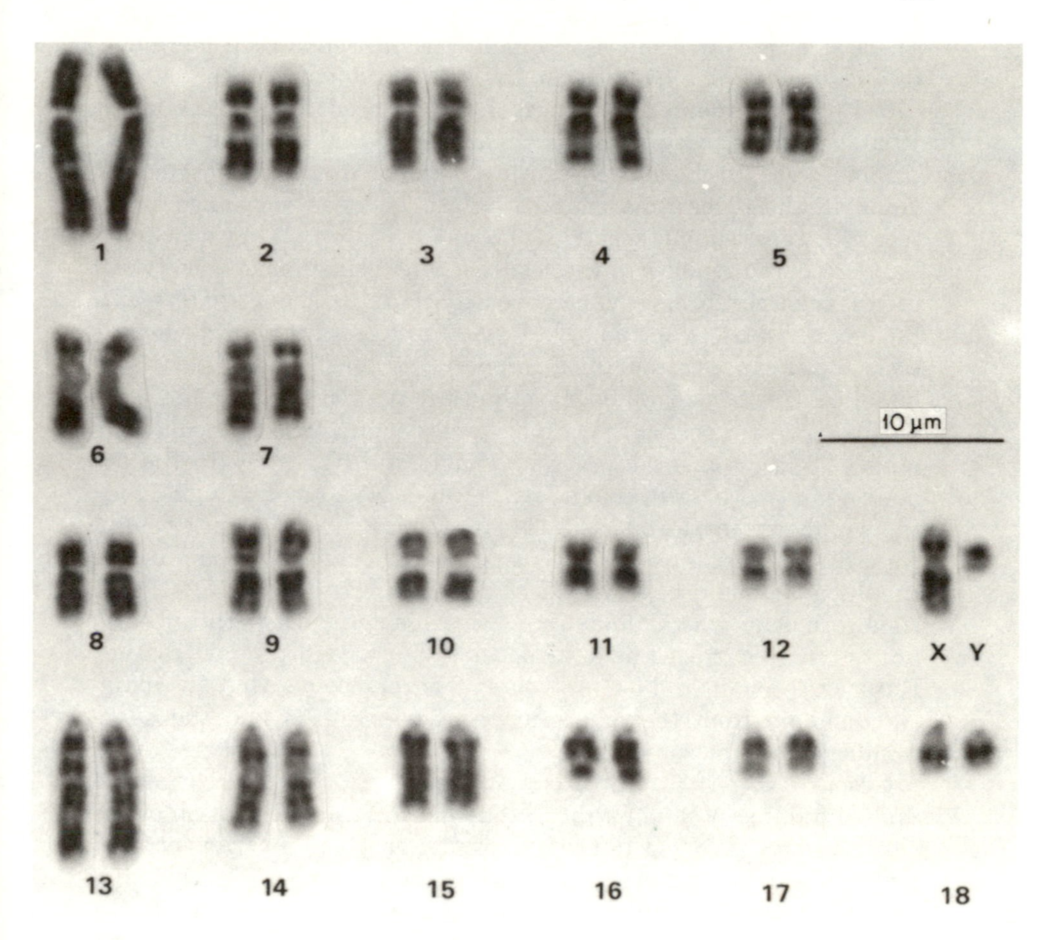

FIG. 13.2. G-banded karyotype of a domestic boar.
Photograph by C. C. Lin, courtesy of Hereditas 92, 145–162 (1980).

the paper by Stone was published about two months earlier (see Figs. 13.1 and 13.2).

WILD SWINE

The earliest reference to chromosome studies of European wild pigs (*Sus scrofa*) was by McFee *et al.* (1966A) and Rary *et al.* (1968).

There were pigs in Tennessee which were descendants from importations of wild pigs from Germany in 1912. Some of these had escaped into the mountains where they had bred freely since about 1920. Seventy-three percent of them had 36 and the remainder 37 chromosomes. In the karyotype the submetacentric chromosome not found in the domestic pig fit in length between the longest and second-longest submetacentrics of the domestic pig. They concluded that this chromosome was the result of centric fusion between two of the acrocentric chromosomes normally found in domestic swine, a Robertsonian translocation. They also pointed out that if the domestic swine originated from the European wild boar, as has been postulated, central fission would have had to occur in the process.

The earliest report of the chromosomes of one Japanese wild boar female (*Sus vittatus leucomystax* Major) and hybrids with the domestic pig was by Muramoto *et al.* (1965). No source was given for the wild boar female, but it is assumed that it was of Asiatic origin. According to Grzimek (1972), *Sus scrofa leucomystax* Temminck is the Japanese wild boar. *Sus scrofa vitattus* Boie is the Southeast Asia wild boar, which at one time was considered a separate species but is now recognized as a subspecies that interbreeds with other European–Asian wild boars. It had 38 chromosomes that are indistinguishable from the domestic pig. The hybrids had the same number of chromosomes.

Gropp *et al.* (1969) and Rittmannsperger (1971) confirmed the diploid number of 36 for wild European swine. Identification of the chromosomes involved in the apparent Robertsonian translocation which led to the difference in chromosome numbers in wild and domestic pigs was first reported by McFee and Banner (1969). In 1973 Gustavsson *et al.*, using a wild boar which had descended from animals imported earlier into Sweden from Finland (originally from Holland and Belgium), identified the Robertsonian translocation by banding. He concluded that chromosomes 15 and 17 were involved (Figs. 13.3 and 13.4). Comparing his illustrations with those published by McFee and Banner, they appear to be the same, even though the two authors used different schemes for preparing karyotypes.

Tikhonov and Troshina (1974, 1975, 1980A,B) made several reports on the chromosomes of both the European and Asian wild boars in the USSR. They studied four subspecies of the wild European–Asiatic boar *Sus scrofa nigripes, Sus scrofa ussuricus, Sus scrofa attila,* and *Sus scrofa ferus* and their hybrids with domestic swine. In the Far Eastern subspecies, *Sus scrofa attila,* no animals were found with 36 chromosomes. This lends additional support to the hypothesis that domestic swine descended from Asiatic wild swine. Tikhonov *et al.* (1974), through immunogenetic studies, also supported the origin of domestic pigs from Asian wild boars. They used a method for karyotyping in which the chromosomes were divided into four groups: A, large submetacentrics; B, acrocentrics; C, medium-sized submetacentrics; and D, metacentrics. By G-banding they determined that chromosome B5 (number 17 by the Reading Conference) had formed a Robertsonian translocation with B4 (Reading number 16) in one subspecies and with B3 (Reading number 15) in another.

Animals with the two different translocations have been crossed and have produced viable offspring. Matings of the F_1 hybrid have not been reported, but theoretically these F_1 animals would produce six different gametes (not necessarily in equal numbers) if two centromeres went to each pole during meosis.

15 + 16/17	balanced
15 + 16	deficient
15 + 15/17	duplication
16 + 15/17	balanced
15/17 + 16/17	duplication
16 + 16/17	duplication

Three different zygotes with balanced chromosome content could result from a union of such gametes: (15, 15, 16/17, 16/17), (16, 16, 15/17, 15/17), and (15, 16, 15/17, 16/17). All other combinations would have deficiencies or duplications. Further matings are being made to test this theory.

In the meantime, crosses have been made to incorporate the two different translocations into domestic species, and the heterozygosity which has resulted has been found to increase productivity (Tikhonov and Troshina 1980A,B).

BANDING AND KARYOTYPING

Gustavsson *et al.* (1971) did the first banding of pig chromosomes, identifying each individual chromosome by the quinicrine mustard fluorescence (Q-band) technique. They organized the chromosomes

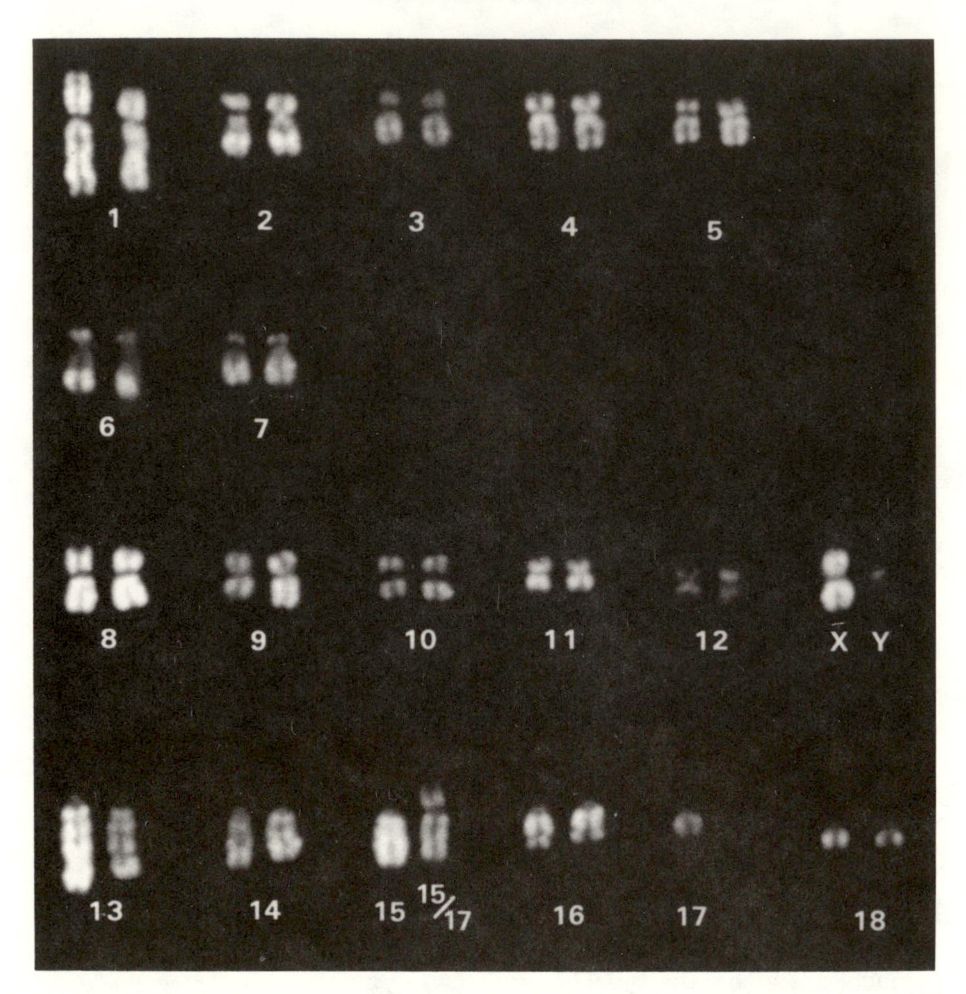

FIG. 13.3. Q-banded karyotype of a hybrid wild swine × domestic swine boar. The chromosome, in heterozygous condition, typical of the wild swine, was 15/17.

Photograph by Gustavsson, courtesy of Hereditas 75, 153–155(1973).

into metacentric–submetacentric group by size, followed by the acrocentrics by size, which has become the more common method for illustrating livestock chromosomes in karyotypes. In the following year Hansen (1972) published another karyotype by Q-banding and identified a chromosome different from Gustavsson's as the X chromosome. He also arranged the karyotype in a different manner. Berger (1972) used 8 *M* urea to band the pig chromosomes. Several other reports were published using the trypsin method from Seabright (1971) to produce G-bands. In 1977 Hansen published a very complete report using Q-, G-, and R-bands, presenting the karyotype in the order agreed upon at the Reading Conference.

Since the X chromosome of the pig is similar in size and form to

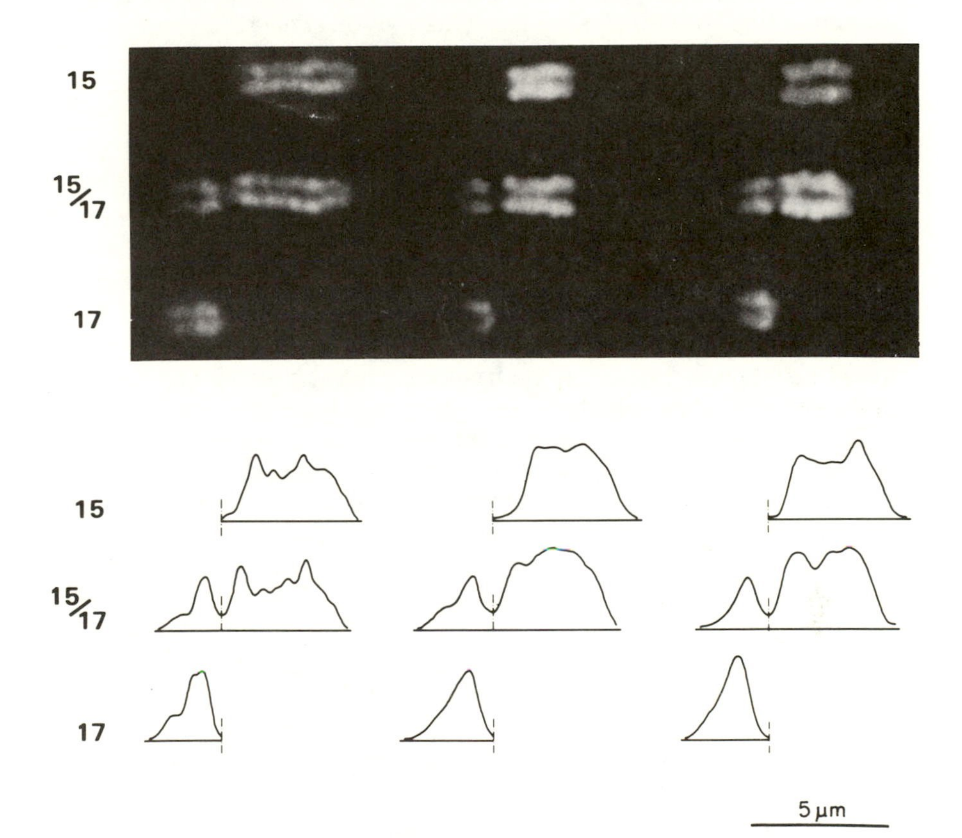

FIG. 13.4. Q-banded chromosomes 15, 17, and 15/17 from three cells and photoelectric recordings. These clearly illustrate the conclusion that the chromosome which distinguishes wild from domestic swine is either a centric fusion of 15/17 in the wild swine or a centric fission of this chromosome in domestic swine.

Photograph by Gustavsson, courtesy of Hereditas 75, 153–155(1973).

some other submetacentric chromosomes, it has been rather difficult to identify. By tritiated thymidine and late labeling, Cornefert-Jensen *et al.* (1968) clearly identified it. In the male only the Y chromosome was found to be late-labeled. Sysa (1977) used BUdR-treated cultures to establish the X chromosome as the one with a fluorescing R-band at the end of the p arm.

C-banding has shown (Christensen and Smedegard 1979) that some polymorphisms exist in pig chromosomes, especially on chromosome 15 by the Gustavsson (see Gustavsson *et al.* 1972) system of karyotyping.

SWINE INTERSEXES

Intersexes in swine are not numerous, but are not rare. Breeuwsma (1969) estimated that in the Netherlands the frequency in the total pig population was .4%, but that this was apparently greater than in Sweden and Germany. Finding a statistical difference among progeny of different boars led to the conclusion that hereditary factors influence the frequency of intersexes. The degree of expression varies widely.

The frequency of occurrence of intersexes has resulted in numerous reports on the chromosomes of intersex swine. Apparently the first was by Makino *et al.* (1962) on a Yorkshire that appeared masculine in face and skeleton with an enlarged clitoris that had penile characteristics. From 40 cells, 36 had the normal diploid 38,XX number. Some concern was expressed about the male characteristics in the absence of a Y chromosome.

Previous to the above report, Cantwell *et al.* (1958) had demonstrated that there was a sex chromatin body in the intersex cells at about the same frequency as in females and much higher than in males. They pointed out the great variation in degree of intersexuality found in swine.

For a review of early studies on the anatomical characteristics of intersex swine, see Hughes (1929). Four pairs of heterosexual twins which were monochorionic were found to have modified gonads, especially in the females. Placental fusion is rare in swine.

The first intersex with other than 38,XX chromosomes was reported by McFee *et al.* (1966B). It had 90% XX and 10% XY cells, a mosaic condition similar to freemartin cattle. Anatomically it was an underdeveloped male. Bruere *et al.* (1968) found a similar case with 42 XY and 4 XX cells and over 80% of the nerve cells with sex chromatin bodies.

Breeuwsma (1968) reported an intersex Dutch Landrace pig with an XXY chromosome constitution. Harvey (1968) found an animal reported as a male with a similar chimerism, but found 6% with two sex chromatin bodies and therefore concluded that it was a mosaic XXY/XXXY.

Lojda (1975) found a sow which when mated to four different boars, produced in each litter, on the average, about half normal and half intersex pigs. The intersex pigs are listed in Table 13.1. It was concluded that the condition was inherited either as a dominant gene or a recessive sex linked gene, similar in effect to the testicular feminization syndrome (tfs) found in humans. The variety of chromosomal conditions were difficult to explain.

A very good review of intersex swine is found in the 1973 paper of Miyake in which he reported 8 more cases of intersexes with variable morphological characteristics. From 30 karyotypes per animal he concluded that all 8 were genetically females.

It is interesting to note that in true hermaphrodites in humans, persons with both ovarian and testicular tissue, Jones *et al.* (1965) found that, of 29 cases in the literature where chromosomes were examined, 24 were chromosomally female, 46,XX. Among the other 5 cases, one was XY and all others were chimeric in various combinations of XO,XY, XX/XY, XX/XXY/XXYYY, and XX–XY.

Intersex swine have been reported to be preponderantly chromosomal females (42 out of 46 in Table 13.1). Three of the 46 were chimeric, which indicates a high probability that placental fusion similar to that found in freemartin cattle and some other species may also cause a small number of intersexes in swine. The predominance of chromosomally female karyotypes among the intersex swine gives some support to the hypothesis that intersexes in swine usually may be the result of some gene causing intersexuality, as proposed by Okamoto (1978). He assembled information on 80 intersex swine, with no reference to cytological findings, and from pedigree analysis suggested a recessive gene affecting only females. His statistical analysis did not give strong support to this concept, but some genetic factors other than gonosomal balance or imbalance appear to be possible.

It is generally recognized that maleness in mammals is caused by

TABLE 13.1. Several Reports on Intersex Swine

Number of animals	Chromosomal description	Reference
2	38,XX	Basrur and Kanagawa (1971)
1	38,XX/38,XY; 80% XY	
1	39,XXY	Breeuwsma (1968)
1	38,XX/38,XY; 91% 38,XY	Bruere *et al.* (1968)
7	Sex chromatin studied, concluded all female	Cantwell *et al.* (1958)
1	All 38,XX	Hard and Eisen (1965)
4	38,XY	Lojda (1975)
7	38,XX/38,XY	
1	38,XX	
1	38,XX/38,XY/37,XO	
15	Intersex, but cytological results not given	
1	Described as also having a reciprocal translocation, t(6pt;14q−)	Madan *et al.* (1978)
1	38,XX	Makino *et al.* (1962)
1	37,XX including a Robertsonian translocation 13/17	Masuda *et al.* (1975)
1	38,XX/38,XY; 90% XX	McFee *et al.* (1966B)
2	38,XX	McFeely *et al.* (1967)
7	38,XX	Melander *et al.* (1971)
8	38,XX	Miyake (1973)
12	38,XX[a]	Okamoto and Masuda (1977)
1	38,XX	Vogt (1966)

Totals: 43 38,XX including 1 with t(6pt;14q−) and 1 37,XX with the 13/17 translocation; 10 38,XX/38,XY chimeric; 1 39,XXY; 1 38,XX/38,XY/37,XO.
[a] One of the 12 is the one reported by Masuda *et al.* (1975).

the Y chromosome. Since the preponderance of intersexes in pigs is found in animals with female karyotypes, some deviation from this concept must be occurring in this species. The variation in the degree of expression of intersexuality in pigs leads to the hypothesis that there may be several genes or chromosomal conditions that influence maleness in swine. Chimerism in a few cases is one obvious cause. One or more genes located on either the autosomes or the X chromosome may influence sex expression. Further cytogenetic studies on swine intersexes should help clarify our knowledge about the cause of intersexes in swine and may provide information on the general topic of sex determination.

EMBRYOLOGICAL STUDIES

Only a few studies have been made on the chromosomes of early embryos of swine. Apparently the first was by McFeely (1967); it was based upon 88 blastocysts from 7 gilts assumed to be not closely related, bred to 2 unrelated Yorkshire boars. He found 10% of the blastocysts to have chromosomal defects. Apparently one abnormal blastocyst was found from each of the gilts. The results were not presented in great detail. The types of abnormalities found were triploid XXX; tetraploid XXYY with one submetacentric chromosome missing; one triploid XYY; one tetraploid XXYY labeled as degenerating; one triploid XXY; and one containing a combination of defects including triploid XXX, diploid XX/triploid XXX, and tetraploid XXXX.

Moon *et al.* (1975) found 4 out of 15 blastocysts apparently abnormal. Since 3 of the 4 were mixoploid ($2n/4n$, $2n/6n$, and $2n/32n$), and he did not report the number of cells of each type, some questions might be raised concerning interpretation. Polyploid cells are found in almost all preparations in cattle. They have been found in meiotic preparations of testicular material and regularly in lymphocyte and primary cell cultures, so the occurrence of a small number in a swine blastocyst culture would not necessarily indicate abnormality.

In a report on 169 blastocysts by Dolch and Chrisman (1981), only diploid cells were found. However, the number of usable blastocysts related to the number of corpora lutea was considerably lower than in McFeely's study. Both were on 10-day-old blastocysts. Further work is needed on normal pigs of different breeds, with lymphocyte cultures on parents, to establish the most frequent pattern of chromosomal aberrations in swine.

Smith and Marlowe (1971), in 25-day-old embryos, found 1 out of 76 with approximately 50% of the cells lacking a number 16 chromosome. It can be assumed that other chromosomal aberrations may have occurred at fertilization but by 25 days had degenerated.

Bouters *et al* (1974) reported a study on the chromosomes of embryos from boars with fertility problems. Some deviations from normal were found.

An excellent study has been made by Popescu and Boscher (1982) on 9- and 10-day-old embryos from matings of 4 translocation, 38,XX or XY,t($4q^+$;$14q^-$), males or females to 4 normal males or females. Twenty-seven embryos were karyotyped out of 47 recovered. Eleven, or 41%, of the 27 karyotyped embryos were unbalanced, 5 had duplications, and 6 were deficient. Sixteen of the 27 were balanced. Presumably the embryos with unbalanced karyotypes would have died as embryos, since no pigs born alive have been found to have unbalanced karyotypes. These data on balanced and unbalanced karyotypes related closely to the reduction in fertility of 43% when the size of litters of the original boar was compared with 95 contemporary litters sired by other boars. The reduction in fertility for this translocation was directly related to the frequency of nondisjunction which resulted in chromosomally unbalanced gametes.

ENVIRONMENTALLY INDUCED CHROMOSOMAL ABERRATIONS

Several treatments and chemical agents have been administered to swine, followed by chromosomal studies. Some of these treatments resulted in chromosomal aberrations; some had little effect.

The effects of irradiation on chromosomes of swine have been studied by two different methods: (a) irradiation of the semen prior to breeding and (b) irradiation of the testicles of the boar followed by natural mating after recovery, a period of 5 to 6 months (Zartman *et al.* 1969; Zartman 1971). The effects of irradiation of the testicles were examined by culturing cells from stillborn pigs. No consistent modifications were found in each of the 108 successful cultures, but pericentric inversions and sister-chromatid fusions occurred with high enough frequency to suggest a delayed effect of irradiation. The irradiation of the semen, however, resulted in four chromosomal aberrations, which were consistently found in all cells of each animal. One was a reciprocal translocation between the short arm of C1 and the long arm of B2. Another was a centric fusion between one of the A2–7 group and D2–3. A third was between A1 and one of the smallest acrocentrics, D5–6. The fourth was an apparent pericentric inversion in the smallest of the A2–7 group. All four of the aberrations occurred in Hampshires whose sire's testicles had been irradiated. A mosaicism was also found in three other animals from females which were from boars with irradiated testicles.

Fries *et al.* (1983) reported that, using irradiated semen, they were able to produce eight different translocations and two inversions in swine. Using these and the Robertsonian translocation 15/17, they were able to identify chromosome 15 for the locus of G-blood groups.

Treatment of swine with chemical agents or vaccines increased the frequency of chromosomal aberrations. These included live vaccine against swine fever (Lojda and Rubes 1977; Manolache *et al.* 1978), aflatoxin B_1 (Lojda and Petrickova 1977; Petrickova *et al.* 1976) in live animals. Hog cholera virus was also found to cause chromosomal aberrations in a cultured pig kidney cell line (Pirtle 1966).

These effects lead to the obvious conclusion that when attempting to establish chromosomal patterns for apparently normal animals, or for phenotypically aberrant animals, careful attention must be given to any treatments or exposure of the animals to environmental agents.

Dolch and Chrisman (1981) treated prepubertal gilts with gonadotropins and studied the chromosomes of the preimplantation blastocysts. In 169 blastocysts, they found only diploid cells, looking at 20 cells per animal. Apparently the gonadotropic treatment had no deleterious effect upon the chromosomes. However, they recovered at 10 days after breeding slightly less than half as many blastocysts as there were corpora lutea. This loss of blastocysts might have covered some effects of the treatment.

CHROMOSOMAL ABERRATIONS OCCURRING NATURALLY

Thirteen reciprocal translocations and one Robertsonian translocation, different from the ones found in wild swine, have been reported (Table 13.2). Popescu (1982) described the effects of these translocations on performance. Six of these translocations were reported in 1983 from boars that had been identified by siring litters of fewer pigs than normal. Nine of the 13 reciprocal translocations were associated with reduced litter size, which identifies this type of chromosomal aberration as an important cause of lowered fertility in swine. Chromosome 1 was involved in five of the translocations, which may indicate that it has more potential breakage points; the

TABLE 13.2. Naturally Occurring Autosomal Aberrations in Swine

Description of aberration[a]	Other information	Reference
t rcp(1p−;6q+)	Reduced litter size	Locniskar *et al.* (1976)
t rcp(1p−;16p+)		Forster *et al.* (1981)
t rcp(1p+;14q−)	Reduced litter size	Gustavsson and Settergren (1983)
t rcp(1q−;17q+)	Reduced litter size	Gustavsson and Settergren (1983)
t rcp(1p−;8q+)	Reduced litter size	Gustavsson and Settergren (1983)
t rcp(4q+;14q−)	Reduced litter size	Popescu and Legault (1979)
t rcp(6p+;14q−)	Intersex	Madan *et al.* (1978)
t rcp(6p+;15q−)		Bouters *et al.* (1974)
t rcp(7q−;11q+)	Reduced litter size	Gustavsson and Settergren (1983)
t rcp(7q−;15q+)	Reduced litter size	Popescu (1983)
t rcp(9p+;11q−)	Reduced litter size	Gustavsson and Settergren (1983)
t rcp(11p+;15q−)		Hageltorn *et al.* (1973) Henricson and Backstrom (1964)
t rcp(13q−;14q+)	Reduced litter size	Hageltorn (1976)
rob(13/17)	Normal phenotype	Miyake *et al.* (1977)
rob(13/17)	(2) Intersex	Masuda *et al.* (1975)

[a] The numbers are not strictly comparable because different standard karyotype references were used by some.

translocations may also simply be related to the longer length of that chromosome. Since some of the translocations had the chromosomes identified from standards other than the one established by the Reading Conference, the identification of the chromosomes may not be strictly comparable. Two cases have chromosomes near enough the same relative size, t rcp(6p+;14q−) and t rcp(6p+;15q−), however, to lead to the suggestion that these two might be identical. In two cases (Hageltorn 1976; Popescu and Legault 1979), transmission of the translocation chromosome, t($4q^+$;$14q^-$), to offspring was indicated, and approximately 50% received the aberrant chromosome (Fig. 13.5).

Madan *et al.* (1978) found the reciprocal translocation t($6p^+$;$14q^-$) in an intersex pig. The fact that the pig was an intersex was considered to be fortuitous. No relatives or offspring were studied so that transmission was unknown (Figs. 13.6 and 13.7, pp. 236, 237).

The Robertsonian translocation 13/17 has been found twice in Japan, once in four animals of one litter, in which one animal had deformed nostrils (Miyake *et al.* 1977), and once in an intersex pig (Masuda *et al.* 1975). In neither case was the translocation considered to be the cause of the phenotypic abnormality.

Polymorphism occurs in C-banding of the number 15 chromosome (Christensen and Smedegard 1979). The chromosome with the large C-band was transmitted in a 1 : 1 ratio from a heterozygous boar. Sysa (1978) also found variability in the form and type of chromatin in the centromeric region of chromosomes 13 through 18 and variability in secondary constrictions, NORs, and N-bands (an early banding method for NORs).

CHROMOSOMES OF PHENOTYPICALLY ABNORMAL SWINE

At least four different types of pigs with phenotypic abnormalities have been studied cytologically. Vogt (1967) studied two pigs, both females, affected with atresia ani. Neither of the animals had any deviation from the normal female karyotype.

One porcine cyclops pig (Arakaki and Vogt 1976) was found to have a normal female karyotype. In humans, several chromosomal defects have been found to be associated with similar abnormalities.

Hansen-Melander and Melander (1972) found a female which lacked a vulva, although at slaughter the remaining portions of the genital tract were found to be normal. It had a normal female karyotype, 38,XX.

Miyake *et al.* (1977) found a 13/17 Robertsonian translocation in a piglet whose nostrils opened directly into the oral cavity. Three more pigs in the same litter, two males and one female, were phenotypically normal but had the same chromosomal aberration in their lymphocytes. It was concluded that the phenotypic abnormality was not the result of the translocation.

Since so many phenotypic abnormalities in humans are associated with chromosomal aberrations, it is surprising that so few cases have been found in livestock. Whether this is due to the selection for physical fitness, productivity, and fertility in livestock, as compared with the relatively random mating of humans, or to perhaps a greater human species sensitivity to correlated chromosomal-phenotypic responses is not yet known.

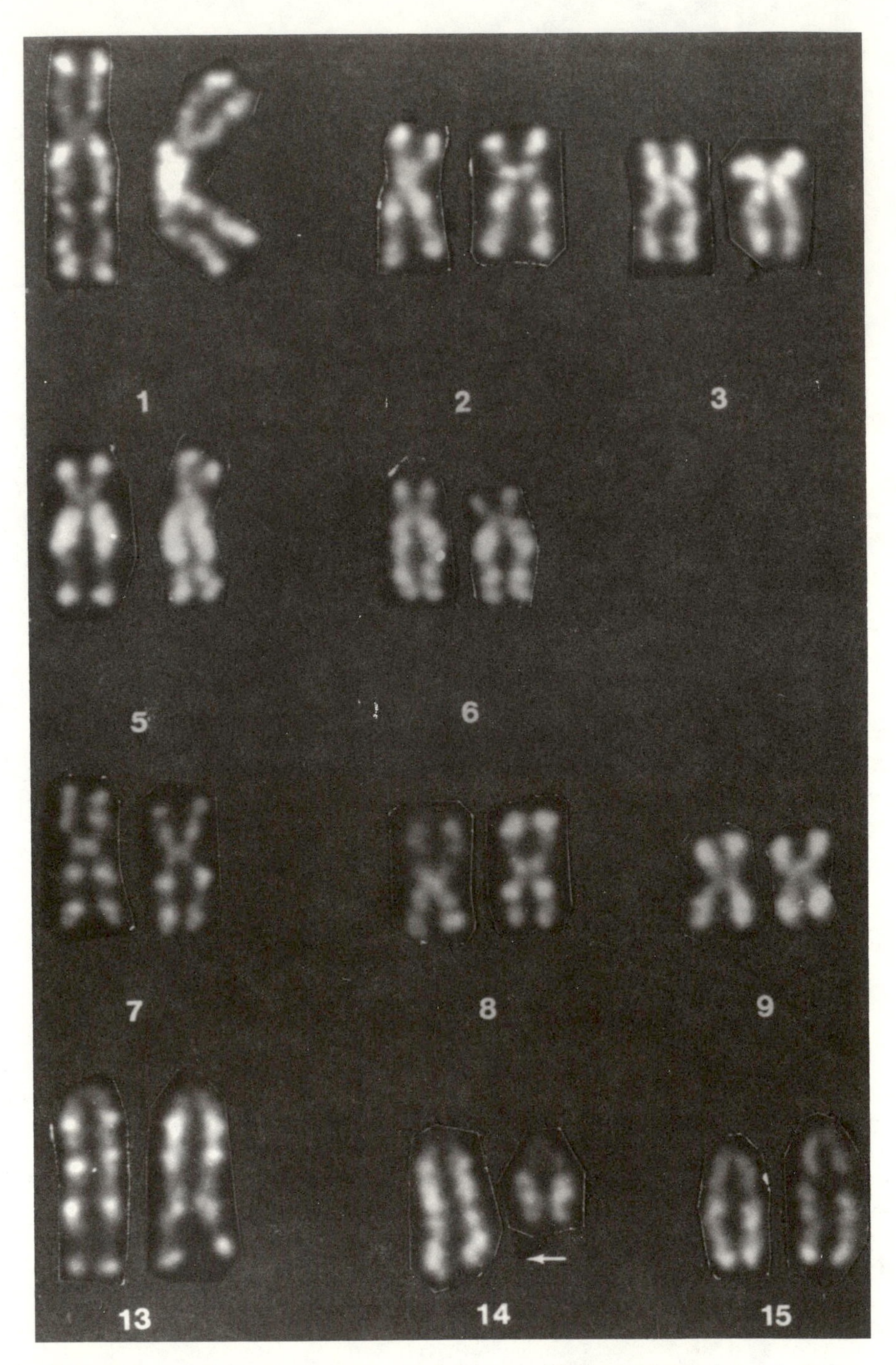

FIG. 13.5. R-banded karyotype from a boar with the reciprocal translocation 38,XY,t(4q^{+}; 14q^{-}). Chromosomes are located over the arrows.

Photograph by Popescu.

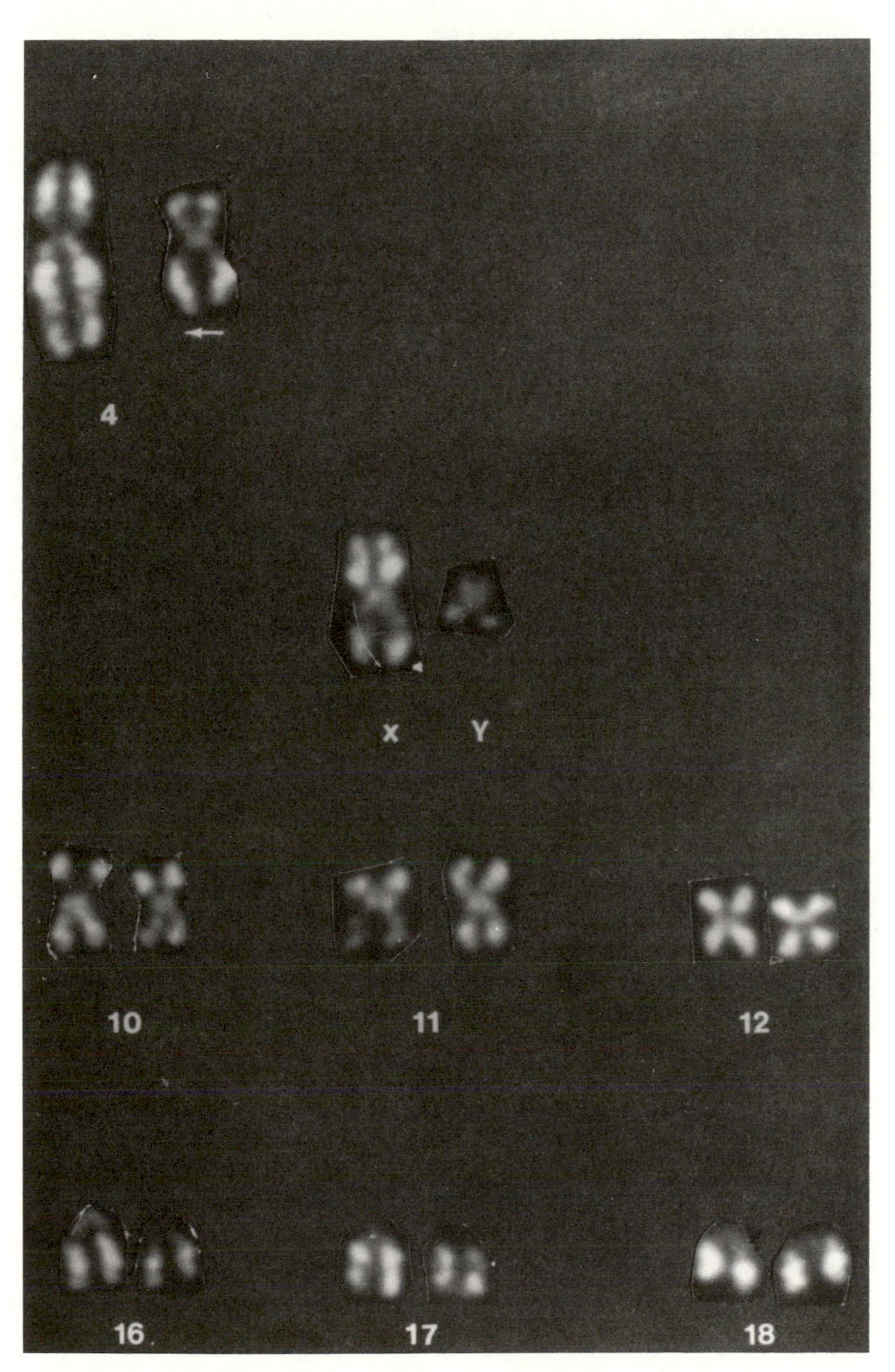

FIG. 13.5. (*Continued*)

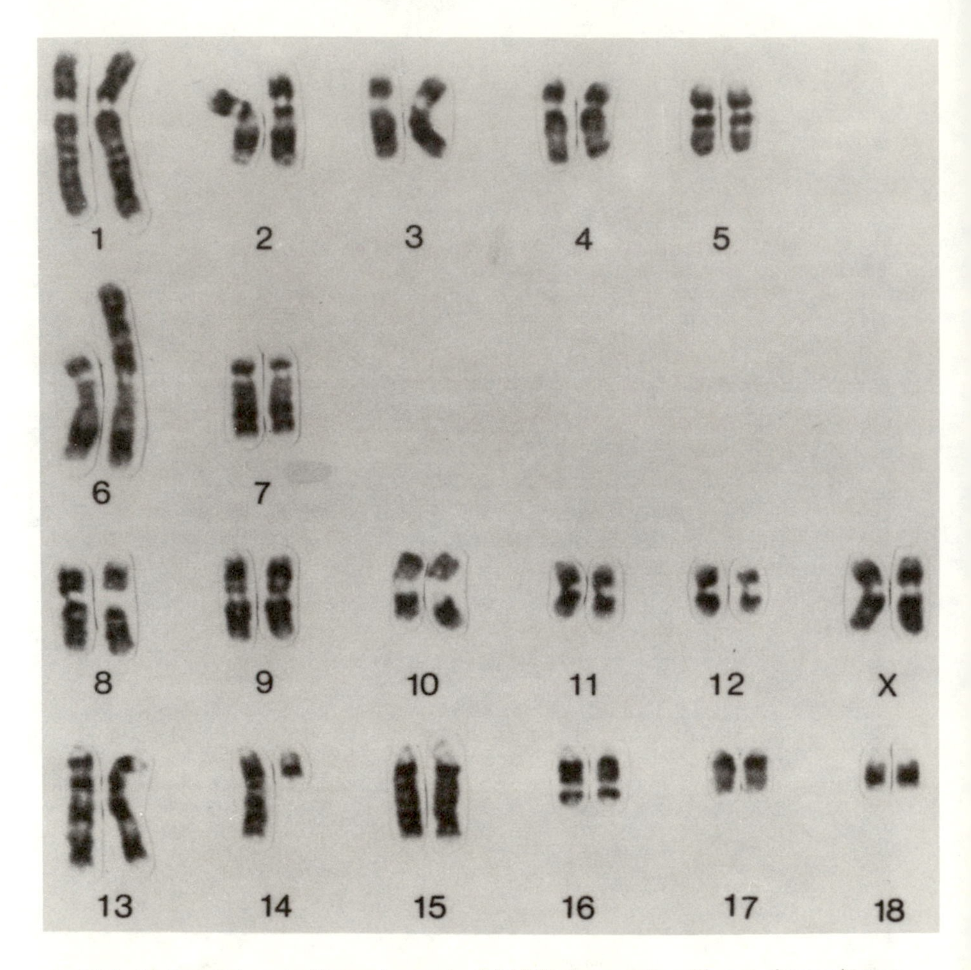

FIG. 13.6. G-banded karyotype of a 38,XX intersex pig with a reciprocal translocation 38,XX,t(6p+; 14q−).

Photograph by Madan et al., courtesy of J. Reprod. Fert. 53, 395–398(1978).

GENERAL

Swine is a species that merits additional attention by livestock cytogeneticists for several reasons. Swine have a relatively small number of chromosomes when compared with other common farm livestock—cattle, sheep, and horses. The chromosomes are more easily identified with standard staining procedures other than the X chromosome. Lymphocyte cultures yield far more usable metaphase spreads, perhaps because of the higher leukocyte count typical of swine. Litters permit the use of greater numbers and more reliable statistics than single-born offspring, and the relatively short generation interval permits more rapid appraisal of transmission of chromosomal aberrations. Further work needs to be done on the chromosomes of blastocysts and oocytes, as well as on meiosis from testicular material.

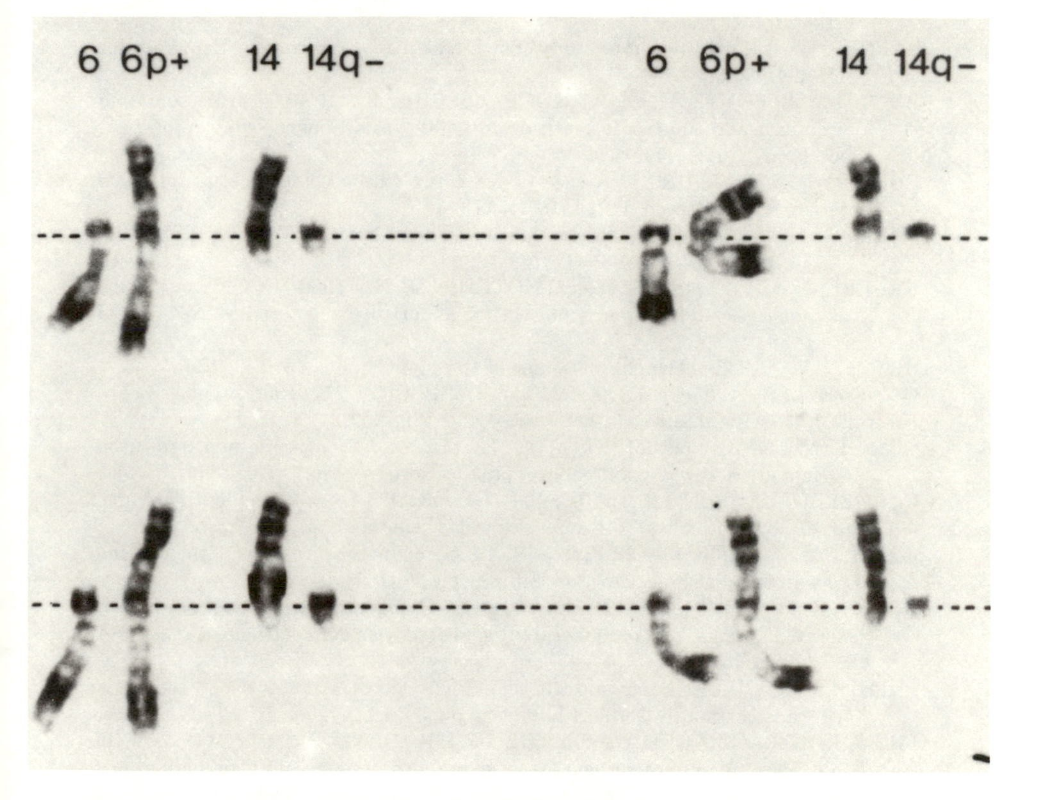

FIG. 13.7. Chromosomes from four cells showing the reciprocal translocation 38,XX,t(6p+; 14q−). The G-banding illustrates quite clearly the portions of chromosomes involved in the translocation.

Photograph by Madan et al., courtesy of J. Reprod. Fert. 53, 395–398(1978).

Although few studies have been made on phenotypic abnormalities, these are not rare in swine. Several of the aberrations reported have resulted from examining the chromosomes of boars which produced smaller than average litters.

REFERENCES

APARICIO, R. D. 1960. Cytogenetic study of spermatogenesis in the pig. Arch. Zootec. (Cordoba) *9*:103. (Anim. Breed. Abstr. *29,* 7, 1961.)

ARAKAKI, D. J., and VOGT, D. W. 1976. A porcine cyclop with normal female karyotype. Am. J. Vet. Res. *37*(1), 95–96.

BASRUR, P. K., and KANAGAWA, H. 1971. Six anomalies in pigs. J. Reprod. Fertil. *26,* 369–371.

BERGER, R. 1972. Karyotype study of swine with a new technique. Exp. Cell Res. *75*(1), 298–300.

BOUTERS, R., BONTE, P., and VANDEPLASSCHE, M. 1974. Chromosomal abnormalities and embryonic death in pigs. 1st World Congr. Genet. Appl. Livestock Prod. *3,* 169–171.

BREEUWSMA, A. J. 1968. A case of XXY sex chromosome constitution of an intersex pig. J. Reprod. Fertil. *16*(1), 119–120.

BREEUWSMA, A. J. 1969. Intersexuality in pigs. Neth. J. Vet. Sci. *94*(7), 493–504.

BRUERE, A. N., FIELDEN, E. D., and HOTCHINGS, H. 1968. XX/XY mosaicism in lymphocyte culture from a pig with freemartin characteristics. N.Z. Vet. J. *16,* 31–38.

BRYDEN, W. 1933. The chromosomes of the pig. Cytologia *5,* 149–153.

CANTWELL, G. E., JOHNSTON, E. F., and ZELLER, J. H. 1958. The sex chromatin of swine intersexes. J. Hered. *49*(5), 199–202.

CHRISTENSEN, K., and SMEDEGARD, K. 1979. Chromosome markers in domestic pigs. A new C-band polymorphism. Hereditas *90*(2), 303–304.

CORNEFERT-JENSEN, FR., HARE, W. C. D., and ABT, D. A. 1968. Identification of the sex chromosomes of the domestic pig. J. Hered. *59*(4), 251–255.

CREW, F. A. E., and KOLLER, P. C. 1939. Cytogenetical analysis of the chromosomes in the pig. Proc. Roy Soc. Edinburgh *59,* 163.

DOLCH, K. M., and CHRISMAN, C. L. 1981. Cytogenetic analysis of preimplantation blastocysts from prepuberal gilts treated with gonadotropins. Am. J. Vet. Res. *42*(2), 344–346.

FORSTER, M., WILLEKE, H., and RICHTER, L. 1981. An autosomal, reciprocal 1/16 translocation in German Landrace pigs. Zuchthygien. *16,* 54–57.

FRIES, R., STRANZINGER, G., and VOEGELI, P. 1983. Is the gene locus of the G-blood group on chromosome 15 in swine? An approach to gene mapping in swine using natural and induced marker chromosomes. J. Dairy Sci. *66* (Suppl. 1), 252.

GIMENEZ-MARTIN, G., LOPEZ-SAEZ, J. F., and MONGE, F. G. 1962. Somatic chromosomes of the pig. J. Hered. *53*(6), 281 and 290.

GROPP, A., GIERS, D., and TATTENBORN, V. 1969. Chromosomes of wild swine (*Sus scrofa*). Experientia *25,* 778.

GRZIMEK, H. C. B. 1972. Animal Life Encyclopedia, Vol. 13. Van Nostrand Reinhold, New York.

GUSTAVSSON, I., and SETTERGREN, I. 1983. Reciprocal chromosome translocations and decreased litter size in the domestic pig. J. Dairy Sci *66* (Suppl. 1), 248.

GUSTAVSSON, I., HAGELTORN, M., JOHANSSON, C., and ZECH, L. 1972. Identification of the pig chromosomes by the quinacrine mustard fluorescence technique. Exp. Cell Res. *70,* 471–474.

GUSTAVSSON, I., HAGELTORN, M., ZECH, L., and REILAND, S. 1973. Identification of the chromosomes in a centric fusion/fission polymorphic system of the pig (*Sus scrofa* L.). Hereditas *75,* 153–155.

HAGELTORN, M. 1976. Detailed analysis of a reciprocal translocation (13q-; 14q+) in the domestic pig by G- and Q-staining techniques. Hereditas *83*(2), 268–272.

HAGELTORN, M., GUSTAVSSON, I., and ZECH, L. 1973. The Q- and G-banding patterns of a t(11p+;15q−)in the domestic pig. Hereditas *75*(1), 147–151.

HANCE, R. T. 1917. The diploid chromosome complexes of the pig (*Sus scrofa*) and their variations. J. Morphol. *30,* 155–222.

HANSEN, K. M. 1972. The karyotype of the pig (*Sus scrofa domestica*), identified by quinacrine mustard staining and fluorescence microscopy. Cytogenetics *11*(4), 286–294.

HANSEN, K. M. 1977. Identification of the chromosomes of the domestic pig (*Sus scrofa domestica*). An identification key and a landmark system. Ann. Genet. Sel. Anim. *9*(4), 517–526.

HANSEN-MELANDER, E., and MELANDER, Y. 1972. A malformed pig with a normal female karyotype. Hereditas *70*(1), 154.

HARD, W. L., and EISEN, J. D. 1965. A phenotypic male swine with a female karyotype. J. Hered. *56,* 254–258.

HARVEY, M. J. A. 1968. A male pig with an XXY/XXXY sex chromosome complement. J. Reprod. Fertil. *17*(2), 319–324.

HENRICSON, B., and BACKSTROM, L. 1964. Translocation heterozygosity in a boar. Hereditas *52*(2), 166–170.

HILLEBRAND, P. 1936. Chromosomal investigation of three different breeds of domestic swine (Deutsches weisses Edelschwein, veredeltes Landschwein und Berkshire). Diss. Phil. Breslau.

HUGHES, W. 1929. The freemartin condition in swine. Anat. Rec. *41,* 213–246.

JONES, H. W., JR., FERGUSON-SMITH, M. A., and HELLER, R. H. 1965. Pathological and cytogenetic findings in true hermaphroditism. Obstet. Gynecol. *25*(4), 435–447.

KRALLINGER, H. F. 1931. Cytological studies on some domestic animals. Arch. Tierernaehr. Tierz. Abt. B, *5*:127–187.

LOCNISKAR, F., GUSTAVSSON, I., HAGELTORN, M., and ZECH, L. 1976. Cytological origin and points of exchange of a reciprocal chromosome translocation (1p−; 6q+) in the domestic pig. Hereditas *83*(2), 272–275.

LOJDA, L. 1975. The cytogenetic pattern in pigs with hereditary intersexuality similar to the syndrome of testicular feminization in man. Doc. Vet., Brno *8,* 71–82.

LOJDA, L., and PETRICKOVA, V. 1977. The effect of feeding various levels of aflatoxin B_1 on the chromosome pattern of rats and pigs. Ann. Genet. Sel. Anim. *9*(4), 539.

LOJDA, L., and RUBES, J. 1977. Chromosome aberrations in pigs after vaccination with living vaccine against swine fever. Ann. Genet. Sel. Anim. *9*(4), 540.

MADAN, K., FORD, C. E., and POLGE, C. 1978. A reciprocal translocation t(6p+;14q−) in the pig. J. Reprod. Fertil. *53*(2), 395–398.

MAKINO, S. 1944. The chromosome complex of the pig (*Sus scrofa*). Cytologia *13,* 170–178.

MAKINO, S., SASAKI, M. S., SOFUNI, T., and ISHIKAWA, T. 1962. Chromosome condition of an intersex swine. Proc. Jpn. Acad. *38*(9), 686–689.

MANOLACHE, M., VOICULESCU, I., MANOLESCU, N., and POPA, M. 1978. Chromosomal damage induced by swine fever virus. 14th Int. Congr. Genet., Moscow.

MASUDA, H., OKAMOTO, A., and WAIDE, Y. 1975. Autosomal abnormality in a swine. Jpn. J. Zootech. Sci. *46*(12), 671–676.

McCONNELL, J., FECHHEIMER, N. S., and GILMORE, L. O. 1963. Somatic chromosomes of the domestic pig. J. Anim. Sci. *22*(2), 374–379.

McFEE, A. F., and BANNER, M. W. 1969. Inheritance of chromosome number in pigs. J. Reprod. Fertil. *18*(1), 9–14.

McFEE, A. F., BANNER M. W., and RARY, J. M. 1966A. Variation in chromosome number among European wild pigs. Cytogenetics *5*(1–2), 75–81.

McFEE, A. F., KNIGHT, M., and BANNER, M. W. 1966B. An intersex pig with XX/XY leucocyte mosaicism. Can. J. Genet. Cytol. *8*(3), 502–505.

McFEELY, R. A. 1967. Chromosome abnormalities in early embryos of the pig. J. Reprod. Fertil. *13*(3), 579–581.

McFEELY, R. A., HARE, W. C. D., AND BIGGERS, J. D. 1967. Chromosome studies in 14 cases of intersex in domestic animals. Cytogenetics *6*(3–4); 242–253.

MELANDER, Y., HANSEN-MELANDER, E., HOLM, L., and SOMLER, B. 1971. Seven swine intersexes with XX chromosome constitution. Hereditas *69,* 51–58.

MIYAKE, Y.-I. 1973. Cytogenetical studies on swine intersexes. Jpn. J. Vet. Res. *21*(3), 41–49.

MIYAKE, Y.-I., KAWATA, K., ISHIKAWA, T., and UMEZA, M. 1977. Translocation heterozygosity in a malformed piglet and its normal littermates. Teratology *16*(2), 163–168.

MOON, R. G., RASHAD, M. N., and MI, M. P. 1975. An example of polyploidy in pig blastocysts (pre-implantation zygotes) J. Reprod. Fertil. *45*(1), 147–149.

MURAMOTO, J., MAKINO, S., and ISHIKAWA, T. 1965. On the chromosomes of the wild boar and the boar-pig hybrids. Proc. Jpn. Acad. *41*(3), 236–239.

OKAMOTO, A. 1978. Genetic studies of the intersex swine. Bull. Coll. Agric. Utsonomiya Univ. *10*(2), 23–36.

OKAMOTO, A., and MASUDA, H. 1977. Cytogenetic studies of intersex swine. Proc. Jpn. Acad. *53*(Ser. B, No. 7), 276–281.

PETRICKOVA, V., LOJDA, L., RUBES, J., and STOVIKOVA, M. 1976. Effect of aflatoxin B_1 on the chromosomal pattern and reproduction of rats and pigs. 8th Int. Congr. Anim. Reprod. Artif. Insem. 199.

PIRTLE, E. C. 1966. Chromosomal variations in a pig kidney cell line persistently infected with hog cholera virus. Am. J. Vet. Res. *27*(118), 737–745.

POPESCU, C. P. 1982. Reciprocal translocations in pigs and their effects on performance. Pig News Inf. 3, 255–257.

POPESCU, C. P. 1983. Cytogenetic evaluation of boars with low prolificacy. J. Dairy Sci. 66 (Suppl. 1), 252.

POPESCU, C. P., and BOSCHER, J. 1982. Cytogenetics of preimplantation embryos produced by pigs heterozygous for the reciprocal translocation (4q+; 14q−). Cytogenet. Cell Genet. *34,* 119–123.

POPESCU, C. P., and LEGAULT, C. 1979. A new reciprocal translocation t(4q+; 14q−) in domestic swine (*Sus scrofa domestica*). Ann. Genet. Sel. Anim. *11*(4), 361–369.

RARY, J. M., HENRY, V. G., MATSCHKE, G. H., and MURPHREE, R. L. 1968. The cytogenetics of swine in the Tellico Wildlife management area, Tenn. J. Hered. *59*(3), 201–204.

READING CONFERENCE 1980. Proceedings of the First International Conference for the Standardisation of Banded Karyotypes of Domestic Animals, 1976. C. E. Ford, D. L. Pollock, and I. Gustavsson, Editors. Hereditas *92,* 145–162.

RITTMANNSPERGER, CH. 1971. Chromosome studies on wild and domestic swine. Ann. Genet. Sel. Anim. *3*(1), 105 (Abstr.).

SACHS, L. 1954. Chromosomal numbers and experimental polyploidy in the pig. J. Hered. *45,* 21–24.

SEABRIGHT, M. 1971. A rapid banding technique for human chromosomes. Lancet *2,* 971–972.

SMITH, J. H., and MARLOWE, J. J. 1971. A chromosomal analysis of 25-day-old pig embryos. Cytogenetics *10*(6), 385–391.

SPALDING, J. F., and BERRY, R. O. 1956. A chromosome study of the wild pig (*Pecari angulatus*) and the domestic pig (*Sus scrofa*). Cytologia *21,* 81–84.

STONE, L. 1963. A chromosome analysis of the domestic pig (*Sus scrofa*) utilizing a peripheral blood culture technique. Can. J. Genet. Cytol. *5*(1), 38–42.

SYSA, P. 1977. Remarks on identification of X chromosome in pig. Ann. Genet. Sel. Anim. *9*(4), 539.

SYSA, P. 1978. Chromatin polymorphism of metaphase chromosomes in the pig (*Sus scrofa* Dom. L.). 14th Int. Congr. Genet., Moscow 278.

TIKHONOV, V. N., and TROSHINA, A. I. 1980A. Marker chromosome translocations Tr 1 (16/17) and Tr 2 (15/17) in development of commercial Landrace × wild boars hybrids and Siberian mini-pigs. Proc. 4th Eur. Colloq. Cytogenet. Domest. Anim. 242–248.

TIKHONOV, V. N., and TROSHINA, A. I. 1974. The identification of chromosome rearrangements of the wild and domestic pigs by the Giemsa banding method. 1st World Congr. Genet. Appl. Livestock Prod. 193–196.

TIKHONOV, V. N., and TROSHINA, A. I. 1975. Chromosome translocations in the karyotypes of wild boars *Sus scrofa* L. of the European and Asian areas of USSR. Theoret. Appl. Genet. *45*(7), 304–308.

TIKHONOV, V. N., TROSHINA, A. I., and GORELOV, I. G. 1974. Immunogenetical studies of European, Asian and American wild suiformes in connection with phylogenesis of domestic pigs. 1st World Congr. Genet. Appl. Livestock Prod. *3,* 197–202.

TROSHINA, A. I., and TIKHONOV, V. N. 1980B. The reproductive features of Landrace × wild boar hybrids with two chromosomal translocations Tr. 1 (16/17) and Tr. 2 (15/17). Proc. 4th Eur. Colloq. Cytogenet. Domest. Anim. 250–261.

VOGT, D. W. 1966. Cytological observations on an intersex pig. J. Anim. Sci. *25,* 252.

VOGT, D. W. 1967. Chromosome condition of two atresia ani pigs. J. Anim. Sci. *26*(5), 1002–1004.

WODSEDALEK, J. E. 1913. Spermatogenesis of the pig with special reference to the accessory chromosomes. Biol. Bull. *25,* 8.

ZARTMAN, D. L. 1971. Chromosomal aberrations in cultured porcine fibroblasts after X-irradiation of male progenitors. J. Anim. Sci. *32*(1), 1–9.

ZARTMAN, D. L., FECHHEIMER, N. S., and BAKER, L. N. 1969. Chromosomal aberrations in cultured leukocytes from pigs derived from X-irradiated semen. Cytogenetics *8,* 355–368.

14

CHROMOSOMES OF HORSES, ASSES, AND MULES

CHROMOSOME NUMBERS

Early reports of the chromosome numbers in horses (*Equus caballus*) showed the total to be less than the number that actually exists. Low numbers of chromosomes were reported for other species of livestock as well. This undoubtedly reflected the limitations imposed by the techniques available at the time. According to Makino (1951), Kirillow first reported, in 1912, that the diploid number in the male was 20 to 32 chromosomes. In 1914 Wodsedalek reported 37 chromosomes with an XO sex chromosome pattern. Painter (1924) found 57 to 60 chromosomes and referred to Masui who reported 33 to 38 chromosomes.

While studying the chromosomes of new cell types in kidney cultures, Rothfels *et al.* (1959) found the chromosome number to be 64 in horses. This is the first report of the correct number.

In recent reports (Eldridge and Blazak 1976; Cribiu and DeGiovanni 1978) the earlier work was reviewed and the numbers and descriptions were substantiated. The horse has 64 chromosomes, with 26 metacentric or submetacentric and 36 acrocentric autosomes. The X chromosome is the second largest submetacentric chromosome and the Y chromosome is the smallest or one of the smallest acrocentrics. Scott and Long (1980) disagreed with the conclusion that the Y chromosome was the smallest chromosome in the horse, but preferred to agree with several other investigators that it is *one* of the smallest. Since the Y chromosome has been found to be polymorphic in most species, these differences in interpretation of its length relative to the other small chromosomes probably reflect differences in the specific animals studied. The breed of horses studied should be included in future reports.

Apparently the results of the earliest studies on ass (*Equus asinus*) chromosomes were published in 1962 by Benirschke *et al.* and by Trujillo *et al.* They were not certain of the number of metacentrics, however. The ass has 62 chromosomes, with 48 metacentric and submetacentric autosomes and 12 acrocentric autosomes (Eldridge and Blazak 1976). The X chromosome is submetacentric (or subtelocentric) and follows chromosome 3 in length. It is very similar to the number 4 pair of autosomes in size and morphology. The Y chromosome appears to be metacentric in the clearest metaphase spreads and is the smallest chromosome, although further detailed studies on animals from other areas may reveal variation in the size of the Y chromosome. Figures 14.1 and 14.2 show horse and ass chromosomes.

Hybrids between horses and asses are called mules or hinnies depending upon the species of the dam. If a jack, the male ass, is mated to a mare, the female horse, the resulting offspring is a mule. If a stallion, the male horse, is bred to a jenny, the female ass, the the offspring is called a hinny. Apparently the first report on the chromosomes of mules and hinnies was also made in 1962 by Benirschke *et al.* and by Trujillo *et al.* As expected, these hybrids had 63 chromosomes. There is a wide discrepancy in the morphological structure of the chromosomes of horses and asses (Fig. 14.3). The fact that the fertility of mares bred to jacks is little different from mares bred to stallions is both unusual and unexpected when chromosomal deviations between the two species are considered. According to Gray (1972), most of the species within the genus *Equus* have been reported to have produced offspring when mated with other species. In most cases the hybrids are sterile.

The Przewalski horse, *Equus przewalskii* Poliakov, has been found to have 66 chromosomes (Short *et al.* 1974). The hybrids between it and the horse have 65 chromosomes, including one unpaired metacentric chromosome, and both males and females are fertile. Evolutionary development may have been through the oc-

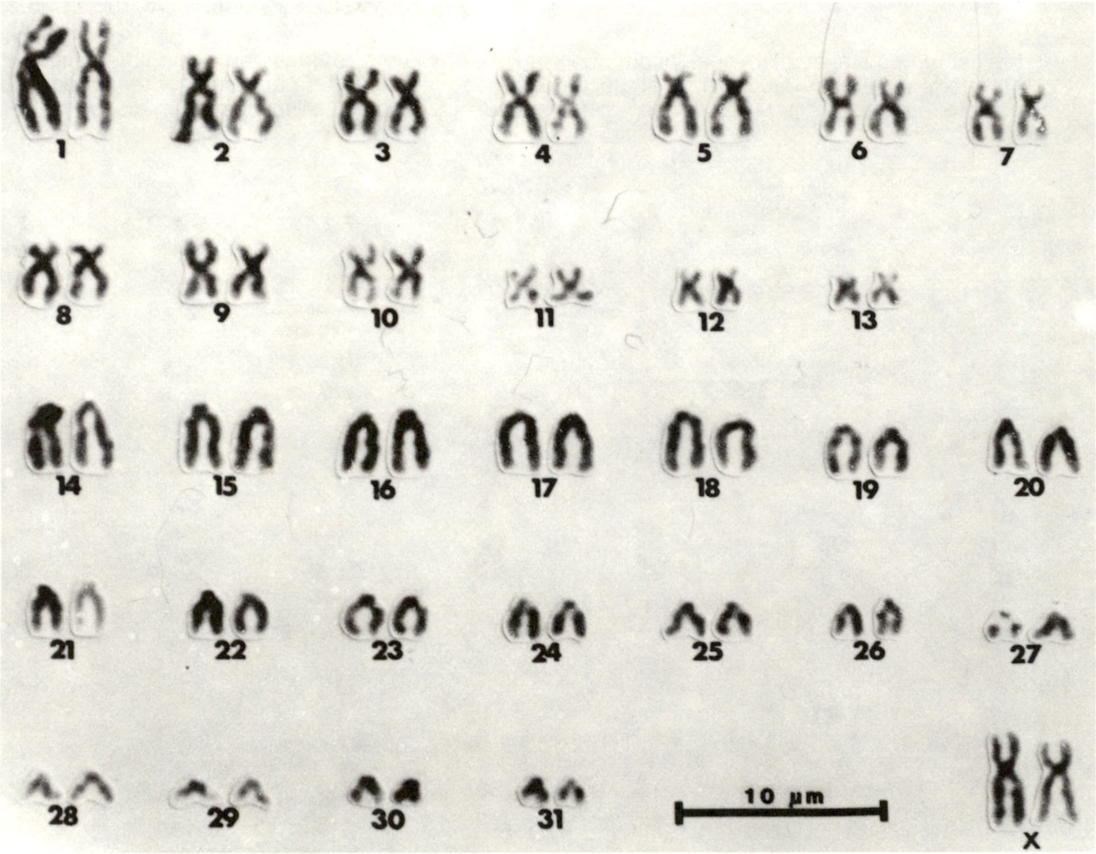

FIG. 14.1. Giemsa-stained karyotype of a mare, female horse. There are 13 pairs of metacentric and submetacentric autosomes and 18 acrocentric autosomes. The X chromosome is the second-largest submetacentric. Frequently, number 27 may appear as three dots.

Photograph by Eldridge, courtesy of J. Hered. 67, 361–367(1976).

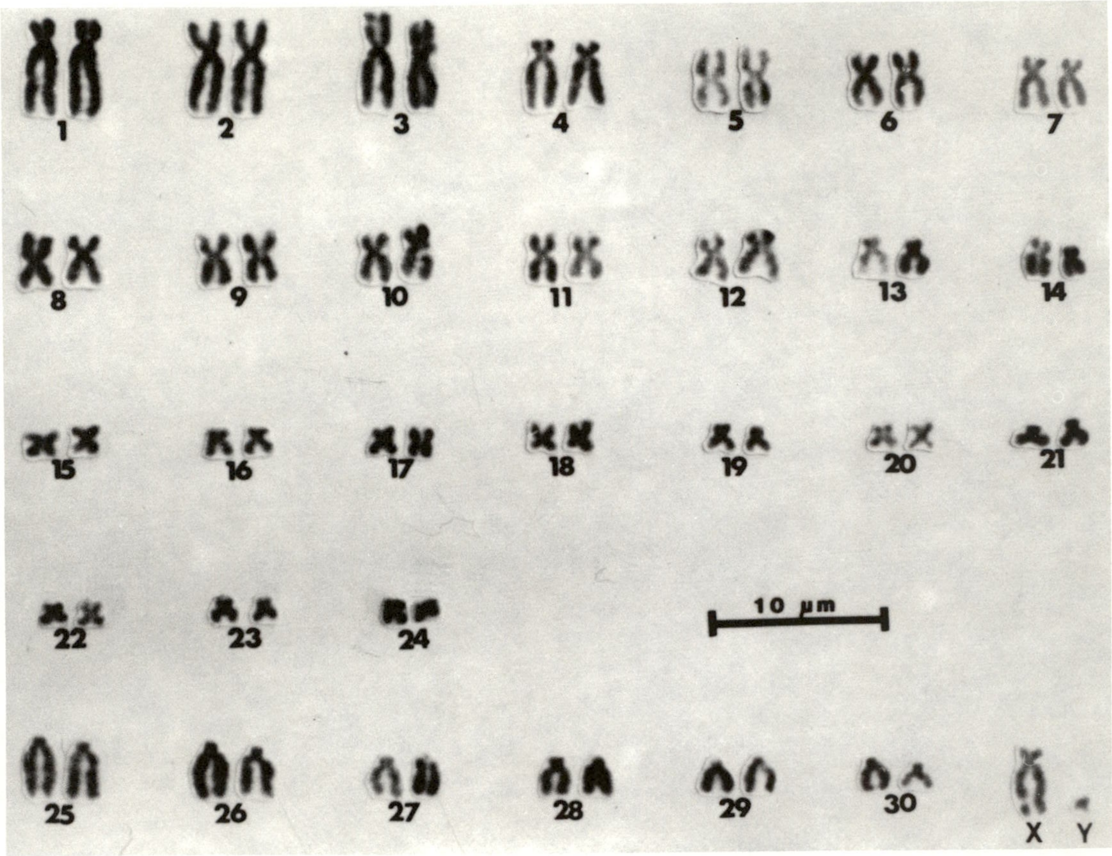

FIG. 14.2. Giemsa-stained karyotype of a jack, the male ass. There are 24 pairs of metacentric and submetacentric autosomes and 6 pairs of acrocentric autosomes. The X chromosome is about the fourth or fifth largest submetacentric chromosome. The Y chromosome is the smallest chromosome, possibly metacentric.

Photograph by Eldridge, courtesy of J. Hered. 67, 361–367(1976).

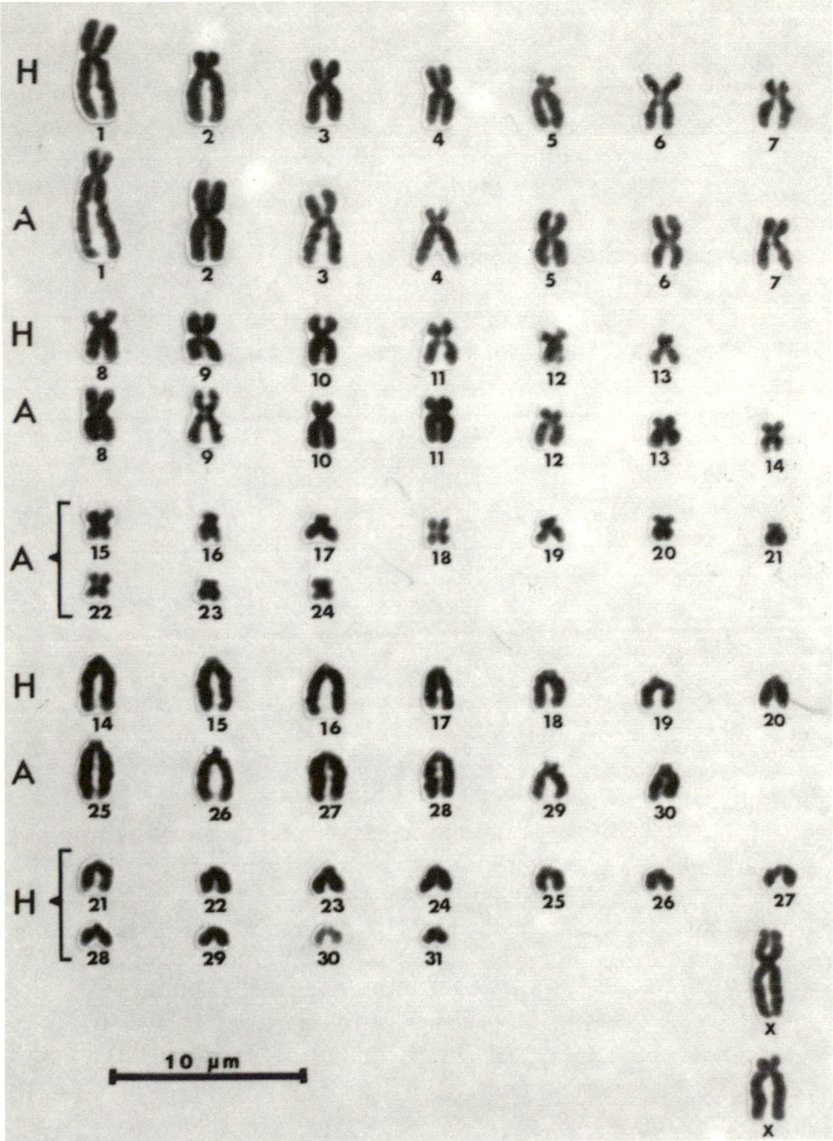

FIG. 14.3. Giemsa-stained karyotype of a mare mule, female. The chromosomes cannot be paired, but are placed in rows marked H for horse and A for ass. Among the metacentric and submetacentric autosomes the first 5 and the 13th were assigned with confidence to the horse parent, and the first 5 and 15 through 24 to the ass. Among the acrocentrics 27 through 31 were assigned with confidence to the horse. The two X chromosomes were also assigned with considerable confidence. All other chromosomes were assigned arbitrarily to the horse or ass parent.

Photograph by Eldridge, courtesy of J. Hered.

currence of a Robertsonian translocation, considering the Przewalski horse as the more primitive ancestor, or by centric fission if the domestic horse was the original ancestor. Unsuccessful attempts have been made to mate Przewalski stallions with female asses, but one hybrid Przewalski × Mongolian horse produced three sterile offspring from matings to jackasses (Gray 1972). There are limited numbers of Przewalski horses alive in parks and zoos.

Three species of zebras, including five subspecies, have been studied chromosomally (Hansen 1975). The diploid chromosome numbers vary from 32 to 46. Numerous cases of hybrids between zebras and other species of *Equus* have been reported (Gray 1972), but little chromosome work has been done with these hybrids.

CHROMOSOME BANDING

Several studies have been made of G-, Q-, C-, and R-banding of horses, asses, mules, and zebras. For the horse, the standard karyotype was adopted at the Reading International Conference in 1976 (Ford *et al.* 1980) (Fig. 14.4). Some further refinements were made by Hansen (1980), identifying landmarks for the identification of horse chromosomes and illustrating the G-bands in considerably more detail. The G-banded karyotype by Maciulus and Bunch (unpublished) shows even more detailed banding (Fig. 14.5); however, they are not in the same order as the Reading Standard. The schematic of the G-bands by Maciulus and Bunch (Fig. 14.6) corresponds to their photographed G-banded karyotype. The discrepancies in numbering, as described by Hansen (1980), will hopefully be resolved in future conferences so that one standard karyotype can be used for reference to chromosome numbers in horses.

The clearest C-banded karyotype of the horse has been published by Melchior and Hohn (1976) (Figs. 14.7 and 14.8). They have shown that one pair that they identify as number 12, a metacentric autosome, is lacking a C-band and that the X chromosome has two distinct C-bands, the second located in the proximal third of the q arm. Cribiu and DeGiovanni (1978) agree in their description, although placing the chromosome without C-banding as number 11. They also reported C-banding in the ass and the mule and found distinct differences between the two species, which made possible a distinction between the two haploid sets in the hybrid.

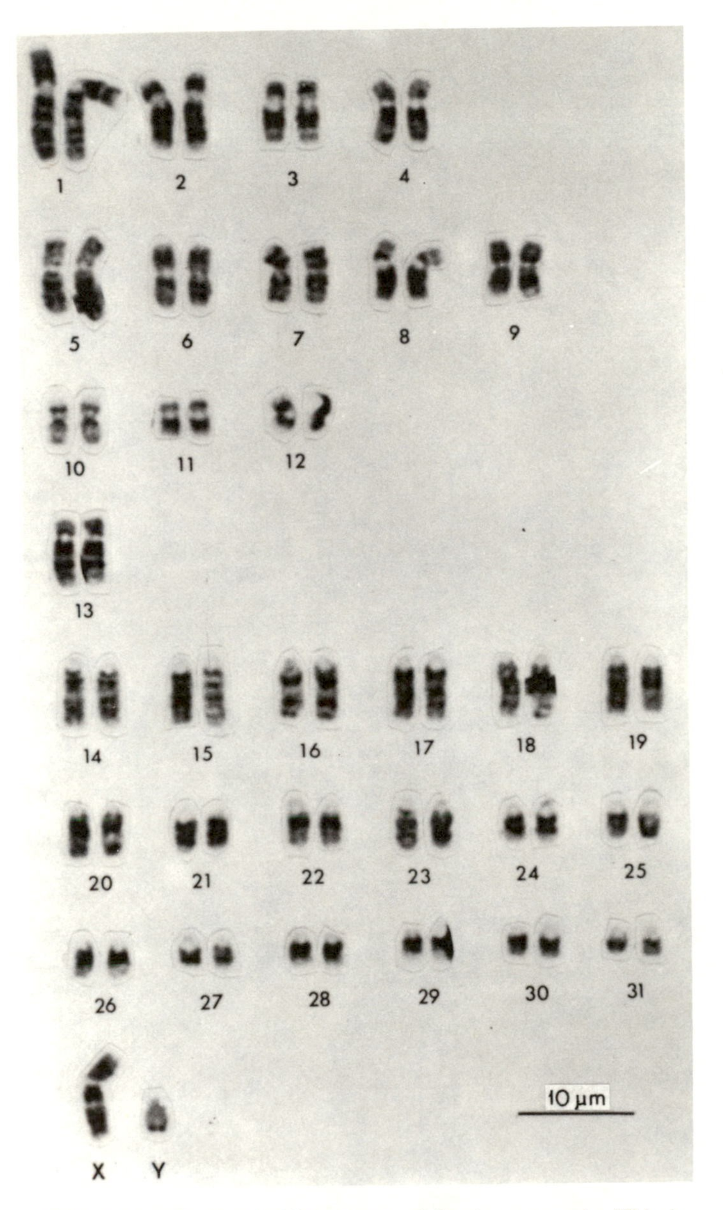

FIG. 14.4. G-banded karyotype of the horse, male. This is the standard adopted by the Reading Conference.
Photograph by R. A. Buckland, courtesy of Hereditas.

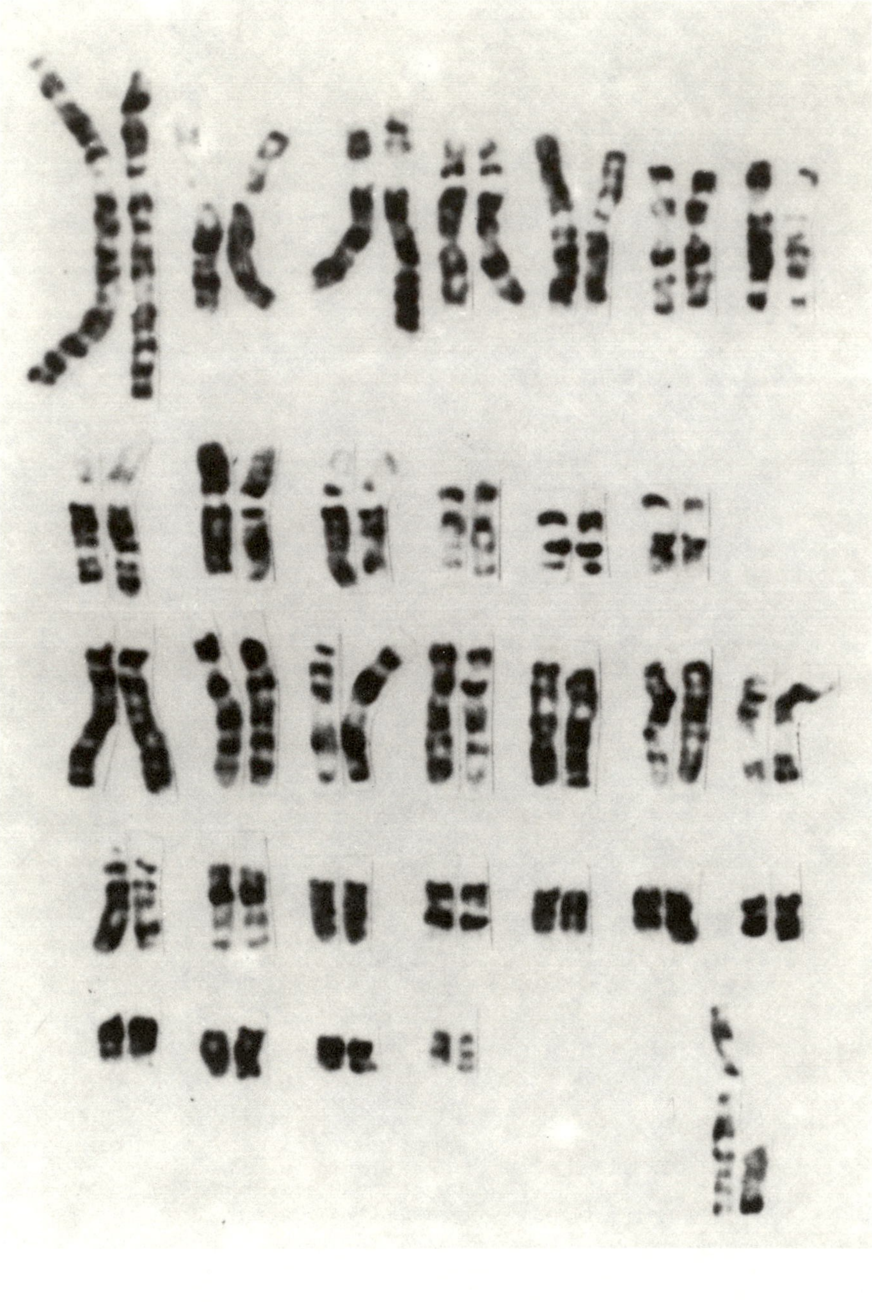

INTERSEXUALITY

Intersexuality in horses, although rare, was recognized and described as early as 1909 according to Bornstein (1967), who first examined two such horses cytologically. He found both to be cryptorchids, each with testes undescended, and both were cytologically female 64,XX. Since then at least nine more cases of horses with abnormal behavior and abnormal external genitalia have been studied cytologically (see Table 14.1, p. 256).

The intersexes are not associated predominantly with any one pattern of chromosomes, as with swine where most intersexes are chromosomally female. Three cases are female, 64,XX, one is 64,XY, and another 66,XXXY. All the rest are mixoploids with some form of chimerism or mosaicism. Only one seems to follow the chimeric pattern of freemartin cattle and is 64,XX/64,XY. The other combinations of chromosomes are 64,XX/64,XY/65,XXY/63,XO; 64,XX/65,XXY; 63,XO(?)/64,XX/65,XXY; 63,XO/65,XYY; and 64,XX (possibly some 64,XY and 65,XXY).

Cryptorchidism appears to be associated consistently with intersex horses, and the anatomical variation that brings these horses to attention is the presence of both a vulva and a penis, but highly variable in expression. The one deviation was the case reported by Kieffer *et al.* (1976), where it was concluded that the behavioral variation was probably the testicular feminizing syndrome. Sexual differentiation in mammals is also influenced by the *tfm* gene, which is usually carried on the X chromosome. This gene, which controls an inherited defect, causes genetic males, XY, to be apparently females which are sterile, and which usually have testes located within the abdomen. It is found in man, mice, rats, and cattle.

The suggestion has been made by some authors that cryptorchid animals should be examined routinely for chromosomal pattern, even though they do not have abnormal external genitalia. Cryptorchidism has a genetic cause, as illustrated by the larger number of cryptorchids in some related groups than in others, but chromosomal aberrations may be the basic reason for the condition more frequently than has been reported to date.

FIG. 14.5. G-banded karyotype of the male horse. The bands are shown in greater detail than the Reading Standard, but are not in the same order.
Photograph courtesy of Alma Maciulus and T. D. Bunch, unpublished.

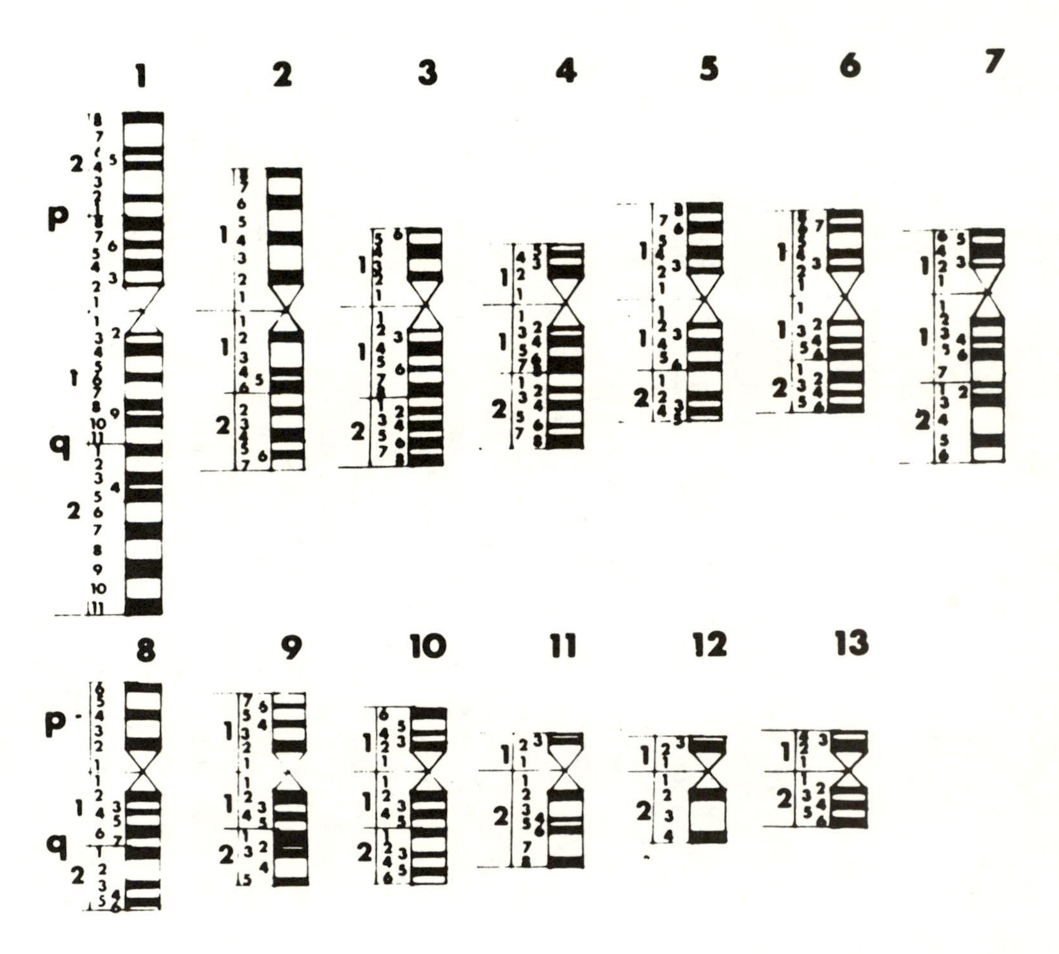
1
2
3
4
5
6
7
8
9
10
11
12
13
p
q

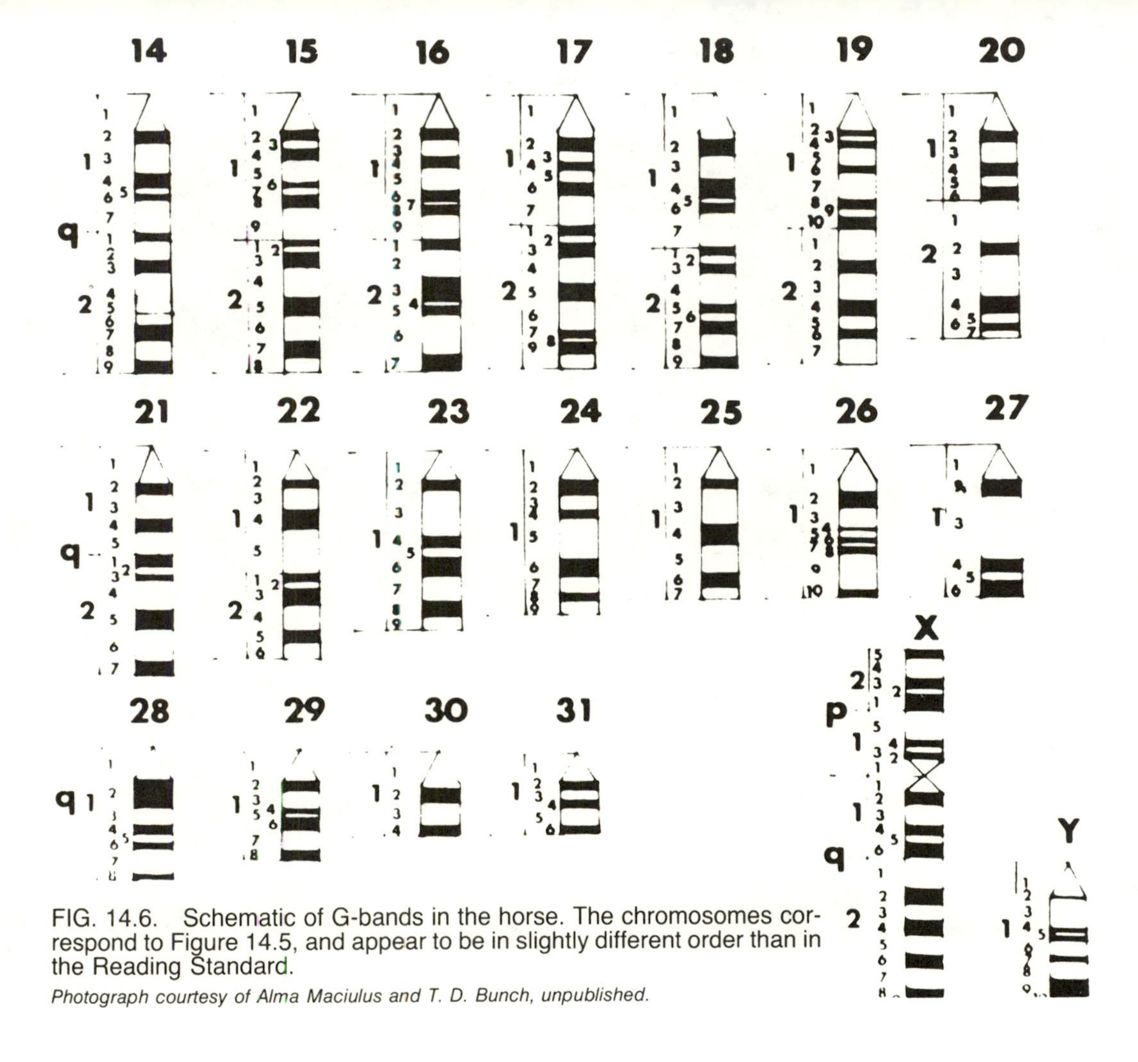

FIG. 14.6. Schematic of G-bands in the horse. The chromosomes correspond to Figure 14.5, and appear to be in slightly different order than in the Reading Standard.

Photograph courtesy of Alma Maciulus and T. D. Bunch, unpublished.

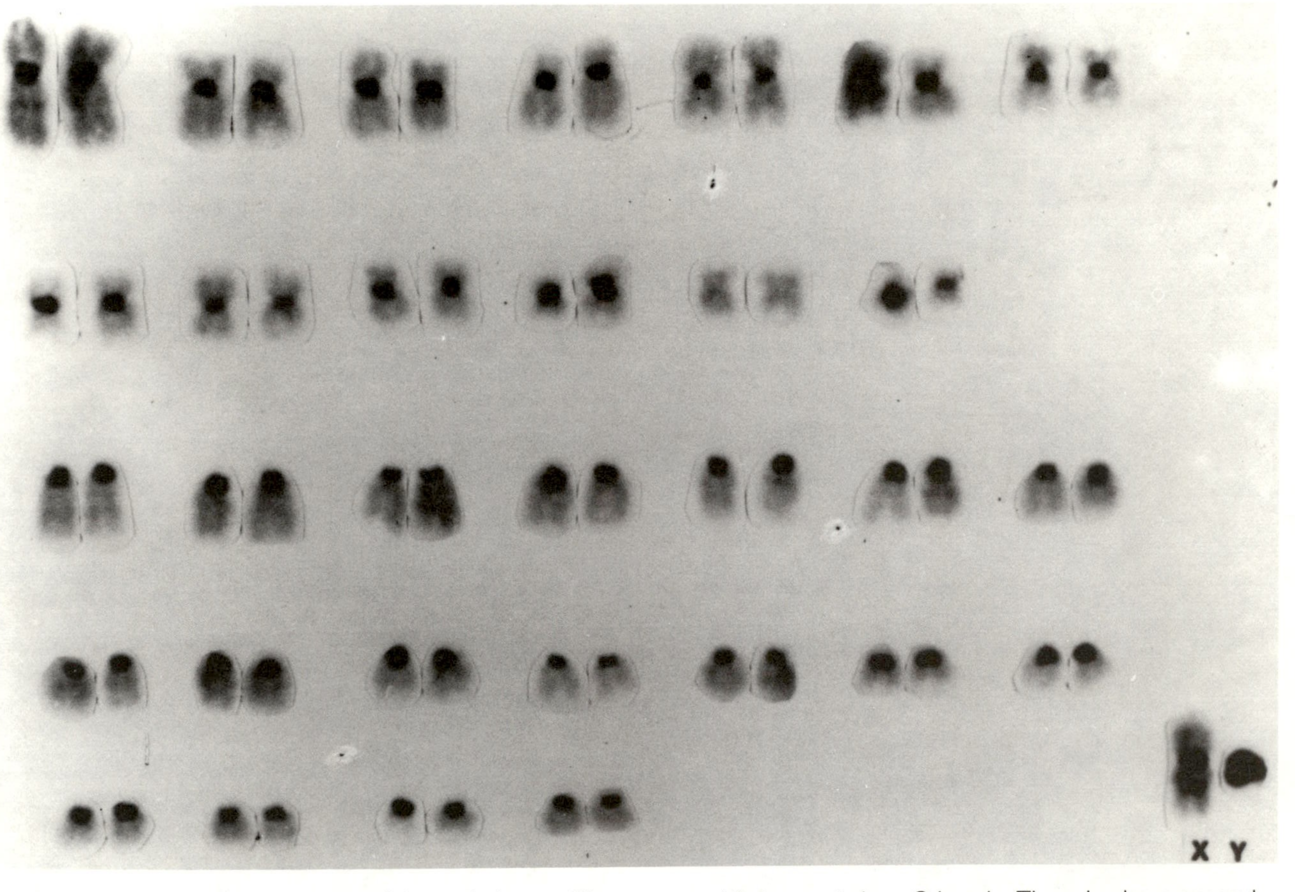

FIG. 14.7. C-banded karyotype of the male horse. Chromosome 12 does not show C-bands. There is also a second C-band in the q arm of the X chromosome, seen more clearly in Figure 14.8.

Photograph by Melchior and Hohn, courtesy of Giessener Beitr. Erbpathol. Zuchthyg. 6, 179–194.

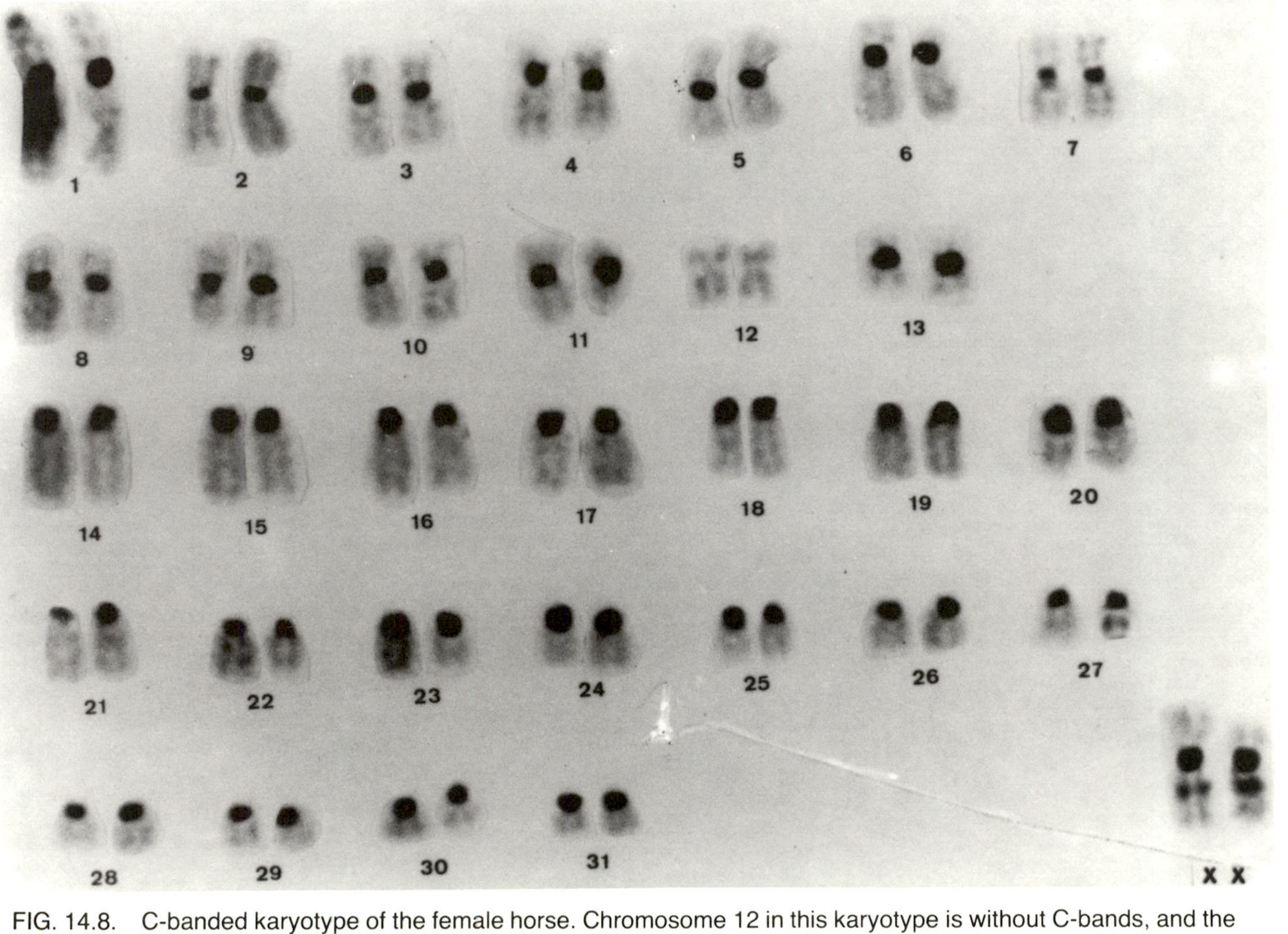

FIG. 14.8. C-banded karyotype of the female horse. Chromosome 12 in this karyotype is without C-bands, and the C-bands in the q arm of the X chromosomes are quite clear.

Photograph by Melchior and Hohn, courtesy of Giessener Beitr. Erbpathol. Zuchthyg. 6, 179–194.

TABLE 14.1. Reports on Intersexuality in Horses

Number of animals	Description	Cytological sex	Reference
2	Cryptorchid, genitalia varied, Ardent and North Swedish breeds	64,XX	Bornstein (1967)
1	Unilateral cryptorchid, internal "testes" without testicular characteristics. Conclude whole-body chimerism Clydesdale, Percheron cross	64,XX/64, XY/65, XXY/63,XO	Basrur *et al.* (1970)
1	Arabian filly, enlarged clitoris, cryptorchid	64,XX	Gerneke and Coubrough (1970)
1	Monius breed, vulvar slit and glans penis, internal genitalia female except for two internal testes	66,XXXY	Gluhovschi *et al.* (1970)
1	Belgian stallion, no uterus, cryptorchid, no germinal cells	64,XX (possibly some 64,XY and 65,XXY)	Dunn *et al.* (1974)
1	Welsh pony, cryptorchid, spermatogonia absent, vascular anastomosis or zygote fusion	64,XX/65,XXY	Bouters *et al.* (1975)
1	Arabian filly, greater distance from anus to ventral commissure than usual, erectile tissue, no uterus palpable, testicles internal, no spermatogenesis	63,XO?/64, XX/65,XXY	Fretz and Hare (1976)
1	Quarter horse, phenotypically female, male-like behavior, testicular feminizing syndrome	64,XY	Kieffer *et al.* (1976)
1	Arabian cross, cryptorchid, normal male penis, propose vascular anastomosis as cause	64,XX/64,XY	McIlwraith *et al.* (1976)
1	Thoroughbred, penis-like phallus in preputial-like vulva, cryptorchid, hypoplastic testicles, male behavior	63,XO/65,XYY	Hohn *et al.* (1980)

CHROMOSOMAL ABERRATIONS ASSOCIATED WITH FERTILITY PROBLEMS

Twenty-five mares with fertility problems, which have been found to be cytologically abnormal, have been reported (Table 14.2). Seventeen of these were found to be 63,XO. Four more were chimeric, with one of the cell lines 63,XO and the other 64,XX or 64,XY. Two were 64,XY. One was 65,XXX and one chimeric animal was 64,XX/65,XX + a fragment.

The large majority of these chromosomally abnormal mares were 63,XO. Variation in anatomy and sexual behavior was found, but in general the animals were smaller than average for the breed; the external genitalia appeared nearly normal; the ovaries were firm, smooth, and without palpable follicles; and if they had estrous cycles at all they were irregular. Frequently the uterus was flaccid and the cervical os was open. These characteristics are similar to those found in Turner's syndrome in humans with some notable exceptions, such as greater development of the gonads. Although some of the 63,XO mares had estrous cycles, some were bred a number of times, and, in one report (Blue *et al.* 1978), some active luteal tissue and some apparently atretic graafian follicles were found, but no case has yet been reported in which a foal was born. This is similar to Turner's syndrome in humans, but differs from mice in which XO females are fertile. Not all infertile mares with small, inactive ovaries are chromosomally abnormal. Blue *et al.* (1978) and Hughes and Trommershausen-Smith (1977) both reported that such animals had been found with normal 64,XX chromosomes. Eldridge (1980) also found one such normal animal with all the other characteristics of a 63,XO mare.

The other two had 64,XY chromosome constitution. To these could be added the male pseudohermaphrodite horse reported by Kieffer *et al.* (1976). These three were probably the result of the testicular feminizing syndrome (TFS), which has been postulated to be controlled by a single gene and cannot be detected chromosomally.

Several suggestions have been given concerning the cause of these chromosomal abnormalities. In the XO type the most simple hypothesis is that nondisjunction during meiosis produced a spermatozoon lacking either gonosome or an ovum without an X chromosome. The peculiar properties of X chromosomes in which one is normally inactive in each cell permit the zygote with only one X to

TABLE 14.2. Cases of Infertile Mares Which Were Examined Cytologically

Number of animals	Description	Cytological sex	Reference
1	Thoroughbred mare, no sex chromatin bodies, bred several times	63,XO with remaining X chromosome modified	Payne *et al.* (1968)
7	All seven mares were sterile, normal-appearing externally (one slightly gelding like); normal vagina, cervix, and vulva; small uterus, except for two normal; gonads small to very small or nonpalpable; 4 with no estrous cycles; 2 with variable cycles; 1 regular. Thoroughbreds (1)(3)(6)(7); Arab (2); Palomino Welsh pony (4); thoroughbred × Anglo Arab (5)	(1) 65,XXX (2) 63,X/64,XX (3) 63,X/64,XX (4) 63,XO (5) 63,XO (6) 64,XX/65,XX plus a fragment (7) 64,XY	Chandley *et al.* (1975)
12	All mares had fertility problems; smaller than average except for pony; external genitalia normal, but small; no clitoris enlargement; uteri small and flaccid; cervical os frequently dilated; ovaries small, firm, no palpable follicles; no cyclic estrous pattern. Arabian (1–7), quarter horse (8–10), Appaloosa (11), pony (12).	(1–9) 63,XO (10) 63,XO/64,XY (11) 63,XO/64,XX (12) 64,XY	Hughes and Trommershausen-Smith (1977)
1	Thoroughbred mare, small, never observed in estrus, but had some active luteal tissue and some apparently atretic graafian follicles	63,XO	Blue *et al.* (1978)
1	Phenotypically normal mare (breed not identified); sterile; genital tract normal; ovaries small, smooth, firm, no follicles; stroma tissue only	63,XO	Metenier *et al.* (1979)
3	All were sterile; treatment with hormones was not effective; ovaries small, nonfunctional. Thoroughbred (1,2), Arabian (3)	63,XO	Miyake *et al.* (1979)

be viable. Monosomy of any of the autosomes usually does not occur in diploid species because of the inability to survive. Fertility of XO females appears at this time to be species specific with the sow, mare, and woman apparently sterile while mice are fertile.

The chimeric condition can be explained by an early mitotic error, by fertilization of an ovum with more than one sperm, or by possible early fusion of two zygotes.

MULES, HINNIES, AND OTHER HYBRIDS

Hybrids between species within the genus *Equus* are well known and numerous (Gray 1972). With the exception of the cross between the domestic horse, *Equus caballus,* and the Przewalski horse, *Equus przewalskii,* all other hybrids appear to be sterile. The diploid chromosome numbers range from 32 in *Equus zebra kartmannae* to 66 in *E. przewalskii,* according to Hsu and Benirschke (1974). The hybrids have the intermediate number as expected wherever the hybrids have been examined cytologically.

Although many cases of apparently fertile mare mules have been reported, since accurate chromosome studies have been possible, no cases have been reported of fertile mules with the chromosome composition proving them to be mules. Benirschke *et al.* (1964) found a fertile mare, which was thought to be a mule, to be chromosomally a donkey. Eldridge and Suzuki (1976), through cytological analysis and blood group and protein systems, found that a mare mule which through lactation had reared an offspring was not the dam, but that a Shetland mare had apparently had twins born 3 weeks apart and the mare mule had adopted the first foal. The mare mule cytologically was a mule but the foal cytologically was a female horse with no ass chromosomes.

Trujillo *et al.* (1969) found a number of tetraploid or near-tetraploid spermatogonia in the hinny testis. This has led to the hypothesis that amphidiploids may occur with homologous horse/horse or donkey/donkey chromosome pairs at meiosis. If tetrapolar spindles were to permit an organization so that these sets were divided into two horse and two donkey gonial cells, in the female or the male, it might be possible for a mule to produce germ cells with a haploid set of strictly horse or ass chromosomes. Rare as it may be, such a development might permit fertilization. The fact that most offspring reported from mare mules have been either typical horses or asses or mules, and not gradations in between, would support this possibility, tenuous as it is.

Mules and hinnies have provided an interesting support to the Lyon hypothesis of X inactivation. Since in a mule the X chromosome of the horse can be distinguished from the X chromosome of the donkey, it is possible to determine the percentage of late-replicating X chromosomes which are from the horse or donkey parent. The glucose-6-phosphate dehydrogenase (G-6-PD) locus is on the X chromosome, and there is a species difference between horses and donkeys. The late-replicating X chromosome has been identified as the inactive chromosome. There is a close correlation between the percentage of inactive X chromosomes and the amount of G-6-PD from the opposite X (Rattazzi and Cohen 1972). Furthermore, clones of cells from single mule cells have 100% of one type of G-6-PD or the other. In mules or hinnies, the X chromosome from the donkey parent is more frequently inactive than the X chromosome from the horse parent (Hamerton *et al.,* 1971). Thus, X inactivation is not a random event in these hybrids.

REFERENCES

BASRUR, P. K., KANAGAWA, H., and PODLIACHOUK, L. 1970. Further studies on the cell population of an intersex horse. Can. J. Comp. Med. *34*(4), 294–298.

BENIRSCHKE, K., BROWNHILL, L. E., and BEATH, M. M. 1962. Somatic chromosomes of the horse, the donkey and their hybrids, the mule and the hinny. J. Reprod. Fertil. *4*(3), 319–326.

BENIRSCHKE, K, LOW, R. J., SULLIVAN, M., and CARTER, R. M. 1964. Chromosome study of an alleged fertile mare mule. J. Hered. *55,* 31–38.

BLUE, M. G., BRUERE, A. N., and DEWES, H. F. 1978. The significance of the XO syndrome in infertility of the mare. N.Z. Vet. J. *26*(6), 137–141.

BORNSTEIN, S. 1967. The genetic sex of two intersexual horses and some notes on the karyotype of normal horses. Acta Vet. Scand. *8*(4), 291–300.

BOUTERS, R., VANDEPLASSCHE, M., and DE MOOR, A. 1975. An intersex (male hermaphrodite) horse with 64XX/65XY mosaicism. J. Reprod. Fertil., Suppl. *23,* 375–376.

CHANDLEY, A. C., FLETCHER, J., ROSSDALE, P. D., PEACE, C. K., RICKETTS, S. W., McENERY, R. J., THORNE, J. P., SHORT, R. V., and ALLEN, W. R. 1975. Chromosome abnormalities as a cause of infertility in mares. J. Reprod. Fertil., Suppl. *23,* 377–383.

CRIBIU, E. P., and DEGIOVANNI, A. 1978. Karyotype of the domestic horse

(*Equus caballus*), the ass (*Equus asinus*) and the mule by the C-band method. Ann. Genet. Sel. Anim. *10*(2), 161–170.

DUNN, H. O., VAUGHAN, J. T., and McENTEE, K. 1974. Bilaterally cryptorchid stallion with female karyotype. Cornell Vet. *64*, 265–275.

ELDRIDGE, F. E. 1980. Unpublished data.

ELDRIDGE, F., and BLAZAK, W. F. 1976. Horse, ass and mule chromosomes. J. Hered. *67*(6), 361–367. [Note the correct karyotype for Figure 3 is in *68*(1), 58.]

ELDRIDGE, F. E., and SUZUKI, Y. 1976. A mare mule—dam or foster mother? J. Hered. *67*, 353–360.

FORD, C. E., POLLOCK, D. L., and GUSTAVSSON, I. 1980. Proceedings of the First International Conference for the Standardization of Banded Karyotypes of Domestic Animals. Hereditas *92*(2), 145–162.

FRETZ, P. B., and HARE, W. C. D. 1976. A male pseudohermaphrodite horse with 63XO?/64XX/65XXY mixoploidy. Equine Vet. J. *8*(3), 130–132.

GERNEKE, W H., and COUBROUGH, R. I. 1970. Intersexuality in the horse. Onderstepoort J. Vet. Res. *37*(4), 211–215.

GLUHOVSCHI, N., BISTRICEANU, M., SUCIU, A., and BRATU, M. 1970. A case of intersexuality in the horse with type 2A + XXXY chromosome formula. Br. Vet. J. *126*(10), 522–525.

GRAY, A. P. 1972. Mammalian hybrids. Commonwealth Bur. Anim. Breed. Genet., Edinburgh, Tech. Commun. *10*.

HAMERTON, J. L., RICHARDSON, B. J., GEE, P. A., ALLEN, W. R. and SHORT, R. V. 1971. Non-random X chromosome expression in female mules and hinnies. Nature (London) *232*, 312–315.

HANSEN, K. M. 1975. The G- and Q-band karyotype of Bohm's or Grant's zebra (*Equus burchelli bohmi*). Hereditas *81*(2), 133–140.

HANSEN, K. M. 1980. The G-band karyotype of the domestic horse (*Equus caballus*). Proc. 4th Eur. Colloq. Cytogenet. Domest. Anim. 386–389.

HOHN, H., KLUG, E., and RIECK, G. W. 1980. A 63,XO/65,XYY mosaic in a case of questionable equine male pseudohermaphroditism. Proc. 4th Eur. Colloq. Cytogenet. Domest. Anim. 82–92.

HSU, T. C., and BENIRSCHKE, K. 1974. An Atlas of Mammalian Chromosomes. Springer-Verlag, New York.

HUGHES, J. P., and TROMMERSHAUSEN-SMITH, A. 1977. Infertility in the horse associated with chromosomal abnormalities. Aust. Vet. J. *53*(6), 253–257.

KIEFFER, N. M., BURNS, S. J., and JUDGE, N. G. 1976. Male pseudohermaphroditism of the testicular feminizing type in a horse. Equine Vet. J. *8*(1), 38–41.

MAKINO, S. 1951. Chromosome Numbers in Animals, 2nd Edition. Iowa State College Press, Ames, IA.

McILWRAITH, C. W., OWEN, R. A. P. R., and BASRUR, P. K. 1976. An equine cryptorchid with testicular and ovarian tissue. Equine Vet. J. *8*(4), 156–160.

MELCHIOR, I., and HOHN, H. 1976. Karyotype of the horse (*Equus caballus*) made with G- and C-banding technique. Giessener Beitr. Erbpathol. Zuchthyg. *6*(3), 179–194.

METENIER, L., DRIANCOURT, M. A., and CRIBIU, E. P. 1979. An XO chromosome constitution in a sterile mare (*Equus caballus*). Ann. Genet. Sel. Anim. *11*(2), 161–163.

MIYAKE, Y.-I., ISHIKAWA, T., and KAWATA, K. 1979. Three cases of mare sterility with sex-chromosomal abnormality (63,X). Zuchthygiene *14*(4), 145–150.

PAINTER, T. S. 1924. Studies in mammalian spermatogenesis. V. The chromosomes of the horse. J. Exp. Zool. *39*(2), 229–248.

PAYNE, H. W., ELLSWORTH, K., and DE GROOT, A. 1968. Aneuploidy in an infertile mare. J. Am. Vet. Med. Assoc. *153*(10), 1293–1299.

RATTAZZI, M. C., and COHEN, M. M. 1972. Further proof of genetic inactivation of the X chromosome in the female mule. Nature (London) *237*, 393–396.

ROTHFELS, K. H., AXELRAD, A. A., SIMINOVITCH, L., McCULLOCH, E. A., and PARKER, R. C. 1959. The origin of altered cell lines from mouse, and man, as indicated by chromosome and transplantation studies. *In* Proceedings of the Third Canadian Cancer Conference, pp. 189–214. Academic Press, New York.

SCOTT, I. S., and LONG, SE. 1980. An examination of chromosomes in the stallion (*Equus caballus*) during meiosis. Cytogenet. Cell Genet. *26*, 7–13.

SHORT, R. V., CHANDLEY, A. C., JONES, R. C., and ALLEN, W. R. 1974. Meiosis in interspecific equine hybrids. II. The Przewalski horse/domestic horse hybrid. Cytogenet. Cell Genet. *13*(5), 465–478.

TRUJILLO, J. M., STENIUS, C., CHRISTIAN, L. C., and OHNO, S. 1962. Chromosomes of the horse, the donkey and the mule. Chromosoma *13*, 243–248.

TRUJILLO, J. M., OHNO, S., JARDINE, J. H., and ATKINS, N. B. 1969. Spermatogenesis in a male hinny: Histological and cytological studies. J. Hered. *60*, 79–84.

WODSEDALEK, J. E. 1914. Spermatogenesis of the horse with special reference to the accessory chromosome and the chromatoid body. Biol. Bull. *27*(6), 295–325.

15

Bird Cytogenetics*

Birds share many common attributes, such as feathers, nucleated red blood cells, and egg laying. Their chromosomes also have similar features, as all karyotyped species show two more or less distinct classes, macrochromosomes and microchromosomes. In general, the chromosomes of the *Saurpsida* group are characterized by a macro-micro chromosome configuration with less DNA per nucleus than mammals. Estimates of the DNA content of bird chromosomes range from 1.7 to 3.6 picograms per nucleus, which is about 50% of the nuclear DNA of mammals. Repetitive DNA is also less, being in the neighborhood of 17–25% of the genome, compared to 50% or more in mammals.

The common techniques for securing chromosome preparations from bone marrow, peripheral blood leukocyte culture, and other cell cultures have been adapted successfully by cytologists working with avian material except that an incubation temperature close to 40°C is optimum for cell growth. The feather pulp technique first proposed by Sandnes (1954) and refined further by Shoffner *et al.* (1967) is useful and it is widely used for birds. The growing point of the young pinfeather has a high mitotic index and provides excellent material for squash chromosome preparations. Its principal advantage is that metaphase chromosomes can be secured within a few hours from unharmed specimens. The oviparous nature of birds makes it relatively easy to obtain fertilized eggs for incubation.

*This chapter was written by Dr. R. N. Shoffner, Department of Animal Science, University of Minnesota, St. Paul, MN 55108.

Embryos at 16–72 hours of development exhibit a very high mitotic index which results in excellent chromosome preparations. Thousands of early chicken embryos have been screened cytologically for frequency of chromosomal abnormalities, study of induced rearrangements, and linkage. A considerable amount of avian cytogenetic knowledge, aside from karyotype documentation and cytotaxonomic considerations, has resulted from study of the chicken, *Gallus domesticus*. Consequently, a considerable portion of the discussion in this chapter originates from research in this species.

AUTOSOMES

Macrochromosomes

The macrochromosomes are those which can be readily identified and usually comprise the six to ten largest pairs. In spite of the limited number of recognizable macrochromosomes, each species, except for some closely related ones, has one or more marker chromosomes for a unique karyotype. The partial chromosome karyotypes of the five common species shown in Figure 15.1 illustrate the general chromosome configuration found in birds. Consequently, only a few of the larger chromosomes can be described in the conventional metacentric, submetacentric, subtelocentric, acrocentric, and telocentric terms. There are a few exceptions, such as the secretary bird, *Sagittarius serpentarius,* where nearly 60% of the $2n$ complement have distinct biarmed chromosomes (DeBoer 1976). The decorative Mandarin duck, *Aix galericulata,* on the other hand, is at the opposite extreme, as all chromosomes are acrocentric, leaving length as the only discriminating feature.

Microchromosomes

The majority of birds have a $2n$ number of about 80 chromosomes; the smaller, individually unidentifiable microchromosomes outnumber the larger ones by about 3 to 1, but account for only 15–20% of the nuclear chromatin. The microchromosomes are true chromosomes, although at one time cytologists suspected aneuploidy and other irregular properties of these small elements. Because they are very small (Fig. 15.2) and may be lost or covered over in prepara-

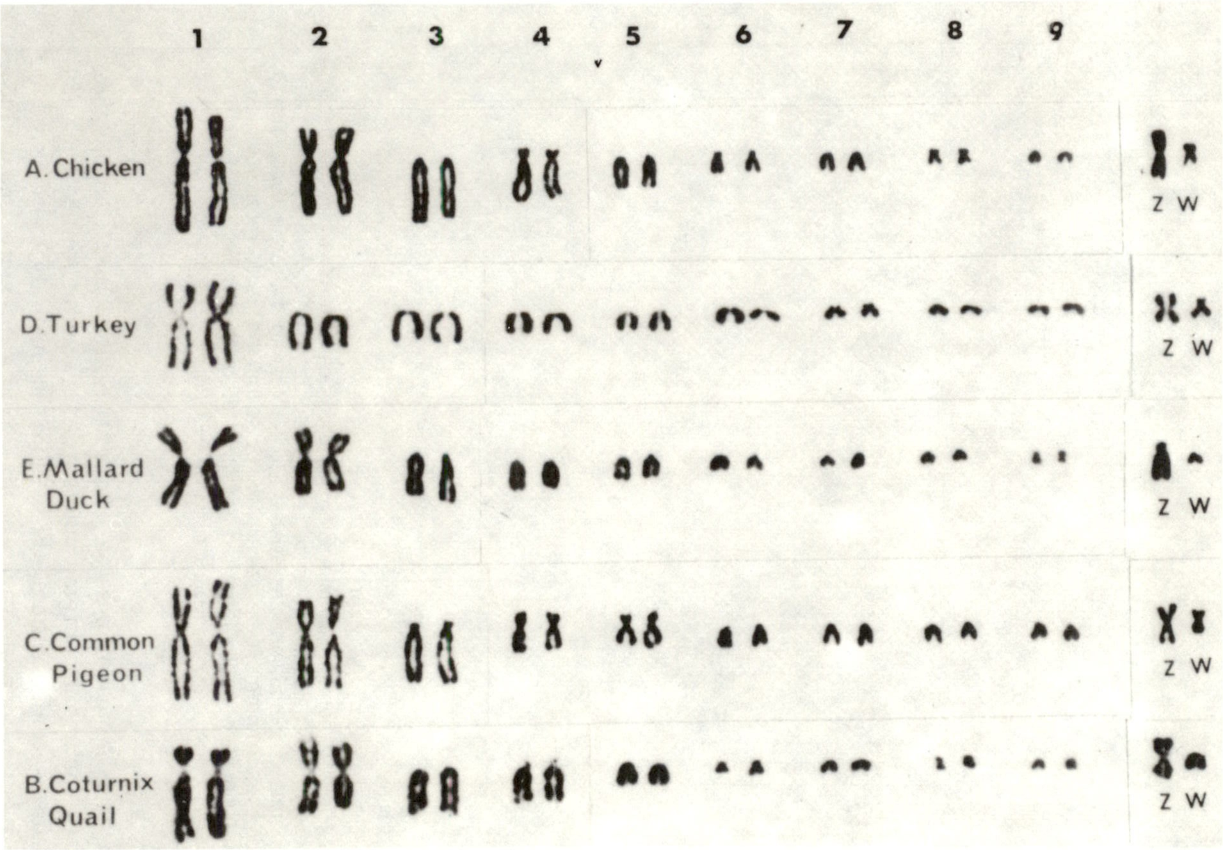

FIG. 15.1. The partial karyotypes of these species illustrate the common macrochromosome and microchromosome configuration commonly found in birds. The Z sex chromosome is usually about fourth largest in size, while the W sex chromosome varies considerably in both size and morphology.

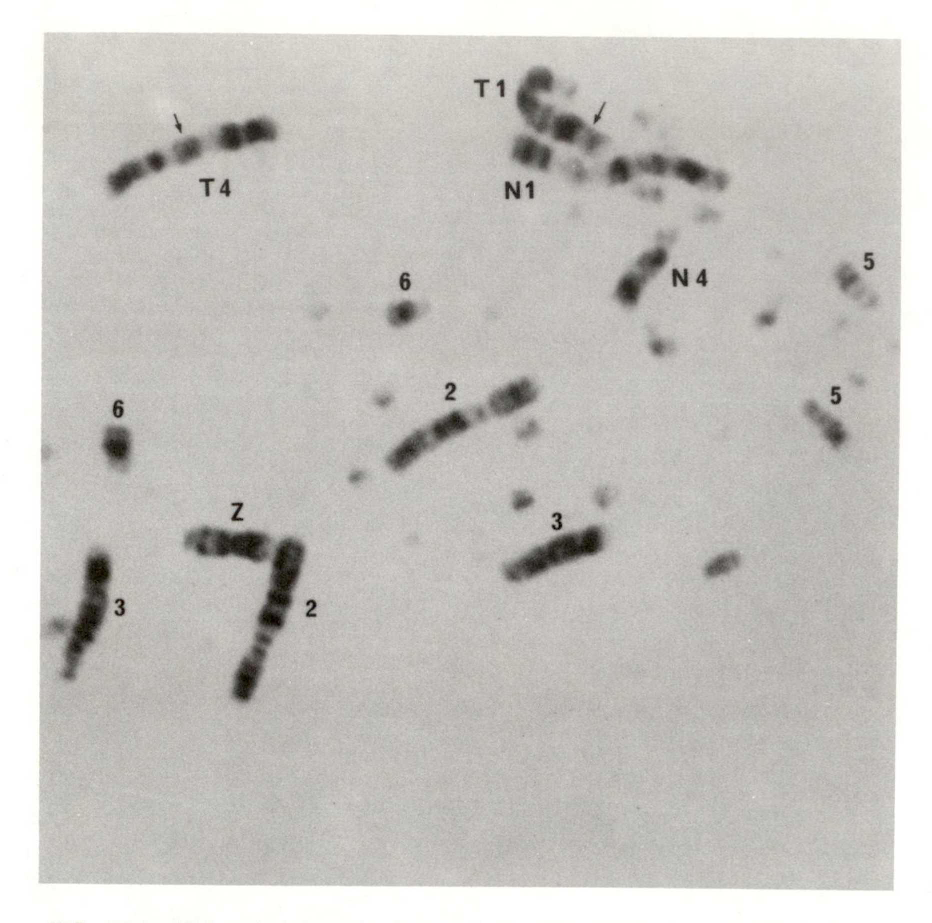

FIG. 15.2. G-banded (trypsin–Giemsa) chromosomes of a chicken heterozygous for a translocation between chromosomes 1 and 4. The translocation and normal chromosomes are labeled T1, T4, N1, and N4, respectively. The G-bands of each chromosome pair are unique so that the break–reunion region of the translocation can be identified, as indicated by the arrows.

Courtesy of J. J. Bitgood, Department of Animal Science, University of Minnesota.

tion, it is almost impossible to identify all elements in a single mitotic metaphase chromosome spread. Counts obtained from M1 diakinesis configurations (Fig. 15.3) are usually more reliable. There is considerable variation in staining intensity among these elements from very light to very dark within the same preparation.

It is not known whether this is artifactual or due to some intrinsic property of these small chromosomes. There are conflicting reports on the onset of DNA synthesis for the large and small chromosomes during the S period of the cell replication cycle as estimated by the administration of tritiated thymidine (see review by Shoffner 1974). Thus, it is not certain if there are actual differences in structure or function of these elements or if the reported variation originates from technique. There is, however, a distinct variation in the amount and distribution of constitutive heterochromatin in the microchromosomes of different species of birds as shown by Stefos and

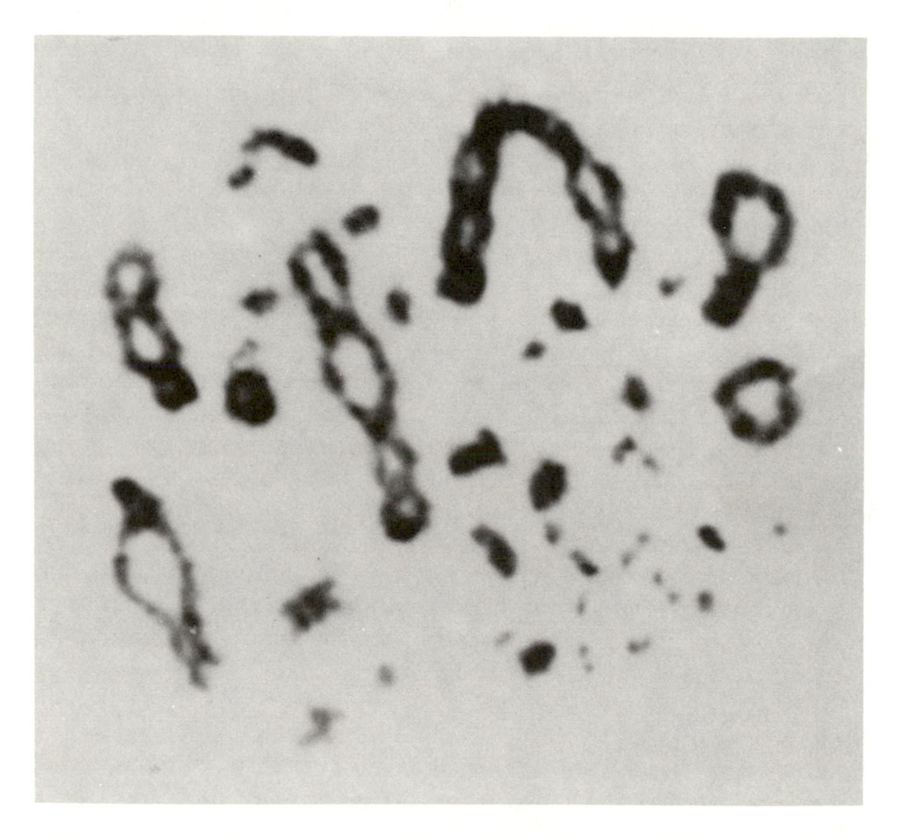

FIG. 15.3. The diakinesis stage of spermatogenesis in the chicken shows pronounced chiasmata in the larger chromosomes. Counts of microchromosomes are more accurate in this stage than are those in mitotic metaphase because there are only half as many elements.

Arrighi (1974) for the parakeet (*Melopsittacus undulatus*), bobwhite quail (*Colinus virginianus*), chukar (*Alectoris graeca*), chicken (*Gallus domesticus*), ring-necked pheasant (*Phasianus colchicus*), and pigeon (*Columba livia*). These findings indicate that certain of these small elements have a large proportion of highly repetitive sequences of DNA and that the frequencies of these types of chromosomes vary between species.

GONOSOMES

The Z Chromosome

Sokolow and Trofimow (1933) reported that the chicken male had two Z chromosomes and the female only one. Subsequent karyotyping of other species has established the homogametic nature of the avian male. The Z sex chromosome comprises about one tenth of the haploid genome and ranks approximately fourth or fifth in size for almost all karyotyped bird species. Facultative heterochromatin, observed as a prominent dark body in the nucleus, is an inactive X chromosome in the mammalian (XX) female but is not present in the homogametic (ZZ) male bird. Thus, it appears that there is not a corresponding inactivation of one Z chromosome as a dosage-compensation mechanism in birds. In mammals, the inactivation of one X in the female has been interpreted as a mechanism for preventing the doubling of the effect of the X chromosome, or a dosage-compensation mechanism. There are Z-linked genes which have a definite dosage effect, for example, barring. Cock (1964) reviewed the literature regarding sex-linked genes in birds and found no evidence for dosage compensation. Schmid (1962) determined that the Z chromosomes of the male and female chicken replicate at the same time as the autosomes.

The sex-linked traits in chickens comprise the largest known linkage group, with 15 traits assigned to the Z chromosome and the relative map distances determined (Somes 1978). Sex-linked traits have also been found for the turkey, domestic duck, and pigeon. The Mendelian traits on the Z chromosome range from lethals to non-detrimental phenotypic variants. There are also major genes affecting quantitative traits such as sexual maturity and rate of feathering. A number of single genes with considerable commercial value, such as barring (*B*), silver (*S*), and rapid feathering (*K*), are utilized in the sex determination of baby chicks. The sex-linked recessive dwarf gene (*dw*) is also economically important, as reduction in body size improves feed efficiency in egg layers and broiler breeder females.

The W Chromosome

It appears that all female birds have a W chromosome, although for some time it was believed that the sex chromosome makeup was ZO instead of the now known ZW composition. Even today, occasional references cite a ZO composition rather than ZW. The W chromosome varies considerably between species in size, ranging from as little as one eighth of the size of the Z chromosome of some species to as much as one third in other species. It is readily recognized in some species, whereas in others it is difficult to distinguish from the larger microchromosomes except by elective staining for constitutive heterochromatin (C-banding). The W chromosome is composed mainly of highly repetitive DNA sequences which stain darkly heterochromatic when subjected to selective cytochemical C-banding procedures. The W chromosome can be used to identify sex in species that do not exhibit sexual dimorphism. The presence of the W in a metaphase chromosome configuration of cells secured from the specimen in question identifies the sex. The C-banding method of Stefos and Arrighi (1971) and the G-band trypsin modification of Wang and Shoffner (1974) are discriminatory even with poor-quality preparations.

A distinct, positive heteropycnotic (tightly coiled and, therefore, darkly staining) body has been observed in the interphase nucleus of the female parakeet (*Melopsittacus undulatus*) and with a lesser intensity and regularity in the female chicken (Stefos and Arrighi 1971). This dark staining body produced by C-banding cytochemistry is believed to be the W chromosome. If so, this explains the conflicting reports about the sex-associated dimorphism for the heterochromatic body observed in the interphase nuclei of females in several species of birds (Shoffner 1974).

The W chromosome does have some unique DNA, as a histocompatibility locus was identified by the response of reciprocal male–female skin grafts (Gilmour 1967; Bacon and Craig 1969). Some compatibility factor of the W is involved in the development of the

ZW females in the chicken × coturnix quail hybrids, as they die in early embryogenesis; on the other hand, the ZZ males are developmentally compatible. The only difference between the chromosome constitutions of the male and female hybrids is the W chromosome. The double dose of the W in the 3A-ZWW triploid embryos is also lethal, whereas the 3A-ZZW and 3A-ZZZ genotypes hatch and develop into adults.

Sex Determination

In the chicken, individuals with the chromosome constitution 2A-ZZ or 3A-ZZZ are males and have been found invariably to possess bilateral spermatozoa-producing testes (Shoffner 1974). The female chicken, on the other hand, is potentially bisexual, as 2A-ZW and 3A-ZZW females may produce either female or male gametes in either the right or left gonad. When the left ovaries of the AA-ZW females are removed prior to 30 days of age, the phenotypes of these females range from a few with strong femininity through many with intersexuality, to a few with strong masculinity. Triploid female chickens with the genotype of 3A-ZZW have an intersex external phenotype. On examination of more than 30 3A-ZZW triploid females it was found that spermatozoa may be produced by either the right or left gonad and ova from the left ovary depending upon the sexual development of the specimen. Consequently, individuals without the W chromosome, that is, with only the Z sex chromosomes, produce male gametes, whereas individuals with a W chromosome may produce either ova or spermatozoa. While this evidence does not rule out the speculation by Abdel-Hameed and Shoffner (1971) that a balance situation between autosomes and the W chromosome is involved in sex determination, neither does it support the balance hypothesis. For additional information about the W chromosome of birds, see the review by Bloom (1974B).

MEIOTIC CHROMOSOMES

Gametogenesis in birds follows the same pattern as that in other vertebrates. In the chicken female both the right and left ovaries of the 18- to 21-day embryo exhibit the typical leptotene, zygotene, and pachytene stages of prophase I. The right ovary regresses shortly after hatching and the functioning left ovary contains the dictyotene-stage germ cells. These are located in the cortex of the left ovary, and as the bird matures they are gradually surrounded by prefollicular cells, with a subsequent enlargement of the oocyte, accumulation of yolk material, and formation of the follicle. Lampbrush chromosomes, which have the characteristic appearance of lateral loops radiating from a central stem, are prominent in the nucleus of the dictyotene-stage cells of birds.

The onset and duration of spermatogenesis vary considerably among species of birds. It may begin as early as 12 weeks of age in some strains of chickens and as late as the second or third year of life in some of the large birds such as the Canada goose. The diplotene and diakinesis stages in the M-I reduction division stage exhibit a high frequency of chiasmata, often with open ends rather than closed at the telomeres. As can be seen in Figure 15.3, chiasmata are numerous in the large chromosomes. Frequency of crossing-over is high in the chicken. For example, the locus for barring (*B*) is on the p arm of the Z chromosome and the locus for silver (*S*) is on the opposite arm, and in genetic tests these two loci appear to be independent (Bitgood *et al.* 1980).

CHROMOSOMAL ANALYSIS

Cytochemical Differentiation

Avian chromosomes can be characterized by differential reaction to cytochemical treatment, such as C-banding, which defines regions rich in constitutive (highly repetitive) DNA. In birds there is considerable heterogeneity in the distribution of heterochromatin between telomeric and centromeric regions (Raman *et al.* 1978; Stefos and Arrighi 1974; Stock *et al.* 1974). Also, the localization of C-bands is restricted to a limited number of the chromosomes. C-band polymorphisms have been identified in the chicken by Pollock and Fechheimer (1981).

The G-banding technique has proven to be exceedingly useful for identification of individual bird chromosomes since it specifically marks chromosomal regions with alternate dark, gray, and light bands (see Fig. 15.3). Also, break and reunion points of chromosome rearrangements can be identified (Wang and Shoffner 1974). Takagi and Sasaki (1974) found that the G-band patterns for the large chromosomes of several bird species were almost identical. This lack of interspecies polymorphism severely limits this cytochemical technique for discriminatory comparison between species. However, G-banding polymorphic patterns may be a characteristic variation within a species, as Mateescu *et al.* (1974) have identified a G-band polymorphism in the large chromosome number 1 of the chicken.

Linkage

Only a few species of birds have recognized linkage groups, and several procedures have been used to localize them to their respective chromosomes. In addition to the traits known to be on the Z chromosome of the chicken, there are nine autosomal linkage groups, only three of which have been located definitely on specific chromosomes; three that reside on microchromosomes are as yet unidentified. Zartman (1973) found evidence that the pea comb gene was located on chromosome 1 and this finding was later confirmed by Bitgood *et al.* (1980) by localizing the gene to the p arm of this large chromosome using marker chromosomes and the principle of segregating homologues, in which homologous chromosomes segregate as Mendel indicated that genes segregate. In the cytogenetic analysis of a specimen that exhibited a 3 allele phenotype for one of the *B* blood group alleles, Bloom *et al.* (1978) established the location of the chicken nucleolar organizer region (NOR) to one of the larger microchromosomes in the size range 15–18. Utilizing the same trisomic stock, Bloom and Cole (1978) determined that the locus of the *Ea-B* blood group was linked to the NOR chromosome.

The somatic cell hybridization technique was used by Kao (1973) to determine that the loci for the biosynthesis of the nucleotide adenine, *Ade-a* and *Ade-b,* are located on chromosomes 6 and 7, respectively. In the same way, Leung *et al.* (1975) located the gene for the production of endogenous cystosol thymidine kinase to one of the microchromosomes.

Padgett *et al.* (1977) separated by centrifugation the larger chicken chromosomes from the smaller ones and with appropriate RNA–DNA hybridization probes determined that the pro-virus site for the endogenous RNA tumor virus (RAV-O) is located on one of the macrochromosomes and the transforming gene for avian sarcoma virus is located on one of the microchromosomes. They also found that the chick ovalbumin gene is on one of the macrochromosomes.

CHROMOSOME VARIATION

Cytotaxonomy

Ohno *et al.* (1964) postulated that birds have a uniform and conservative karyotype. This opinion was based on observations of several species of birds, all of which possessed the characteristic macrochromosome and microchromosome configuration.

Shoffner (1974) summarized the karyotypic information on the larger chromosomes of about 100 bird species and found that approximately 90% had a large metacentric chromosome 1 and almost as many possessed an acrocentric chromosome 3, suggesting a conservative morphology for these two chromosomes. There is also a very close similarity in the G-banding patterns of the larger chromosomes of several avian and reptilian species (Takagi and Sasaki 1974). Takagi and Sasaki suggest that this configuration represents a primitive karyotype which has retained homology over a very long time. However, even with this apparent linage conservatism there are some striking chromosomal variations between species of birds, and nearly all of them have a unique karyotype.

One example among the several published, karyotypic comparisons with cytotaxonomic significance is that by DeBoer (1976), who found a wide variation among the karyotypes of 16 species of *Falconiformes*. There are four representative types based on the ratio of large to small chromosomes and the percentage of bi-armed to acrocentric chromosomes. Certain of the hawks and eagles have a distinct karyotype, characterized by a greater frequency of large bi-armed macrochromosomes as contrasted with other taxa which had a high percentage of acrocentric chromosomes and many microchromosomes.

A phylogenetically important study was reported by Takagi *et al.* (1972) on the chromosome karyotype of four geographically separated species of *Ratitae:* ostrich (*Struthio camelus*), cassowary (*Casuarius casuarius*), emu (*Dromicieus novaehollandiae*), and rhea (*Rhea americana*). Their karyotypes have a distinctly similar acrocentric chromosome complement which is representative of this group and is also distinguished by having in common homomorphic Z and W sex chromosomes. This cytotaxonomic evidence suggests a monophyletic rather than a polyphyletic origin for these large flightless land birds.

Another approach in cytotaxonomic analysis was reported by Shoffner *et al.* (1979) for the chromosome homology between the Ross's goose (*Anser rossii*) and the emperor goose (*Anser canagicus*). There is an obvious difference in morphology between chromosomes 1 and 3, whereas the other large chromosomes are morphologically identical. The cytogenetic evidence from F_1 male meiotic configurations, G-banding analysis, trisomic frequency in the F_2, and F_1 semi-sterility is a strong indication that the karyotypic difference is due to a translocation involving these two chromosomes.

It is worth noting that precise karyotypic comparisons can be made from hybrids between two species as one complement from each parent is present in the same chromosome preparation. Differences between homoelogous chromosomes (chromosomes from different species which appear to be homologous, but for which no real proof exists as to complete homology) have been determined with confidence in the hybrids of chicken × coturnix quail (Bammi and Shoffner 1966), chicken × pheasant (Basrur 1969), and muscovy duck × domestic Peking duck (Mott *et al.* 1968). Similar results have been obtained for several interspecific and intergeneric bird hybrids karyotyped in the University of Minnesota Animal Science Cytogenetics Laboratory.

Polymorphisms in Natural Populations

Both transitory and persistent structural variations of chromosomes are found in interbreeding populations. Reciprocal translocations, inversions, duplications, or deficiencies are usually transitory, as there is considerable selective disadvantage because of the semisterility resulting from the duplication–deficiency gametes of individuals with heteromorphic chromosomes. Persistent chromosome variants may be possible because of special meiotic mechanisms that evade the unbalanced gamete problem, such as that for pericentric inversions in drosophila or the fusion–fission (Robertsonian) chromosomes found in mammals.

Inversions

The first persistent heteromorphic chromosome situation reported in birds was for the white-throated sparrow (*Zonotrichia albicollis*), which has a nesting range in southern Canada and northern United States (Thorneycroft 1966, 1968). This population was found to be polymorphic for pericentric inversions involving chromosome pairs 2 and 3, with six combinations of standard and inversion chromosomes in about 400 specimens examined. Shields (1973) discovered pericentric inversions for chromosomes 2 and 5 in the slate-colored junco (*Junco hyemalis*). The same inversions were found in four other geographically distributed species of the junco. The extensive chromosomal polymorphism described for this genus suggests that the inversions occurred in a single population which has since radiated geographically. Inversion polymorphisms have been found in the common green pigeon (*Treron phoenicoptera*) by Ansari and Kaul (1979B) and in the European green finch *Chloris chloris* (now *Carduelis chloris*) by Hammar and Herlin (1975).

The frequency of inversions in these bird populations has raised some interesting questions about meiotic behavior and the role of these rearrangements in chromosomal evolution and speciation. The inversion heterozygotes may be expected to produce 50% duplication–deficiency gametes from crossing-over during meiosis, which would put a tremendous selective pressure against retention of the inversion, even it if had counteracting adaptive value. In the diakinesis configurations of the male white-throated sparrow there is a distinct, open, nonpaired region for the inversion chromosomes (Thorneycroft 1966). A similar configuration was reported by Shields (1976) in the diakinesis configuration of the pericentric heterozygote male dark-eyed junco (*Junco hyemalis*). A pericentric inversion induced by X-ray irradiation of spermatozoa in chromosome 2—inv(2)—of the chicken did not result in reduced fertility of the heterozygote, and the diakinesis configurations have a decided nonpaired region (Pollock and Fechheimer 1977). In contrast to the inv(2) described by Pollock and Fechheimer (1977), an induced inversion in chromosome 1—inv(1)—of the chicken under study in the Minnesota laboratory has semi-sterility, and duplication–deficiency chromosomes have been observed in mitotic chromosomes of early embryos (Bitgood 1980). Those inversions which persist in populations are most likely to be those which do not pair in meiosis and hence do not produce aberrant gametes. While this mechanism avoids the usual duplication–deficiency gamete semi-sterility problem, it does result in a "locked-in" linkage block encompassing the whole of the inverted segment. The inclusion of advantageous or disadvantageous genes from the original chromosome kept in this region will greatly influence the selective value of the inversion chromosome.

Translocations

A spontaneous translocation involving chromosomes 2 and 3 was reported by Ryan and Bernier (1968) in a White Leghorn male chicken. Since the translocation heterozygotes were semi-sterile, it is probable that this rearrangement was a recent one in chickens. Ansari and Kaul (1978, 1979A) reported translocation heterozygosity in an Indian species of birds, *Lonchura punctulata,* and in the black-headed oriole, *Oriolus xanthornus.* Misra and Srivastsva (1976) also reported a case of translocation heterozygosity in the cattle-egret (*Bubulcus ibis*). It is not known whether the translocations found in these Indian species are transitory of if there is some compensatory mechanism for avoiding the semi-sterility resulting from duplication–deficiency gametes in the translocation heterozygote. The possibility of a fusion–fission (Robertsonian) type of translocation cannot be overlooked, although none of this type has been previously reported in birds.

Induced Rearrangements

Mature spermatozoa of the chicken are susceptible to mutagenic effects of X-ray irradiation (Zartman 1973; Wooster *et al.* 1977). Shoffner (1972, 1975) recovered progeny with chromosomal rearrangements (8 reciprocal translocations and 1 inversion) from mature chicken males injected intraperitoneally with triethylene melamine (TEM) and ethyl methanesulfonate (EMS). Fertility and hatchability returned to normal levels 12 days following injections, indicating that only mature sperm and possibly late spermatids are affected by these mutagenic chemicals.

Altogether, about 20 translocations involving chromosomes 1, 2, 3, 4, 5, and Z and several microchromosomes have been studied extensively for meiotic behavior and used as marker chromosomes in linkage determinations. The general phenotype of the translocation birds is indistinguishable from that of standard chromosome birds. Fertilizing capacity of spermatozoa from both heterozygous and homozygous carriers is the same as for those with standard chromosomes, indicating that the genetic imbalance in the nucleus does not reduce fertilization. The semi-sterility of the heterozygous translocation carriers results from early embryonic death.

PLOIDY

A variety of euploids and aneuploids, including haploids, $2n$ parthenogens, triploids, tetraploids, mixoploids, and trisomics, have been identified in the chicken. These aberrant chromosome situations, which result from abnormal processes in gametogenesis, fertilization, or development, almost invariably terminate in early embryological death, as indicated from extensive studies by Dr. N. S. Fechheimer and colleagues at Ohio State University and by Dr. Stephen E. Bloom at Cornell University. The frequency of these aberrants are genetically influenced, as they vary in stocks of different origins (Bloom 1972; Fechheimer *et al.* 1972).

Haploid chick embryos have been described by Fechheimer *et al.* (1968), Bloom (1969, 1970, 1972), and Miller *et al.* (1971). These embryos are abnormal and succumb early in development, as no late embryos or post-hatched haploid chicks have ever been identified. Fechheimer and Jaap (1978), utilizing chromosome markers in the male parent, found that haploid embryos were largely, if not entirely, of androgenetic origin.

The $2n$ parthenogenetic embryos observed in chickens and turkeys result from either suppression or the reentry of the second polar body and subsequent fusion of the two haploid nuclei. Parthenogenetic embryos are slow in initial organization and development and are often composed of varying frequencies of haploid, diploid, tetraploid, and mixed ploidy levels (Sarvella 1970). Parthenogenetic development is under genetic influence, as shown by Olsen (1966), who increased the frequency in Beltsville Small White turkeys by selection to a level of 42%, some of which developed into mature turkeys. All embryos which hatched were males, the homogametic sex in turkeys. Parthenogenetic development in chickens is largely influenced by a single locus, inherited as an autosomal recessive (Olsen *et al.* 1968).

Spontaneous triploidy occurs with regularity in chickens. Since the first viable triploid chicken with a 3A-ZZW genotype was reported by Ohno *et al.* (1963), more than 50 post-hatched chicks, juveniles, and adult chickens from a variety of genetic sources have been identified in the Minnesota laboratory. The triploid syndrome explains some of the "sex-reversed" chickens that have been observed in poultry flocks. Except for instances in a few fishes, amphibians, and lizards, the triploid chicken is the only other known polyploid vertebrate. Triploid embryos with 3A-ZZZ and 3A-ZZW genotypes survive to hatching time and develop into mature individuals. The 3A-ZWW genotypes have never been recovered in post-hatched chicks, suggesting that the ZWW state is not developmentally compatible. The 3A-ZZZ males are phenotypically indistinguishable from diploids, with active testes that produce genetically imbalanced and morphologically abnormal spermatozoa. The 3A-ZZW specimens exhibit a wide range of sexuality for femaleness and maleness. The vast majority are hermaphroditic with both male and female germinal tissue. Triploidy could arise either as effective diploid gametry or through polyspermy. The evidence indicates that the majority arise through failure of meiosis II during oogenesis (Mong *et al.* 1974). Nondisjunction in meiosis II would account for the 3A-ZZW genotypes and possibly the 3A-ZZZ types, but the 3A-ZWW type would be expected to arise from nondisjunction at meiosis I. Wang and Shoffner (1979) provided evidence that triploids may arise from nondisjunction at either meiosis I or II inhibited by timed intraperitoneal injections of the alkaloid chemical Colcemid (Ciba) 2 to 5 hours prior to ovulation in laying hens.

There are no known tetraploid species of birds, nor have viable ones been identified in the chicken. However, chick embryos of 4A-ZZZZ and 4A-ZZWW have been observed by Bloom (1974A), and mosaic tetraploid embryos have been described by Miller *et al.* (1971).

A considerable number of mixoploid individuals with both euploid and aneuploid cell populations have been found in early chicken embryos. The majority of these abnormal chromosome situations lead to aberrant development and embryonic death. However, two adult mixoploids, a diploid male–female AA-ZZ/AA-ZW and a diploid female–triploid male AA-ZW/AAA-ZZZ were identified in the

Minnesota laboratory (Abdel-Hameed and Shoffner 1971). Both of these individuals had an intersex phenotype and nonfunctional gonads.

A variety of aneuploid types, the most common of which are trisomics involving the macrochromosomes, have been identified in early chick embryos. Two post-hatched aneuploid types have been described, one of these in the chicken which is trisomic for one of the microchromosomes in the range 15–18 (Bloom *et al.* 1978). The trisomic chickens appear to reproduce with little inhibition from the aneuploidic condition. Another post-hatched trisomic was identified in an F_2 hybrid 10 months of age between the Ross's and the emperor goose which was trisomic for the large chromosome 1 (Shoffner *et al.* 1973).

SUMMARY

Birds have a macro–micro chromosome configuration, and the male is homomorphic with two Z sex chromosomes, while the female is heteromorphic with a Z and a W sex chromosome. The W chromosome is composed almost entirely of constitutive heterochromatin and has a sex-limited immunogenetic locus. Chromosome karyotypes have been documented for about 200 species of the more than 8000 that exist in various parts of the world. Several cytotaxonomic investigations reveal that chromosome karyotype is a trait useful for systematics of birds. Structural chromosome polymorphisms have been documented in several natural breeding populations of birds. Some pericentric inversions appear not to pair during meiosis, thereby circumventing the duplication–deficiency gametes normally arising in heterozygous inversion carriers. Translocations, at least those observed in the chicken, consistently produce duplication–deficiency gametes resulting in 50% semi-sterility from aberrant zygote development. Numerical chromosomal aberrants occur with regularity in chickens, almost all of which cause death in early embryogenesis. Some few types such as triploidy 3A-ZZZ and 3A-ZZW in the chicken and trisomy-(15–18) in the chicken and trisomy-1 in the Ross's–emperor hybrid goose are developmentally compatible. Rearrangements which are recognizable in mitotic chromosome preparations are very useful chromosome markers for cytogenetic studies. Cytochemical differentiation maps, such as G-bands, C-bands, and nucleolar organizer regions, have been developed which mark specific chromosomes and regions of chicken chromosomes. Much has been accomplished in avian cytogenetics in the past two decades, and undoubtedly progress will accelerate as application is utilized in many branches of biology and in conjunction with breeding for genetic improvement.

REFERENCES

ABDEL-HAMEED, F., and SHOFFNER, R. N. 1971. Intersexes and sex determination in chickens. Science *1972,* 962–964.

ANSARI, H. A., and KAUL, D. 1978. Translocation heterozygosity in the bird, *Lonchura punctula* (Linn.) (Plonceidae: Aves). Natl. Acad. Sci. Lett. *1*(2), 83–85.

ANSARI, H. A., and KAUL, D. 1979A. Somatic chromosomes of black-head oriole, *Oriolus xanthornus* (Linn.): A probable case of translocation heterozygosity. Separatum Experientia *35,* 740.

ANSARI, H. A., and KAUL, D. 1979B. Inversion polymorphism in common green pigeon, *Threron phenicoptera* (Latham) (Aves). Jpn. J. Genet. *54*(3), 197–202.

BACON, L. D., and CRAIG, J. V. 1969. Variability of response to the female histocompatibility antigen in chickens. Transplantation *7,* 387–393.

BAMMI, R. K., and SHOFFNER, R. N., 1966. Sex ratios and karyotype in the chicken–coturnix quail hybrid. Can. J. Genet. Cytol. *8,* 533–536.

BASRUR, P. K. 1969. Hybrid sterility. *In* Comparative Mammalian Cytogenetics, K. Benirschke, Editor, pp. 107–131. Springer-Verlag, New York.

BITGOOD, J. J. 1980. Ph.D. Thesis, University of Minnesota Library.

BITGOOD, J. J., SHOFFNER, R. N., OTIS, J. S., and BRILES, W. E. 1980. Mapping of the genes for pea comb, blue egg, barring, silver and blood groups S, E, H and P in the domestic fowl. Poult. Sci. *59,* 1686–1693.

BLOOM, S. E. 1969. Chromosome abnormalities in early chicken (*Gallus domesticus*) embryos. Chromosoma (Berl.) *28,* 357–369.

BLOOM, S. E. 1970. Trisomy-3,4 and triploidy (3A-ZZW) in chick embryos: Autosomal and sex chromosomal nondisjunction in meiosis. Science *170,* 457–458.

BLOOM, S. E. 1972. Chromosome abnormalities in chicken (*Gallus domesticus*) embryos: Types, frequencies and phenotypic effects. Chromosoma *37,* 309–326.

BLOOM, S. E. 1974A. The origins and phenotypic effects of chromosome abnormalities in avian embryos. Proc. 15th World Poultry Congr. New Orleans, 316–320.

BLOOM, S. E. 1974B. Current knowledge about the avian W chromosome. Bioscience *24*(6), 340–344.

BLOOM, S. E., and COLE, R. K. 1978. Chromosomal localization of the major histocompatibility (B) locus in the chicken. Poult. Sci. *57,* 1119.

BLOOM, S. E., SHALIT, P., and BACON, L. D. 1978. Chromosomal localization of nucleolus organizers in the chicken. Genetics *88* (Suppl.), 13.

COCK, A. G. 1964. Dosage compensation and sex chromatin in non-mammals. Genet. Res. (Cambr.) *5,* 354–365.

DEBOER, L. E. M. 1976. The somatic chromosome complements of 16 species of *Falconiformes* (*Aves*) and karyological relationship of the order. Genetica (The Hague) *46*(1), 77–113.

FECHHEIMER, N. S., and JAAP, R. G. 1978. The parental source of heteroploidy in chick embryos determined with chromosomally marked gametes. J. Reprod. Fertil. *52,* 141–146.

FECHHEIMER, N. S., ZARTMAN, D. L., and JAAP, R. G. 1968. Trisomic and haploid embryos of the chick (*Gallus domesticus*). J. Reprod. Fertil. *17,* 215–217.

FECHHEIMER, N. S., MILLER, R. C., and BOERGER, K. P. 1972. Genetic influence on incidence and type of chromosome aberrations in early chicken embryos. Proc. 7th Intl. Congr. Anim. Reprod. Artif. Insem., Munich. *2,* 1129–1133.

GILMOUR, D. G. 1967. Histocompatibility antigens in the heterogametic sex in the chicken. Transplantation *50,* 699–706.

HAMMAR, B., and HERLIN, M. 1975. Karyotypes of four bird species of the order Passiformes. Hereditas *80,* 177–184.

KAO, F. 1973. Identification of chick chromosomes in cell hybrids formed between chick erythrocytes and adenine-requiring mutants of Chinese hamster cells. Proc. Natl. Acad. Sci. U.S.A. *70,* 2893–2898.

LUENG, W. C., CHEN, T. R., DUBBS, D. R., and KIT, S. 1975. Identification of chick thymidine kinase determinant in somatic cell hybrids of chick erythrocytes and thymidine kinase-deficient mouse cells. Exp. Cell Res. *95,* 320–326.

MATEESCU, V., STEFANESCU, M., and STAVAR, P. 1974. G-banding patterns in *Gallus domesticus* macrochromosomes. Proc. 15th World Poultry Congr., New Orleans 321–322.

MILLER, R. C., FECHHEIMER, N. S., and JAAP, R. G. 1971. Distribution of karyotype abnormalities in chick embryo sibships. Biol. Reprod. *14,* 549–560.

MISRA, M., and SRIVASTSVA, M. D. L. 1976. Somatic chromosomes of *Bubulcus ibis* (L.) (Cattle-egret): A case of reciprocal translocation. Genetica *46,* 155–160.

MONG, S. J., SNYDER, M. D., FECHHEIMER, N. S., and JAAP, R. G. 1974. The origin of triploidy in chick (*Gallus domesticus*) embryos. Can. J. Genet. Cytol. *16,* 317–322.

MOTT, C. L., LOCKHART, L. H., and RIGDON, R. H. 1968. Chromosomes of the sterile hybrid duck. Cytogenetics *7,* 403–412.

OHNO, S., KITTRELL, W. A., CHRISTIAN, L. C., STEINHAUS, C., and WITT, G. A. 1963. An adult triploid chicken (*Gallus domesticus*) with a left ovotestis. Cytogenetics *2,* 42–29.

OHNO, S., STENIUS, C., CHRISTIANS, L. C., BECAK, N., and BECAK, M. L. 1964. Chromosomal uniformity in the avian subclass Carinatae. Chromosoma *15,* 280–288.

OLSEN, M. W. 1966. Frequency of parthenogenesis in chicken eggs. J. Hered. *57,* 23–25.

OLSEN, M. W., WILSON, P. P., and MARKS, H. L. 1968. Genetic control of parthenogenesis in chickens. J. Hered. *59,* 41–42.

PADGETT, T. G., STUBBLEFIELD, E. and VARMUS, H. E. 1977. Chicken macrochromosomes contain an endogenous provirus and microchromosomes contain sequences related to the transforming gene of ASV. Cell *10,* 649–657.

POLLOCK, B. J., and FECHHEIMER, H. S. 1977. An autosomal pericentive inversion in *Gallus domesticus* (3rd Colloq. Cytogenet. Domest. Anim., Jouy-en-Josas). Ann. Genet. Sel. Anim. *9,* 453–541.

POLLOCK, B. J., and FECHHEIMER, N. S. 1981. Variable C-banding patterns and a proposed C-band karyotype in *Gallus domesticus.* Genetica *54,* 273–279.

RAMAN, R., JACOB, M. and SHARMA, T. 1978. Heterozygosity in distribution of constitutive heterochromatin in four species of birds. Genetica *48,* 61–65.

RYAN, W. C., and BERNIER, P. E. 1968. Cytological evidence for a spontaneous chromosome translocation in the domestic fowl. Experientia *24,* 623–624.

SANDNES, G. C. 1954. A new technique for the study of avian chromosomes. Science *119,* 508.

SARVELLA, P. 1970. Sporadic occurrence of parthenogenesis in poultry. J. Hered. *61,* 215–219.

SCHMID, W. 1962. DNA replication patterns of the heterochromosomes in *Gallus domesticus.* Cytogenetics. *1,* 344–352.

SHIELDS, G. F. 1973. Chromosomal polymorphism common to several species of *Junco* (Aves). Can. J. Genet. Cytol. *4,* 461–471.

SHIELDS, G. F. 1976. Meiotic evidence for pericentric inversion polymorphism in *Junco* (Aves). Can. J. Genet. Cytol. *18,* 747–751.

SHOFFNER, R. N. 1972. Mutagenic effects of triethylene melamine (TEM) and ethyl methanesulfonate (EMS) in the chicken. Poult. Sci. *51,* 1865.

SHOFFNER, R. N. 1974. Chromosomes of birds. *In* The Cell Nucleus, E. H. Busch, Editor, Vol. 2, pp. 223–261. Springer-Verlag, New York.

SHOFFNER, R. N. 1975. Heteroploidy and chromosomal rearrangements in *G. domesticus. In* Genetics Lectures, Vol. 4, pp. 177–188. Oregon State University Press, Corvallis.

SHOFFNER, R. N., KRISAN, A., and HAIDEN, G. J. 1967. Avian chromosome methodology. Poult. Sci. *46*(2), 334–344.

SHOFFNER, R. N., OTIS, J. S., and LEE, F. 1973. Trisomy and abnormal segregation in Ross's × Emperor goose hybrids. Genetics *74,* 2253 (Proc. 13th Intl. Congr. Genet., Berkeley).

SHOFFNER, R. N., WANG, N., LEE, F., KING, R., and OTIS, J. S. 1979. Chromosome homology between the Ross's and Emperor goose. J. Hered. *70,* 395–400.

SOKOLOW, N. N., and TROFIMOW, I. E. 1933. Individualität der Chromosomen und Geschlechtsbestimmung beim Haushuhn. Z. Ind. Abst. Vererb. *65,* 327–352. (Quoted by Hutt, F. B. 1949 in Genetics of the Fowl, McGraw-Hill, New York.)

SOMES, R. G. 1978. New linkage and revised chromosome map of the domestic fowl. J. Hered. *69,* 401–403.

STEFOS, K., and ARRIGHI, F. E. 1971. Heterochromatic nature of W chromosome in birds. Exp. Cell Res. *68,* 229–231.

STEFOS, K., and ARRIGHI, F. E. 1974. Repetitive DNA of *Gallus domesticus* and its cytological locations. Exp. Cell Res. *83,* 9–14.

STOCK, A. D., ARRIGHI, F. E., and STEFOS, K. 1974. Chromosome homology in birds: Banding patterns of the chromosomes of the domestic chicken, ring-necked dove and domestic pigeon. Cytogenet. Cell Genet. *13*(5), 410–418.

TAKAGI, N., and SASAKI, M. 1974. A phylogenetic study of bird karyotypes. Chromosoma *46,* 91–120.

TAKAGI, N., ITOH, M., and SASAKI, M. 1972. Chromosome studies in four species of Ratitae (Aves). Chromosoma (Berl.) *36,* 281–291.

THORNEYCROFT, H. B. 1966. Chromosomal polymorphism in the white-throated sparrow *Zontrichia albicollis* (Gmelin). Science *154,* 1571–1572.

THORNEYCROFT, H. B. 1968. A cytogenetic study of the white-throated sparrow, *Zonotrichia albicollis* (Gmelin). Ph. D. Thesis, University of Toronto.

WANG, N. and SHOFFNER, R. N. 1974. Trypsin G- and C-banding for interchange analysis and sex identification in the chicken. Chromosoma *47*(1), 61–69.

WANG, N., and SHOFFNER, R. N. 1979. Induction of heteroploidy in *Gallus domesticus.* Mutation Res. *69,* 263–273.

WOOSTER, W. E., FECHHEIMER, N. S., and JAAP, R. G. 1977. Structural rearrangements of chromosomes in the domestic chicken: Experimental production by X-irradiation of spermatozoa. Can. J. Genet. Cytol. *19,* 437–446.

ZARTMAN, D. L. 1973. Location of the pea comb gene. Poult. Sci. *52,* 1455–1462.

16

Application of Cytogenetics to Livestock Improvement

The rate of progress in the field of cytogenetics of livestock today is comparable to that of genetics between 1910 and 1920. Research results in cytogenetics are being published at the rate of several each month compared to only a very few per year 20 years ago. Many of these papers describe new chromosomal aberrations, or estimate their frequency in different populations of livestock. Other papers are reports of the relationship between the aberration itself and some anatomical or physiological characteristic in the animals. Most frequently however, no relationship has been established, owing perhaps to the fact that the cytological observations made by different researchers do not have the same degree of refinement as yet.

Each segment of scientific research must go through a descriptive stage, the fact-finding stage. When sufficient numbers of basic or fundamental facts or repeatable observations have been described, relationships can be established and then the newly acquired knowledge can be applied. For example, freemartins have been recognized in cattle for centuries, and the classic work of Lillie in 1916 provided us with knowledge of the role that placental anastomosis played in the development of this form of female sterility. The discovery, through blood typing, that each of the individuals in most cattle twins possessed the set of antigens for its twin as well as itself, was further evidence of the continued effect of the sharing of a common blood supply in utero and the transfer of hematopoietic cells from each twin to the other. Finding sex chromosome chimerism in each of the cattle twins from most heterosexual pairs added another piece of knowledge to the understanding of freemartinism. The practical application of this knowledge was the availability of a new technique, karyotyping, that could be used to separate the freemartins from the fertile females born twin to bulls.

When the first single-born, sex chromosome chimeric animals were found, the most logical and least complex explanation for their occurrence was that there must have been a twin present in early pregnancy that established the opposite sex chromosome cell line, that subsequently it had died and was aborted or resorbed. The study of chromosomes of early embryos by Fechheimer and Harper (1980) indicated that some mechanism other than the presence of a twin of the other sex could cause chimerism. From this study it appears that fertilization of two maternal pronuclei by two sperm, or the fusion of two zygotes, may occur more frequently than has been recognized. The phenotype of male × male or female × female chimeras might be affected by this chimerism, but such animals, in the absence of a marker chromosome such as a Robertsonian translocation, could not be detected by chromosome analysis. If the phenotypic response to such chimerism were large, and if the chimerism were also present in the gonads, then the occurrence of such animals probably would add to the error variance in quantitative analysis. Identification of such animals could lead to refinement of quantitative analyses. If the animal were a true chimera in all tissues including the gonads, then more than two alleles could be present. Therefore any one gamete might carry any one of four alleles, leading to a frequency of less than one half for any one allele. The total frequency of such chimeras in cattle was postulated to be between 4% and 5% by Fechheimer and Harper (1980).

The unexpectedly large number of meiotic errors found by Koenig *et al.* (1983) in slaughter heifers could have been caused by either feed additives or heredity. If a significant variation among strains of cattle does exist in the frequency of abnormal chromosomes in oocytes, then selection against these meiotic errors could be used to increase fertility in cattle. If, on the other hand, the pre-slaughter treatment caused these meiotic errors, then selection would be ineffective. However, such errors could be used in diagnostic analysis of pre-slaughter treatments.

The prevailing attitude of researchers has been that selection could be directed against any chromosomal abnormalities. A good example has been the decisions of some countries to eliminate the 1/29 Robertsonian translocation whenever it is found, since heterozygous females are usually found to have a fertility level 5–10% below average. No positive effect of this aberration has been clearly identified. However, the fact that this aberration has been found so widespread throughout the world, and with such a high frequency (up to 40%) in some breeds, could lead to the conclusion that it may have some positive benefits. Bruere and Ellis (1979) have clearly stated a position that seems reasonable with respect to the attitude that should be taken concerning Robertsonian translocations in livestock, especially in cattle, sheep, and goats: "It now seems clear that the presence of at least 3 stable centric fusions (Robertsonian translocations) in domestic sheep, even in triply homozygous animals ($2n = 48$), is in no way detrimental to either the individual or the population. The mere persistence of such a polymorphic system both naturally for probably 100 years or more and experimentally gives weight to the argument that future investigation of such systems in domestic animals should be to determine their genetic meaning. In no way should they be eliminated arbitrarily from our animal populations on limited veterinary hypothesis." Since it is postulated that the decrease in fertility is a result of nondisjunction and aneuploid gamete formation in heterozygotes, it is also logical to assume that homozygotes would have fully restored fertility. This latter assumption is supported by the work of Bruere and Ellis (1979) in which homozygous rams have been found to have no aneuploid secondary spermatocytes. It is possible, therefore, that 58-chromosome cattle homozygous for the 1/29 translocation might not have reduced fertility and could have some selective advantage over the normal 60-chromosome animals. It is generally agreed that new species have developed naturally by chromosomal aberrations leading to sexual incompatibility and additional gene mutation and chromosomal modifications. Could there be an advantage in smaller numbers of chromosomes? Although selection for such a trait temporarily would reduce the selection pressure for economically important characteristics, the chromosomal variation could be very rapidly fixed in a population and selection for economically important traits could be resumed.

Since the generally accepted view is to select against chromosomal aberrations, the identification of important breeding animals carrying these aberrations becomes highly desirable. Bulls being used in artifical insemination, and potentially capable of producing many thousands of offspring, should be routinely examined for chromosomal variations. The cost of such an examination is relatively low, generally only 1/1000th of the total investment an artificial insemination organization puts into each dairy bull by the time it has been tested for production and type-transmitting ability. It would not appear unreasonable to make chromosomal analysis mandatory for all young sires which are to be used in artificial insemination.

Several chromosomal aberrations in swine have been found to cause small litters, from four to six pigs per litter. These anomalies can be detected by karyotype analysis so that abnormal boars or gilts, from a desirable strain of swine, can be identified and only the animals free from the anomalies would be used for breeding.

It has been assumed that most single-gene-caused differences in animals, such as black vs. red, polled vs. horned, solid color vs. spotted, in cattle (and presumably genes causing quantitative effects on milk production or rate of gain, etc.), are the result of modifications in the DNA that cannot be detected by chromosomal analysis, even with the most detailed banding procedures. However, in several types of defective animals and humans in which the defect originally had been thought to be the result of dominant or recessive inheritance, the cause of the defect has more recently been found to be a cytological aberration (Hozier *et al.* 1982; Donahue *et al.* 1968; Ellsworth 1979; Zellweger 1979; Ledbetter *et al.* 1982). Therefore, when abnormal characteristics in livestock have been explained by dominant or recessive inheritance, they may be the result of some chromosomal aberrations rather than point mutations. More refined methods of G-banding may permit the identification of some chromosomal changes, which then could be used to identify carriers of the "genes."

Sexing of embryos for embryo transfer in cattle can be done by cytological observation of a small amount of material from the embryo since the X and Y chromosomes are easily and accurately identifiable. The feasibility of sexing embryos in other species is dependent upon the ease of identifying X and Y chromosomes. The cost of embryo transfer places limits on its use to date. As the cost decreases it may become economically desirable for dairymen to nearly double the numbers of heifer calves that are produced each year. Twins from sexed embryo transfers would eliminate the occurrence of freemartins, and twinning could become a more desirable characteristic in embryo transfer. The increase in heifer calves each year would permit higher selection pressure, as well as produce more calves from the top females in the herd. More efficient techniques for freezing embryos for transfer would increase the adaptability of the procedure for use in smaller herds of cattle. Better cytogenetic methods for sexing could improve the frequency of successful transfer after sexing compared to present methods. Sexing of embryos for the production of bulls from highly superior cows could also be done, but heifers from these matings would also be so desirable that seldom would the female embryos be deliberately discarded.

Genetic engineering, narrowly defined as the incorporation of desirable DNA segments into the genome, could be a major contribution to livestock improvement in the future. A large amount of cytogenetic work needs to be done with livestock in order to accomplish this. First, the desirable DNA segments need to be identified and located on the chromosomes of livestock. Some syntenic groups of enzymes have been identified, such as LDHB (lactate dehydrogenase B), TPI (triose phosphate isomerase), and PEPB (peptidase B) as one group in cattle. This same syntenic group has also been found in sheep and in rabbits, and on chromosome 12 in humans. This group of enzymes is not syntenic in swine. PGD (phosphogluconate dehydrogenase) and ENO1 (enolase 1) are syntenic in cattle but are not syntenic with PGM1 (phosphoglucomutase 1) as on human chromosome 1. Nine enzymes have been found to be located on nine different chromosomes in cattle, or are nonsyntenic: LDHA (lactate dehydrogenase A), MDH2 (malate dehydrogenase 2), IDH1 (isocitrate dehydrogenase 1), SOD1 (superoxide dismutase 1), PKM2 (pyruvate kinase M2), PGM1 (phosphoglucomutase 1), PGM2 (phosphoglucomutase 2), AK1 (adenylate kinase 1), and MPI (mannose phosphate isomerase). Similar identifications have been made for sheep, swine, and rabbits. As these studies are extended and marker chromosomes (or G-banded chromosomes) are identified, livestock chromosome maps can be prepared (Echard *et al.* 1980).

Most economic traits of livestock are influenced by multiple genes, or loci, and improvement through selection can be accomplished most readily by the application of quantitative genetics methods. Reproductive traits, however, usually have a low heritability, and improvement has been very slow. Several chromosomal aberrations, including aneuploidy in oocytes, have been found to affect level of fertility. Studies of meiotic chromosomes have been neglected in recent years since improved methods for studying mitotic chromosomes were found. Further meiotic studies of both males and females might provide information which could be used to improve fertility.

It must be recognized that cytogenetics of livestock is still a relatively new branch of biological science. Many more basic facts may need to be discovered before application of this information can be used directly in the economic improvement of livestock production.

REFERENCES

BRUERE, A. N., and ELLIS, P. M. 1979. Cytogenetics and reproduction of sheep with multiple centric fusions (Robertsonian translocations). J. Reprod. Fertil. *57,* 363–375.

DONAHUE, R. P., BEAS, W. B., RENWICK, J. H., and McKUSICK, V. A. 1968. Probable assignment of the Duffy blood group locus to chromosome 1 in man. Proc. Natl. Acad. Sci. U.S.A. *61,* 949–955.

ECHARD, G., HORS-CAYLA, M. C., GELLIN, J., SAIDI-MEHTAR, N., HEURTZ, S., BENNE, F., and GILLOIS, M. 1980. The status of the gene map of domestic animals (cattle, sheep, pigs and rabbits). Proc. 4th Eur. Colloq. Cytogenet. Domest. Anim., Uppsala, Sweden 332–343.

ELLSWORTH, R. M. 1979. Retinoblastoma. *In* Birth Defects Compendium, D. Bergsma, Editor, pp. 937–938. National Foundation–March of Dimes, Alan R. Liss, New York.

FECHHEIMER, N. S., and HARPER, R. L. 1980. Karyological examination of bovine fetuses collected at an abattoir. Proc. 4th Eur. Colloq. Cytogenet. Domest. Anim. 194–199.

HOZIER, J., SAWYER, J., CLIVE, D., and MOORE, M. 1982. Cytogenetic distinction between the TK^+ and TK^- chromosomes in the L5178Y $TK^{+/-}$ 3.7.2C mouse–lymphoma cell line. Mutation Res. *105,* 451–456.

KOENIG, J. F. L., ELDRIDGE, F. E., and HARRIS, N. 1983. A cytogenetic analysis of bovine oocytes cultured in vitro. J. Dairy Sci. *66* (Suppl. 1), 253.

LEDBETTER, D. H., RICCARDI, V. M., AIRHART, S. D., STROBEL, R. J., KEENAN, B. S., and CRAWFORD, J. D. 1982. Deletion of chromosome 15 as a cause of the Präder-Willi syndrome. New Engl. J. Med. *304,* 325–329.

LILLIE, F. R. 1916. The theory of the freemartin. Science *43,* 611–613.

STRANZINGER, G. 1980. Application of cytogenetics to the breeding of useful animals (in German). Jahrb. Schweiz. Naturforschenden Ges. *3,* 68–79.

ZELLWEGER, H. 1979. Präder-Willi syndrome. *In* Birth Defects Compendium, D. Bergsma Editor, pp. 883–884. National Foundation–March of Dimes, Alan R. Liss, New York.

Index

A

Aberrations, chromosomal
  birds, 264, 274–279
  cattle, 122–163
  fertility, 78, 86–88
  goats, 207
  mutation, 5
  selection, 286, 288
  sheep, 195–197, 200, 204
  swine, 231–233
Acrocentric chromosomes, 20, 23, 24, 25
  asses, 244
  birds, 264
  goats, 204, 205, 209
  horses, 243
  sheep, 194, 209
  swine, 219
  Y, in cattle, 115, 160
Acrosome, 61
Adjacent segregation, 43
Allophenic, 106
Alternate segregation, 43
Amniocentesis, 178
Anaphase, 12, 14, 18
Androgenesis, 63
Aneuploidy, 5, 31, 78, *see also* Robertsonian translocation
  birds, 277
  cattle, 86, 153–158
  horses, 257, 258
  sheep, 200, 204
  spermatozoa, 87
  swine, 227
  trisomy, 85, 153–158
  variation, 29
Anidian monsters
  cattle, 153
  sheep, 200
Asses, 248
  chromosome banding, 248
  chromosome number, 244
  early study, 243
  karyotype, 246
  meiosis, 259
Atretic follicles, 178
Autologous plasma, 101

B

Banding, 51
  C, 52, 122, 248
  fluorescent, 51
  G, 52, 122, 127, 204, 248
  NOR, 26, 59
  Q, 51, 52, 122, 127, 204, 248
  R, 52, 127, 204, 248
  T, 52
Barr body, 66
Birds
  aneuploidy, 277
  chimeric, 278

Birds (*cont'd.*)
  chromosomal aberrations, 274–279
  chromosome banding, 271
  chromosome number, 273
  G-banded, 266, 271
  gametogenesis, 270
  gonosomes, 268
  induced aberrations, 276
  inversions, 275
  karyotype, 265
  linkage, 272
  macrochromosomes, 263, 264
  meiotic chromosomes, 267, 270
  microchromosomes, 263, 264
  mosaic, 278
  parthenogenetic embryos, 277
  ploidy, 277
  polyploid, 277
  repetitive DNA, 268
  sex determination, 268–270
  tetraploid, 277
  translocation, 276
  triploid, 277, 278
  trisomic, 277, 279
  W chromosome, 269
  Z chromosome, 268
Bivalents, 19
Blastocoele, 64
Blastocyst, 64, 107–108
  cattle, 125, 170
  swine, 229
Blastomeres, 64
Bos indicus, 160
Bos taurus, 160

C

Cats
  calico, 67
  tortoiseshell, 67
Cattle
  chiasma frequency, 172
  chromosomal aberrations, 122–163
  early studies, 115–122
  hybrids, 164
  intersex, 162, 83–85
  karyotype, 49, 54–55, 57, 117, 118–119, 120–121, 128, 130, 135, 136, 137, 140, 142, 148–149, 154, 156, 166
  meiosis, 172
  mitochondria, 180
  oocyte studies, 173–178
  polyploid, 172
  sex chromosome, 116
  sex identification, 178
  sister-chromatid exchange, 179
  triploid blastocyst cells, 170
  tritiated thymidine, 147
Cattle chromosomes
  C-banded, 122
  G-banded, 118–121, 122, 127, 130–131, 136, 142, 149, 154, 156, 164
  Q-banded, 122
  61,XXX female, 158
C-banding, 52–56, 122
C_d-banding, 56
Cell
  cumulus, 109
  environmental variables, 11
  established lines, 106
  nutrient availability, 11
  temperature, 11
Cell culture, 96
Cell division
  interphase, 9
  metabolic activity, 9
  resting stage, 9
Centric fusion, 31, 35
Centrioles, 2, 13, 14, 62
Centromere, 11, 14, 24
Centrosome, 13, 62
Chiasmata, 19, 172–173
  cattle, 173
Chimerism, 62, 67, *see also* Mosaicism
  cattle, 79, 83, 88, 161–163, 167, 172, 286
  goats, 208
  horses, 251–259
  sheep, 197–200
Chromatids, 10, 11, 13
  breaks, 88
Chromomeres, 18
Chromosome, 1, 2, 10, 12, 13, 14, 18, 29
  acrocentric, 20, 23, 24, 25, 160
  biochemical nature, 6

blastocyst, 107, 170
breaks, 31–44, 169
DNA structure, 6
first observation, 4
haploid set, 22
homologous, 5, 19
length, 27
metacentric, 20, 24, 25
name, 4
replication, 17
Robertsonian translocation, 132–143
selective loss, 72
submetacentric, 25, 160
subtelocentric, 24
telocentric, 24
telomeric end, 26
Chromosome aberrations, *see* Aberrations
Chromosome banding, 3, 122
asses, 248
birds, 271
cattle, *see* Cattle chromosomes
horses, 248
mules, 248
swine, 223
Chromosome maps, 46
linkage, 5
somatic cell hybrid, 72
syntenic groups, 289
Chromosome morphology, 23–27
Chromosome numbers, 23
asses, early work, 243
birds, 273
cattle, 116
goats, 204–206
horses, early work, 243
mules, early work, 243
sheep, 189
swine, early work, 219
Cleavage, 64
Colcemid, 53, 102
Colchicine, 31, 53, 101, 102
Congenital abnormalities, 169
Contamination, 95
Culture media, 12
Crossing-over, 19
Cytogenetics
application, 285–290
definition, 1
Cytokinesis, 14
Cytology, 1, 5

D

Deficiencies, 36
Deletion, 36, 78
Deoxyribonucleic acid (DNA), 6, 11, 12
Diakinesis, 18, 19, 21
Dicentric chromosome, 32, 38
Dicentric translocations, cattle, 143, 146
Diestrus, 21
Digyny, 62
Diploid ($2n$), 11, 17
Diplotene, 18, 19, 21
DNA, *see* Deoxyribonucleic acid
Drumstick, 66
Duplication, 36
Dyad, 20

E

Embryonic loss, 78
Embryo sexing, 288
Epididymis, 21
Equational division, 20
Equatorial plate, 13
Euchromatin, 56
Euploid, 29, 30
Extrachromosomal inheritance, mitochondrial, 2, 180

F

Facultative heterochromatin, 66, 268
Fallopian tube, 64
Fertility, 35, 39, 43, 75–89, 170, 290
cattle, 77, 139, 141, 146, 150, 173
sheep, 196, 200
sheep × goat hybrid, 213
translocation, 124–127, 287
Fertilization, 61
Fetal calf serum, 101
Fibroblasts, 96

Fixative, 103
Fluorescent banding, 51
Follicular cells, 61
Freemartinism, 79–82, 161, 289
  birds, 278
  horses, 251, 256, 257
  sheep, 197
  swine, 226, 228

G

Gamete, 21–22, 61–62
  cattle, 115–116, 172–178
  female, 17
  male, 17
  unbalanced, 39
Gametogenesis, 21–22
  birds, 270
G-banding, 52, 122, 204, 248
  birds, 266, 271
  cattle, 118–121, 122, 127, 130–131, 136, 142, 149, 154, 156, 164
  goat, 206, 207
  horses, 248–250, 252
  identification, 127
  sheep, 183, 196, 198–202
  swine, 221, 223–226, 236–237
Genetic engineering, 289
Germinal cells, 18
Germinal vesicle, 22
Goats
  banding, G, Q, R, 204
  chromosomal aberration, 207
  early study, 204–206
  G-banded, 206, 207
  infertile male, 208
  intersex, 207
  polled intersex, 208
  pseudohermaphrodite, 208
  Robertsonian translocation, 207
  sex chimera XX/XY, 208
  sex chromosome, 204, 207–209
  sheep × goat hybrid, 209–215
Graafian follicle, 22
Gynogenesis, 63

H

Haploid (*n*), 17, 21, 30
Heparin, 99, 101
Heritability, 75–76
Hermaphrodite, 83, *see also* Intersex
  cattle, 162
  swine, 227
Heterochromatin, 56, 144
  birds, 268, 269
  cattle, 143
Heterogametic sex, 64
  birds, 269
Heteroploid, 30, 78
Histone, 12
Homology
  cattle, sheep, goats, 163–164
  NF (*nombre fondamentale*), 163
Horses, 243–259
  chimerism, 251, 256, 257
  chromosome banding, 248
  early study, 243
  G-banded, 248–250, 252
  hybrids, 244
  infertility, 257–259
  intersexuality, 251–256
  karyotype, 245, 249, 250, 252–253, 254, 255
  meiosis, 259
  Przewalski, 244
  Robertsonian translocation, 248
  sex chromosome, 243, 257
Hyaluronidase, 61
H-Y antigen, 65
  cattle, 85
Hybrids
  asses, 244
  birds, 274
  cattle, 164–167
  Equus, 259–260
  goat × ibex, 215
  horses, 244–248
  sheep × goat, 209–215
  swine, 222–223
Hybrid mammalian cells, 71
Hypotonic solution, 102, 108

I

Idiogram, 50
Implantation, 64
Infertility, *see also* Fertility
  cattle, 79, 124, 125, 150, 152, 156–160, 162, 178, 286
  goats, 208
  horses, 257–259
  hybrid, 164–167, 259
  swine, 238, 231–233
Interkinesis, 18, 20
Interphase, 10
  S period, 11
Intersex, 83–85
  cattle, 162
  goats, 207
  horses, 251–256
  polled gene, effect, 207
  swine, 226–228
Inversion, 37, 39, 46
  birds, 275
  cattle, 158–160
Isochromosomes, 37

J

Jersey Y chromosome, 161

K

Karyotype, 25, 46–59, *see* individual species
Klinefelter's syndrome, 200
  61,XXY male cattle, 156, 158
Laboratory procedure, 93–112
  blastocyst chromosome, 107
  established cell line, 106
  feather pulp, 263
  fibroblast culture, 96
  incubation temperature, 97
  lymphocyte culture, 99–105
  meiosis, 108
  squash, 110, 263
  tissue culture, 96
Leptotene, 18
Lethal brachygnathia trisomy syndrome (LBTS), 153
Luteinizing hormone (LH), 22
Lymphocyte cultures, 99–105

M

Meiosis, 2, 4, 5, 17–22, 29
  asses, 259
  birds, 267
  cattle, 172
  horses, 259
  mules, 259
  reduction division, 17
  Robertsonian translocations, 32–35
  swine, 237
Meiotic errors, 78
  cattle, 174–178, 286
Meiotic studies, 108
  cattle oocytes, 173–178
  chiasmata, 172
  chromosome, 27, 115, 126, 174–177, 267, 270
Mendel's law, 5
Metacentric chromosomes, 20, 24, 25, 115, 160–161
  asses, 244
  aoudads, 213
  birds, 264
  goats, 209
  horses, 243
Metaphase, 12, 13, 18
  mitotic, 23
Microscopy
  electron, 2, 46, 93
  fluorescent, 3
  interference, 3
  light, 2, 4, 19, 93
  phase-contrast, 3
Microtome, 93
Microtubules, 2, 14
  effect of colchicine, 53, 102
  meiosis, 20
Mitochondria, 2, 180

Mitosis, 2, 9, 10, 11, 27
 definition, 4, 17
 stages, 13–14
Monolayer, 96
Morphology, 23–28, 45
Morula, 64
Mosaic, 79
 birds, 278
 swine, 226
Mosaicism, 67, 79–85, *see also* Chimerism
 birds, 278
 cattle, 141, 158, 161–163
 swine, 226–228
Mules
 chromosome banding, 248
 early study, 243
 karyotype, 247
 meiosis, 259
 spermatogonia 259
 tetraploid, 259

N

n, see Haploid (*n*)
Nondisjunction, 34, 86, 125
NOR, *see* Nucleolus organizer region
Nuclear membrane, 14
Nucleolus, 2, 20, 26
Nucleolus organizer region (NOR), 26, 59
Nucleotides, 10
Nucleus, 2, 13, 14

O

Oocytes, 21, 109
 cattle, 173–178
Oogonia, 21
Ovary, 18, 109
Ovulation, 61
Ovum, 61

P

Pachytene, 18, 19, 21
Paracentric inversion, 37
Parthenogenesis, 63
 birds, 277
Perforatorium, 61
Pericentric inversion, 39
 cattle, 158
Perivitelline space, 62
pH, 97
Phytohemagglutinin (PHA), 98, 99
Polar body, 18, 22
Polyethylene glycol (PEG), 71
Polyploidy, 5, 30, 31
 birds, 277
 cattle, 78, 170–172
 swine, 229
Polyspermy, 62
Population studies, cattle, 167
Preleptotene, 18
Primary cell lines, 106
Primary constriction, 26, 45
Primary cultures, 105
Pronase, 109
Pronucleus, 22, 62, 106
Prophase, 12, 13, 18, 20, 21
Pseudohermaphrodite, 208, *see also* Hermaphrodite

Q

Quadrivalent, 43
Q-banding, 51, 52, 122, 204, 248
 identification, 127

R

R-banding, 52, 204, 248
 identification, 127
Reductional division, 20
Ribosomes, 26
Robertsonian translocation, 31–35, 86
 1/29, cattle, 122–133
 1/29, frequency, 123, 167, 168
 1/29, transmitted as a dominant, 124
 chromosomes involved, 132–143
 goats, 207
 horses, 248

level of fertility, 196
origin, 1/29, 127
phenotypic characteristics, 123
sheep, 195–197
swine, 222

S

Salivary gland chromosomes, 36, 46
Satellite, 26
Secondary constriction, 26, 27, 45
Seminiferous tubules, 21
Sendai virus, 71
Senescence, 106
Sex chromatin, 66
Sex chromosome, 286
birds, 268–270
cattle, 116
freemartins, 79–82
goats, 204, 207–209
horses, 243, 257
intersex, 83–85
sheep, 190–195, 197–200
swine, 224–229
translocation, 147–150
trisomy, 67, 85, 156–158, 162, 169, 200, 227, 229, 257
Sex determination, 5, 64–68, *see also* Sex chromosome
H-Y antigen, 65
Sheep
chromosomal aberrations, 195–197, 200–204
early studies, 189
freemartinism, 197
G-banded, 183, 196, 198–203
karyotype, 191, 192, 193, 194, 201–203, 205, 206
Robertsonian translocation, 195–197
sex chromosome, 190–195, 197–200
sheep × goat hybrid, 209–215
Sheep × goat hybrid, 209–215
karyotype, 212, 214
Somatic cell, 17, 18
hybrid, *see* Chromosome maps
Somatic cell genetics, 71–73, 106
Spermatids, 21
Spermatocyte, 21
Spermatozoa, 21, 61
aneuploid, 87
Spermatogenesis, 21
mules, 259
Spermatogonia
type A, 21
type B, 21
Spindle fiber, *see* Microtubules
Squash technique, 110
Sterility, 30
cattle, 156
hybrid, 164–167, 244
mule, 244
Submetacentric, 25
asses, 244
birds, 264
goats, 209
horses, 243
sheep, 189, 190, 191, 209
sheep × goat hybrid, 213
Y, in cattle, 160
Subtelocentric, 24
birds, 264
horses, 244
Superovulation, 62
Swine
blastocysts, 229
chimerism, 226, 228
chromosomal aberrations, 231–233
chromosome irradiation, 230
early studies, 219
embryological studies, 229
G-banded, 221, 223–226, 236–237
induced aberrations, 230
intersex, 226–228
karyotype, 220, 221, 224, 234–235, 236
meiosis, 237
phenotypic abnormalities, 233
reciprocal translocations, 231
Robertsonian translocations, 22, 231
sex chromosome, 224–229
sister-chromatid fusion, 230
tetraploid, 229
triploid, 229

Syntenic groups, 289, *see also* Chromosome maps
Synapsis, 19
Synaptinemal complex, 19
Syngamy, 62
Synthesis, 9–13, 26
  DNA, 11, 18
  G_1 stage, 10
  G_2 stage, 11
Synthetic medium, 94

T

Tandem translocation, cattle, 87, 146, 152
Telocentric, 24
  birds, 264
Telophase, 11, 12, 18, 20
Terminalization, 19
Testis, 18, 108
  birds, 270
  cattle, 156, 163, 172, 173
  goat, 208
  hinnies, 259
  horses, 251, 256
  swine, 230
Tetraploid, 30
  birds, 277
  blastocyst, 229
  cattle, 171
  mules, 259
  swine, 229
Thymidine, 12
  cattle, 147
  radioactive, 12
  tritiated, 12
Tissue culture, *see* Laboratory procedure
Transformation, 106
Translocation
  balanced, 31
  reciprocal, 40
  Robertsonian, *see* Robertsonian translocation
  X-autosome, 147
Triploid, 30
  birds, 277, 278
  bovine blastocyst, 170
  bovine hermaphrodite, 162
  swine blastocyst, 229
Trisomy
  birds, 277, 279
  brachygnathia, 153
  cattle, 153, 162, 169
Tritiated thymidine, 12, 66
  cattle, 147

V

Virilization, 81
Vitellus, 61

W

W chromosome, 64, 269

X

X-autosome translocation, cattle, 147
X chromosome, 66, 260, *see also* Sex chromosome
X inactivation, 260, *see also* Tritiated thymidine

Y

Y-chromosome polymorphism, 160, 164–165

Z

Z chromosome, 64, 268
Zona pellucida, 22, 61
Zygote, 22
Zygotene, 18, 19